HANDBOOK OF
METALLONUTRACEUTICALS

Nutraceuticals: Basic Research/Clinical Applications
Series Editor: Yashwant Pathak, PhD

Handbook of Metallonutraceuticals
Yashwant V. Pathak and Jayant N. Lokhande

Nutraceuticals and Health: Review of Human Evidence
Somdat Mahabir and Yashwant V. Pathak

Marine Nutraceuticals: Prospects and Perspectives
Se-Kwon Kim

NUTRACEUTICALS Basic Research/Clinical Applications

HANDBOOK OF
METALLONUTRACEUTICALS

Edited by
Yashwant Vishnupant Pathak
Jayant Nemchand Lokhande

CRC Press
Taylor & Francis Group
Boca Raton   London   New York

CRC Press is an imprint of the
Taylor & Francis Group, an **informa** business

CRC Press
Taylor & Francis Group
6000 Broken Sound Parkway NW, Suite 300
Boca Raton, FL 33487-2742

First issued in paperback 2016

© 2014 by Taylor & Francis Group, LLC
CRC Press is an imprint of Taylor & Francis Group, an Informa business

No claim to original U.S. Government works

Version Date: 20140108

ISBN 13: 978-1-138-19923-1 (pbk)
ISBN 13: 978-1-4398-3698-9 (hbk)

Library of Congress Cataloging-in-Publication Data

Handbook of metallonutraceuticals / editors, Yashwant Vishnupant Pathak, Jayant N. Lokhande.
   p. ; cm. -- (Nutraceuticals)
   Includes bibliographical references and index.
   ISBN 978-1-4398-3698-9 (hardcover : alk. paper)
   I. Pathak, Yashwant, editor of compilation. II. Lokhande, Jayant N., editor of compilation. III. Series: Nutraceuticals (Series)
   [DNLM: 1. Dietary Supplements. 2. Trace Elements. QU 145.5]

RA784
613.2'8--dc23
                                                    2013049506

Visit the Taylor & Francis Web site at
http://www.taylorandfrancis.com

and the CRC Press Web site at
http://www.crcpress.com

This book is dedicated to all the rhushies, sages, shamans, medicine men and women, and people of ancient traditions and cultures who contributed to the development of drugs and nutraceuticals worldwide and who kept the science of health alive for the past several millennia.

We thank our parents, friends, and colleagues, who have helped us to put all the ideas and thoughts together in this book.

Contents

Series Preface

Nutraceuticals: Basic Research and Clinical Applications

Nutraceuticals and functional food industries have grown significantly since the 1990s. Foods that promote health beyond providing basic nutrition are termed *functional foods*. These foods have the potential to promote health in ways not anticipated by traditional nutrition science. The acceptance of these products by larger populations, especially in the West, is ever increasing. The United States has been dominating the nutraceuticals and functional food markets recently.

Nutraceuticals provide health benefits by facilitating the healing process and improve overall health by preventing disease processes. For these reasons, they are attracting a large clientele. Health-conscious citizens, both in the United States and worldwide, are considering nutraceuticals as one of the major options for their health care. Nutraceuticals and alternative medicine also have been incorporated in the curriculum for health care professional education. These factors may have contributed to the overall market growth of nutraceuticals, which is predicted to exceed US$90 billion by the end of 2015 in United States, up to US$180 billion worldwide. This growth has been attributed to the aging population worldwide, increased prevalence of serious diseases due to lifestyle changes, and an enhanced focus on preventive medicines globally.

For nutraceuticals, alternative medicines, dietary supplements, and functional foods, it may be challenging to find proper characterization processes, reproducible activities, and clinical evidence to support the claims of their application in prevention or treatment. Looking ahead, it is likely that tougher guidelines from the U.S. Food and Drug Administration (FDA) and good manufacturing practices will soon enter the field. In the near future, the FDA may ask for clinical evidence to allow nutraceuticals and functional food products in the market.

This new series, *Nutraceuticals: Basic Research/Clinical Applications*, will include a range of books edited by distinguished scientists and researchers who have significant experience in scientific pursuit and critical analysis. The series will address various aspects of nutraceuticals, including historical perspectives, the traditional knowledge base, analytical evaluations, green food processing techniques, and applications. The series will be valuable reference books for researchers, academicians, and personnel in the nutraceuticals and food industries.

Here, in the third book of the series, the focus is on metallonutraceuticals. Metals have played a significant role in the treatment of chronic diseases in ancient traditional medicines, such as Ayurveda and Chinese medicines. Some metals are naturally found in the body and are very essential to human health; for example, iron prevents anemia and zinc is a cofactor in more than 100 enzymes. These metals are present in the body at very low concentrations as trace elements. Research in this area—especially the relationship of nutrients in mineral status and behavioral disorders, cardiovascular diseases, and cancer—offers new viewpoints and emerging possibilities. Nanometals have been extensively explored as therapeutic agents. For example, gold nanoparticles (which have been recommended as a treatment in Ayurveda as suvarna bhasma) are now being studied in allopathic medicine for the treatment of cancer. There are many such preparations in Chinese and Ayurvedic medicines.

This book discusses metals from the perspective of nutrients and nutraceuticals. For this purpose, the authors have coined a new term: *metallonutraceuticals*, in which metals are the main constituents. This book may create some controversy, but we hope it leads to a healthy debate about renewing the outlook towards metals and their applications in providing better health care opportunities to the consumer.

Foreword

The *Handbook of Metallonutraceuticals* is not a book; rather, it is an idea bank. Its ideas are based on an incremental and interdisciplinary innovation platform in the nutrition and pharmaceutical industries. Metallonutraceuticals are a new set of specific purpose-built miniaturized nutritional molecules that can carry out new physicochemical reactions in the body for the betterment of health.

Metals in the form of drugs have been used for acute and chronic disorders. Scholars of ancient medicines, such as Ayurveda, have long practiced with such molecules. They have also investigated how natural elemental metals can be converted into bioadaptable and bioacceptable forms by applying special pharmaceutical techniques. In this book, the authors have taken painstaking efforts to explain how metallonutraceuticals can be used in contemporary medicine, and they collaborated with modern pharmaceutical nanoengineers to make the techniques industrially scalable. This book explores such concepts and practices, which can be good stimuli for creating new products in the ever-growing nutritional industry. The authors have also made available rare references and citations of scientific articles to support new approaches, which can be good stepping stones for further applied research.

I was quite surprised to learn that metallonutraceuticals can be integrated with other disciplines, such as biotechnology, biomedical and bioimaging sciences, and medicine, to come up with new smart products for a variety of unresolved issues in drug and medical device development. Because this is a new concept, the authors have also offered new conceptual regulatory frameworks so that the correct placement of products can occur in regulations-driven markets. This book will be a useful reference, as well as a trigger for discussion and dialogue about metals and their applications in health care, for various groups of professionals. The scientific community should welcome this book as an attempt to provide respectability to metals in therapeutic and other applications.

Smita Singhania, PhD
Director
Symbiosis School of Biomedical Sciences
Symbiosis International University
Pune, India

Preface

For many years mankind has known of the various applications of metals in medicine and other fields. The precise nutritional and medicinal values of a few metals, such as zinc, iron, copper, calcium, and sulfur, are known and are currently being practiced. However, it may be possible to use other metals, such as gold, silver, and tin, as essential micronutrients. This is the origin of the concept of metallonutraceuticals. In ancient medicines, such as Ayurveda, these metals are used for *loh siddhi* (transforming lower metals into higher metals) in turn to perform *deh siddhi* (transforming body tissues for longevity). Ancient scholars used metals as lifesaving drugs, for treating chronic disorders, and for specific nutritional enhancement.

Today, the nutraceuticals industry is burgeoning, growing by 20% every year and competing with the pharmaceutical industry. In the past few decades, the nutraceuticals industry has experienced significant product and process innovations. Metallonutraceuticals are well suited for such an innovation-driven nutraceuticals market. However, it is vital to first determine a metallonutraceutical's idealization and novelty, along with industrial scale-up possibilities, applicability, and regulatory frameworks.

The primary challenge of using metals in the human body is identifying the form in which the metals are safe and efficacious. Metals have to be transformed into bioadaptable and bioacceptable molecular forms to possess therapeutic and nutritional properties. Daily recommended values, physiological pathways, site identification for bioactivity, metabolism, and health care applications as nutritional supplements must be determined. Metallonutraceuticals can be conceptually used as smart drug delivery systems, drug adjuvants, and disease-specific medical foods, as well as for some diagnostic and prognostic mapping purposes.

This book discusses all of these basic concepts, product development strategies, and some ready-to-convert ideas for certain nutritional and medicinal conditions to designate further as *condition-specific metallonutraceuticals*. The authors take an interdisciplinary approach for converting such ideas into an industrial product. We hope this book serves as a user-friendly exploratory guide to kickstart new research in the industrial and academic arenas for nutritionists, pharmacologists, pharmaceutical industrialists, nutritional industrialists, research scholars, and academicians.

We welcome comments and suggestions to improve the book and this area of research in the future.

Jayant Lokhande, MD

Yashwant Pathak, PhD

Editors

Yashwant Pathak, PhD, is currently the associate dean for faculty affairs at the newly launched College of Pharmacy, University of South Florida, Tampa, Florida. Pathak earned his MS and PhD degrees in pharmaceutical technology from Nagpur University, Nagpur, India, and EMBA and MS degrees in conflict management from Sullivan University, Louisville, Kentucky. With extensive experience in academia and industry, Pathak has over 120 publications, research papers, abstracts, book chapters, and reviews to his credit. He has presented over 150 presentations, posters, and lectures worldwide in the field of pharmaceuticals, drug delivery systems, and other related topics. He has also coedited six books on nanotechnology and drug delivery systems, two books on nutraceuticals, and several books on cultural studies. Pathak is the holder of two patents. He has travelled extensively to over 75 countries and is actively involved with many pharmacy colleges in different countries.

Jayant Lokhande, MD, obtained his doctoral degree in botanical drugs from University of Mumbai. Subsequently, he completed an MBA in biotechnology (finance and strategy) from University of Pune Campus School. He is an expert in biotechnology business management with integrated economics, catering to the critical business aspects of technology applications. He has worked for several nutraceuticals companies in the United States and India and has formulated successful strategies for marketing nutraceuticals worldwide. He has significant clinical experience in botanical drugs and the application of nutraceuticals in chronic diseases. He recently started his own company in the United States, for which his responsibilities include business analysis, product development, and technical marketing and sales.

Contributors

Vaibhav Alandikar
College of Pharmacy
University of South Florida
Tampa, Florida

Sajid Bashir
Department of Pharmaceutical
    Sciences
Texas A&M Health Science Center
Kingsville, Texas

Amy Broadwater
College of Pharmacy
University of South Florida
Tampa, Florida

Danielle Dantuma
College of Pharmacy
University of South Florida
Tampa, Florida

Pranab Jyoti Das
College of Pharmacy
Dibrugarh University
Dibrugarh, Assam, India

Anastasia Groshev
College of Pharmacy
University of South Florida
Tampa, Florida

Vishal Katariya
Katariya & Associates
Pune, Maharashtra, India

Jingbo Liu
Department of Chemistry
Texas A&M University
College Station, Texas

and

Department of Pharmaceutical
    Sciences
Texas A&M Health Science Center
Kingsville, Texas

Jayant Lokhande
Indus Prowess Inc.
Anaheim, California

Bhaskar Mazumder
College of Pharmacy
University of South Florida
Tampa, Florida

and

College of Pharmacy
Dibrugarh University
Dibrugarh, Assam, India

Srinath Palakurthi
Advanced Light Sources
Lawrence Berkeley National
   Laboratory
Berkeley, California

Ravi Ramesh Pathak
Department of Pathology and Cell
   Biology
University of South Florida
Tampa, Florida

Yashwant Pathak
College of Pharmacy
University of South Florida
Tampa, Florida

Chalres Preuss
Department of Molecular
   Pharmacology and Physiology
University of South Florida
Tampa, Florida

Samuel Rapaka
College of Pharmacy
University of South Florida
Tampa, Florida

Sriniwas Samant
Department of Organic
   Chemistry
Institute of Chemical Technology
Mumbai, Maharashtra, India

Jatin Shah
ViS Research Pvt. Ltd.
Maharashtra, India

Ganesh Shinde
Swami Chikitsalaya
Sindhu Laxmi Niwas
Maharashtra, India

Shivani Soni
Department of Biological Sciences
   and
Center for Nanobiotechnology
   Research
Alabama State University
Montgomery, Alabama

Vijay Sutariya
College of Pharmacy
University of South Florida
Tampa, Florida

Vrinda Vedi
College of Pharmacy
University of South Florida
Tampa, Florida

Komal Vig
Department of Mathematics and
   Science and
Center for Nanobiotechnology
   Research
Alabama State University
Montgomery, Alabama

Jiayou Wang
College of Pharmacy
University of South Florida
Tampa, Florida

and

Guanzhou University of Chinese
   Medicine
Guanzhou City, Guangdong,
   People's Republic of China

Hong-Cai Zhou
Department of Chemistry
Texas A&M University
College Station, Texas

Shufeng Zhou
Department of Pharmaceutical
   Sciences
University of South Florida
Tampa, Florida

Concept, Definition, and Need for Metallonutraceuticals

Jayant Lokhande and Yashwant Pathak

Contents

1.1 Nutraceuticals as an industry

1.1.1 Nutraceuticals definition

Nutrition is a universal phenomenon. It is a science as well as the practice of consuming and using food for living. Food is further composed of an array of nutrients, which exert specific primary and secondary metabolic effects on the body. Nutrients are of basically two types: macronutrients and micronutrients (**Figure 1.1**).

The word *nutraceuticals* (derived from nutrients) is a fairly new concept, but the use of nutraceuticals has been practiced for ages. References can be found in ancient medical treatises from India, China, Egypt, and Rome. The definition of nutraceuticals is subjective because of ethnic societal perceptions and regulatory frameworks. By the mid-nineteenth century, scientists had learned

1

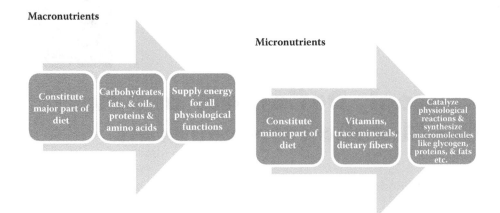

Macronutrients

| Constitute major part of diet | Carbohydrates, fats, & oils, proteins & amino acids | Supply energy for all physiological functions |

Micronutrients

| Constitute minor part of diet | Vitamins, trace minerals, dietary fibers | Catalyze physiological reactions & synthesize macromolecules like glycogen, proteins, & fats etc. |

Figure 1.1 Nutrients from food.

that the primary elements in food are carbon, nitrogen, hydrogen, and oxygen. They divided food constituents into four main types—carbohydrates, fats, protein, and water—but the chemical makeup of the first three classes was unknown (1).

The concept of food as fuel that contains important dietary components was further refined in the United States. Agricultural chemist Wilbur Olin Atwater had spent time in von Voit's laboratory as a postdoctoral researcher, returning to the United States in 1871 to spearhead nutrition science. Atwater spent 5 years in the 1890s building a respiration calorimeter larger than von Voit's, which was able to hold humans for longer than a day. His measurements were so precise that his energy equivalents for protein, fat, and carbohydrate are still used today. Atwater was first to adopt the word *calorie* as an energy unit for food, in which a calorie of food energy is actually equivalent to 1,000 calories of thermal energy (1).

The various synonyms and definitions of nutraceuticals include the following:

- **Dietary supplements:** "A product that is intended to supplement the diet that bears or contains one or more of the following dietary ingredients: a vitamin, a mineral, a herb or other botanical, an amino acid, a dietary substance, for use by man to supplement the diet by increasing the total daily intake, concentrate, metabolite, constituent, extract or combination of these ingredients intended for ingestion in pill, capsule, tablet, or liquid form" (2).
- **Bioactive compounds:** "Naturally occurring chemical compounds contained in or derived from a plant, animal or marine source that exert a desired health/wellness benefit" (3).
- **Novel foods:** "Products that [have] never been used as foods [or] foods that result from [a] process that has not previously been used for food" (4).

- **Nutraceuticals:** "A product isolated and purified from foods that is generally sold in medicinal forms. A nutraceutical is demonstrated to have a physiological effect or protection against chronic disease" (5).
- **Functional ingredients:** "Standardized and characterized preparations, fractions or extracts containing bioactive compounds of varying purity that are used as ingredients by manufacturers of foods, cosmetics and pharmaceuticals" (6).

Regulations of nutraceuticals vary by country (**Figure 1.2**). To consider these regulations, we suggest a modification in the present definition of nutraceuticals, as stated by Dr. Yashwant Pathak:

Nutraceuticals are the products developed from either food or dietary substance, or from traditional herbal or mineral substance or their synthetic derivatives or forms thereof, which are delivered in the pharmaceutical dosage forms such as pills, tablets, capsules, liquid orals, lotions, delivery systems, or other dermal preparations, and are manufactured under strict [current good manufacturing practices]. These are developed according to the pharmaceutical principles and evaluated using one or several parameters and in process controls to ensure the reproducibility and therapeutic efficacy of the product. (7)

1.1.2 Nutraceuticals global market

In a market demand scenario, nutraceuticals are considered to be all commonly understood products falling within a category, such as powder, tablet, or capsule. These products, which can be in natural or synthetic form, are used to improve health conditions, prevent certain diseases, and/or promote energy and wellness (**Figure 1.3**). Nutraceuticals, medical foods, and drugs are compared in **Table 1.1**.

The global latent demand for nutritional supplements in 2009 is shown in **Figure 1.4**. Note that latent demand is sometimes higher than actual industry sales because it is calculated on the basis of potential earnings. The nutraceuticals market has been growing by 5% per year (**Figure 1.5**). The factors influencing the nutraceuticals market could be described by the following marketing fundamentals for functional foods:

- Products with an "easy-to-feel" effect are more successful.
- Products should have innovative packaging that is attractive, functional, easy to carry, and convenient.
- Products are successful when benefits are associated with its relevance and pricing.
- A benefit for a single specific condition and/or multiple benefits for multiple conditions.

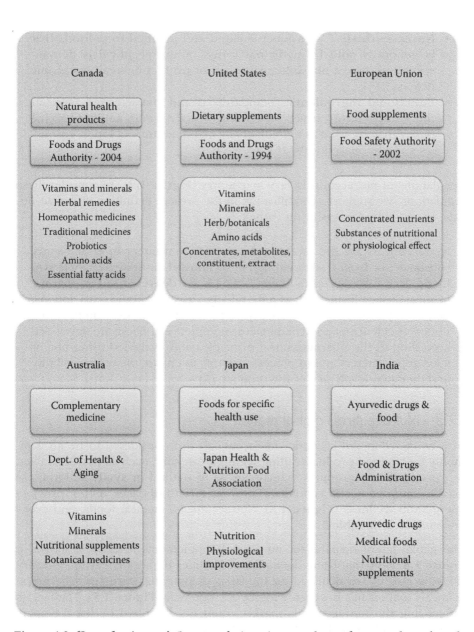

Figure 1.2 *(See color insert.) Country, designation, regulatory framework, and product category.*

There are numerous applications of nutraceutical ingredients in the functional foods industry. A market report by PricewaterhouseCoopers estimated that growth for the functional foods category is between 8.5% and 20% per year (8). Glenn Pappalardo, a director at PricewaterhouseCoopers, said the popularity of functional foods will make a lasting impression on the consumer packaged goods industry: "Fundamentally, this space has a lot of positive opportunities.

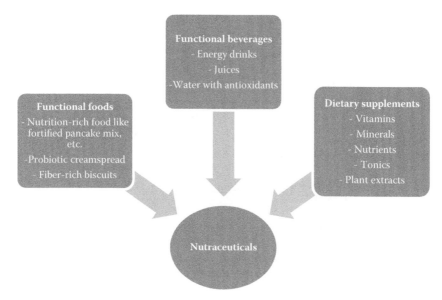

Figure 1.3 Nutraceutical functional product segments.

This is something that is by no means a fad. It will be an ongoing part of our diet and food offerings for many years to come." Consequently, there is a tremendous need for education among consumers, the industry, health care professionals, and interested investors about what functional foods are and where the opportunities lie across segments.

Table 1.1 Distinguishing Factors of Nutraceutical Products

	Nutraceuticals	Medical Foods	Drugs
End use	Energy and vigor boost, fat reduction, liver, heart, kidney disease therapy support, cognitive supplement, hormonal support	Palliative care of a disease or disease condition with distinctive nutritional fulfillment	Treatment of disease and/or signs and symptoms
Consumer end procurement	Consumer desire	Medical supervision	Prescription
Market and sales channel	Supermarkets, health stores, pharmacies, online, grocery retailers	Hospitals, pharmacies, drug stores	Hospitals, pharmacies
Dose	As desired but within product's safe limit	Advised by practitioner	Prescribed by doctor
General governing law	Product claim and labeling verification by U.S. Food and Drug Administration (FDA) and ban on certain ingredients	Product claim and labeling verification by FDA	FDA approval through tier structure

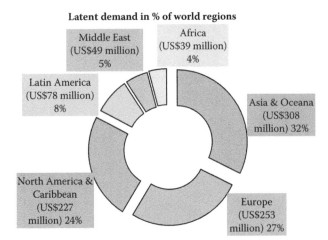

Latent demand in % of world regions

Figure 1.4 Market for nutritional supplements in 2009. (Modified from Parker, P.M. The 2009–2014 World Outlook for Nutritional Supplements. Fountainebleau, France: INSEAD.)

1.1.3 Nutraceuticals adaptation

The basic trends in the nutraceuticals market are given in **Table 1.2**. The use of nutraceuticals is significantly recognized due to changing trends in age demography, purchasing affluence, increased education, and life expectancy. These factors are also contributing to health consciousness. Increased health consciousness is one of the most vital drivers in the nutritional supplements market. Access to information through education and media, along with an understanding of the mode of action and

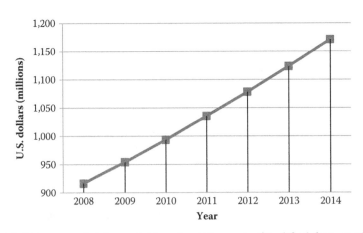

Figure 1.5 Future market for nutritional supplements. (Modified from Parker, P.M. The 2009–2014 World Outlook for Nutritional Supplements. Fountainebleau, France: INSEAD.)

Table 1.2 Trend Interrelations in Healthcare and Nutraceuticals

Healthcare Trend	Nutraceuticals Appropriation	Impact on Market
Personalized medicine approach; drug modification for same disease condition as per race, age, country, etc.	Finely segmented products are likely to lead to greater varieties of functional foods as per unique disease condition profile.	Market fragmentation as per variety and its sustainability
Health care cost upshift so prevention of disease is new mantra	As the burden of health payments shifts to consumers, they may try to limit their exposure to copays and other costs. Nutraceutical consumption can be seen as an alternative affordable way for consumers to better manage their health.	Disease prevention market, which can influence the savings of the consumer

health-promoting effects of nutraceuticals in food and nonfood products, have resulted in a rapidly emerging self-care movement throughout society. When coupled with increased economic prosperity, more people are choosing a proactive approach to managing their own health, rather than using health care services provided by a medical community in response to health problems.

An increase in the aging population (e.g., baby boomers) is precipitating the need for a more responsive and empathetic health care system. This health care system should be able to manage the vast array of age-related maladies faced by older individuals. However, an exponential increase in health care expenditures along with sustainability of the system itself has forced many people to seek out more cost-effective health care alternatives. Therefore, the older population is often attracted to nutritional supplements because of their disease prevention and wellness promotion properties, rather than traditional forms of high-cost health care.

Advances in the areas of food technology, food biochemistry, and the nutritional sciences (e.g., nutritional genomics) are providing access to decipherable benefits of supplements that were not available before. New product development methods being used by the functional food industry to isolate, characterize, extract, and purify nutraceuticals from bacterial, plant, and animal sources are resulting in decreased costs to the industry and are providing new options for the use of functional food products. Changes in global policies and laws governing the distribution and marketing of nutraceuticals have achieved a paradigm shift in attitudes and accountability of governments to the people they represent. The improved communication practices, transportation for marketable nutraceuticals, and appreciation for proprietary patented products are resulting in the provision of a wide variety of products to many people at a given point in time.

1.1.4 Futuristic nutraceuticals

Because nutraceuticals can be used for preventive health care, consumers may more willingly use them to avoid potentially serious health conditions rather than eventually needing much costlier prescription drugs and medical treatments. For older individuals, there is great proliferation of age-related condition-specific supplements, such as products for joint and bone health, eye health, heart health, immune support, and increased energy. Some nutritional supplement companies, such as Solgar and GeneLink, are even using personalized approaches to health management to offer condition-specific prevention by genetically guided nutritional supplements (9). Another example of futuristic nutraceuticals could be for children with autism, who may benefit from a modified diet that improves digestion, putting them in a better frame of mind to learn, think, and interact (10).

The market opportunities are summarized in **Table 1.3**. The market for value-added nutraceutical ingredients is expected to be €2.73 billion at the end of 2016.

1.2 Challenges for nutraceuticals

1.2.1 Science-based evidence for nutraceuticals

In the case of nutritional supplements, such as botanicals, the health-promoting effects are likely due to a complex mix of biochemical and cellular interactions. These biometabolites may function as substrates in metabolic reactions or cofactors of key metabolic enzymes as ligands, which promote or compete with biochemical interactions at the cell surface, or with intercellular receptors, which can enhance absorption and assimilation of important macronutrients

Table 1.3 Opportunity and Related Business Strategy

Opportunity	Technology Development	Health Care Market Strategy	Nutraceuticals Market Strategy	Examples
New market target	Effective food delivery with benefits	Novel benefit of already known product; novel benefit for novel product	Novel benefit of already known product; novel benefit for novel product	Off-patent drugs become part of a medical food or nutraceuticals Consumption of some nutraceuticals is mandatory before taking principal treatment
Research target	New approach pathways to disease	Novel benefit of already known product; novel benefit for novel product	Target-specific food	Genetically modified plants and animals with added functional properties

and micronutrients. In addition, these metabolites may act as enzyme inhibitors, absorbents, or toxicant scavengers that can associate with and help remove damaging substances or toxins from the body. Therefore, there is a critical need for establishment of an efficient process pathway for such metabolites and their final products. In addition, there is a need for scientific evidence in terms of clinical research to support nutritional and medicinal claims being made for botanical metabolites products within industrial environments.

1.2.2 Supplement efficacy still being challenged

The acceptance of nutritional supplements rises and falls according to the changing tides of public perception regarding product safety and effectiveness. Indeed, one of the biggest challenges for nutritional supplements is the enormous amount of constantly changing (sometimes conflicting) advice from countless medical and dietary entities. In the past, negative publicity often focused on the potential dangers of supplements, such as the interactions between prescription and over-the-counter drugs and herbals/supplements, as well as products having unsubstantiated claims.

For example, in summer 2008, Dr. Eric Rimm, an associate professor in the departments of epidemiology and nutrition at Harvard University, told ABC News that dietary supplements cannot provide a nutritional boost to a poor diet, claiming that only a change in diet and lifestyle can boost nutritional efficacy. Dr. Rimm stated, "A supplement is called a supplement because it's supposed to be supplementing a healthy lifestyle" (11). He also claimed that the abundance of vitamin C and folic acid through fortified foods in the diet makes supplementation with these vitamins unnecessary, even for less healthy members of society.

1.2.3 Regulation

Unlike pharmaceutical drugs, nutraceutical products are widely available and are not monitored with the same level of scrutiny by food and drug authorities of many countries. There are still broad-based definitions of nutritional supplements. In addition, undistinguishing standards and functions between "nutraceuticals" and "drugs" create inconsistent credibility for nutritional supplements. Therefore, legitimate companies manufacturing these supplements provide credible scientific research to substantiate their manufacturing standards, products, and consumer benefits and differentiate their products from "nutraceuticals." This lack of regulatory harmonization is resulting in the compromised safety and effectiveness of products.

1.2.4 Bioavailability and the placebo effect

Bioavailability, which can be thought of as the absorption rate of a supplement product, is one of the main challenges in nutraceuticals' product effects.

Nutraceuticals with poor absorption rates result in nutrients being disposed from the body without providing any nutritional or medicinal benefit. Similar to pharmaceuticals, part of the effectiveness of nutraceuticals may be attributed to the placebo effect. People using nutraceuticals may inaccurately credit their use of nutraceuticals for healing illnesses, when in fact the body is often able to recover on its own. Thus, the efficacy and clearly established and sustainable use of nutraceuticals are being questioned.

1.3 Metallonutraceuticals

1.3.1 Nutraceuticals with respect to minerals

Minerals (bioacceptable elemental metals or mineral complexes) include active minerals with established recommended dietary allowances (RDAs), as well as trace minerals or elements. Active minerals, which are required by the body in daily quantities of more than 100 µg, include calcium, chloride, chromium, copper, fluoride, iodine, iron, magnesium, manganese, molybdenum, phosphorus, potassium, and zinc. Trace elements, which are required in amounts less than 100 µg, include arsenic, boron, nickel, and silicon. Eighteen minerals and trace elements are known to be necessary to human health, but the National Academy of Sciences has set RDAs for only seven: calcium, copper, iodine, iron, magnesium, phosphorus, and zinc.

Minerals as nutraceuticals are used in either single-element or multivitamin/ mineral combinations. Single-element vitamins or minerals are most often used to address a specific condition, such as vitamin C to help fight a cold or calcium to help prevent or treat osteoporosis. Multivitamins, which often also include minerals, are typically used as a general dietary complement, with products for children falling into this segment.

The minerals in nutraceuticals can be sourced naturally or synthetically. Mineral nutraceuticals from natural sources include bone meal for calcium. Naturally sourced minerals are probably more bioavailable than synthetically sourced minerals, but scientific studies have shown dichotomous results to some degree. For example, natural vitamin E is more biologically active than synthetic vitamin E and is retained twice as much by the human body. On the other hand, folic acid is absorbed by the body better in synthetic form rather than in its natural form.

Mineral nutritional supplements are consumed in the form of pills, including coated or uncoated tablets, time-release tablets, hard gelatin capsules, and soft-gel capsules ("gelcaps"). They are also available in liquid suspensions, which allow for easy adjustment of dosage and may be placed under the tongue (i.e., sublingual) for faster absorption. Powders, which can be added to liquids or foods, also allow for easy adjustment of dosage. Nasal gels and oral sprays are suited for certain products, such as B vitamins mixed with minerals. Food and beverage formats, such as candies (especially "gummies") and flavored effervescent tablets, are designed to be added to water or other beverages.

1.3.2 Metallonutraceuticals as a novel concept

Metals have played an important role in medicine for centuries. Many metals are essential in our diets in varying quantities, although people have only recently realized their significance due to increased awareness of personal health and media involvement. However, at the other extreme, certain metals remain toxic in trace amounts. These metals can enter the body via a variety of routes and often cannot be excreted, leading to metal toxicity. Despite the dangers metals can pose to the human body, they are also used in the medical profession as invaluable diagnostic tools. These include radioactive and magnetic techniques using a variety of elements to explore the inner structures of the human body without the need for invasive procedures (**Table 1.4**).

Inorganic chemistry and the use of metals in therapeutic drugs have become increasingly important over the last couple of decades, resulting in a variety of exciting and valuable drugs, such as cisplatin for cancer treatment. The metal complexes are amenable to combinatorial synthetic methods, and an immense diversity of structural scaffolds can be achieved. Metal centers are capable of organizing surrounding atoms to achieve pharmacophore geometries that are not readily achieved by other means. Additionally, the effects of metals can be highly specific and can be modulated by recruiting cellular processes that recognize specific types of metal–macromolecule interactions. Metals can be useful probes of cellular functions.

Table 1.4 Metals and Some of Their Known Medical Uses

Metal	Therapeutic/Medical Use
Gold	Arthritis, gout
Silver	Antiseptic agent
Arsenic and antimony	Bactericides
Bismuth	Skin injuries, diarrhea, alimentary diseases
Cobalt	Vitamin B12
Copper	Algaecide, fungicide, insecticide
Gallium	Antitumor agent
Mercury	Antiseptic
Lithium	Manic depression
Manganese	Fungicide, Parkinson disease
Osmium	Antiarthritis
Platinum	Antitumor agents
Rubidium	Substitute for potassium in muscular dystrophy; protective agent against adverse effects of heart drugs
Rhodium	Experimental antitumor agents
Tin	Bactericide, fungicide
Zinc	Fungicide

Metallopharmacology is an area of growing interest, as is evident from the clinical trials that are being conducted worldwide for the use of metals in therapeutics. For example, clinical trials for silver biotics have been carried out to assess its efficacy for a wide variety of human problems, including malaria; upper respiratory tract infections; urinary tract infections; sinusitis; vaginal yeast infections; eye, nose, and ear infections; cuts and fungal skin infections; and even for sexually transmitted diseases, such as gonorrhea. Therefore, it may serve well as an antibiotic alternative at a convenient dosage.

1.3.3 Metallonutraceuticals definition

Metal deficiency diseases have long been known. Diets deficient in Ca, Cu, Zn, and Fe are known to cause serious health problems. More recently, selenium deficiency is of increasing concern. The research into metal metabolism will provide new insights into the essential roles of metals in human health and the effective delivery of trace metals to treat deficiency diseases. Although many metals, such as Cu, Zn, and Fe, are essential for human health and protect against many diseases, they are also implicated in many degenerative diseases, including atherosclerosis (heart disease and stroke), degenerative brain disorders, arthritis, and cancer. Normally, human metabolic processes tightly control the distributions and concentrations of such metals. The causes of these metal imbalances, which may be an important factor in all degenerative diseases, are uncertain.

In the thirteenth and fourteenth centuries, the ancient medical treatise of *Ras Shastra* (the Indian science of alchemy and iatrochemistry) explained the concept of *dehvedha* (the transformation of the body for the prevention of aging and maintenance of positive health), which is originated from *Ayurved* (science of life) in 3000 B.C. *Ras shastra* emphasized more of the *rasayana* (anti-aging and rejuvenation) concept, which indicates the means by which one can achieve the best quality of *rasadi dhatus* (anabolic elements and physiological reactions thereof), which in turn provide positive health to the body. Thus, it prevents diseases and aging, sustaining the body in a healthy and youthful state. From the eighteenth century, a number of metallic and mineral preparations evolved from this concept and were recognized to have the *rasayana* property.

Ras Shastra clearly explained the physiological citation of various metals and minerals:

> *Rasanat sarvadhatunam ras itiabhidiyate, jara ruk mrutyunashay rasyat va raso mata* (**Table 1.5**). (12)

In this verse, *ras* is a Sanskrit term used for "minerals and metals" and *jara ruk mrutyunashay* means they are used for curing disease and anti-aging. In addition, the *Ras Shastra* states:

> *Iti dhansharirbhogan matvanityan sadaiv yataniyam mukta ucchaab-hyasat sa cha sthire dehe.* (13)

Table 1.5 Metals Cited in the Human Body per *Ayurved* (Science of Life)

Metals	Location in Human Body per Indian Science of Alchemy and Iatrochemistry
Gold	Blood, semen, eyes, heart, upper layer of skin and intestine
Silver	Bone marrow, upper layer of bones, gallbladder, pancreas, inner layer of the skin, lungs, muscles, blood vessels, meninges (membrane that surrounds the brain), audio-receptive glands and septum of nose
Copper	Upper and inner layer of skin, mucosa of soft tissue, large glands, eye pupil, hair, pleura and pericardium, skeletal system
Iron	Blood, villi of the intestine, eye pupil, hair, and in small quantity in all tissues of the body
Tin	Abdominal muscles, blood and blood vessels, synovial membrane, outer layer of uterus
Lead	Blood, lymph
Zinc	Blood, brain, sensory tissues, muscles

In this verse, the term *sthire dehe* (sound body) means that the health acquired by use of metals and minerals is important to for one to enjoy *dhansharirbhogan* (prosperity). Thus, metallonutraceuticals can be defined as "essential and bioadaptable single/multiple metal complexes that have a significant impact on physiological and metabolic reactions through the mediation of enzymes and energy-carrier molecules to maintain health and/or cure various disease conditions."

1.4 Conclusion

The importance of nutraceuticals is growing in the arena of health care. Disease prevention and increased health consciousness are the most influencing factors for the use of nutraceuticals. There are certain challenges facing nutraceutical products, including efficacy establishment, standardization, quality, and regulatory harmonization but these challenges can be dealt with by adopting certain measures in manufacturing and efficacy trials.

Along with existing product categories, metals or minerals could be worth exploring to develop new nutraceuticals by considering their past history of medical usage. Although the micronutritional functions of certain metals and minerals are already known, the mapping of their biological pathways can further facilitate the use of unknown metals and minerals as nutritional supplements.

References

1. Outlook Nutrigenomics. 2010. The changing notion of food. *Nature* 468:S17.
2. http://www.fda.gov/food/dietarysupplements/qadietarysupplements/ucm191930.htm

3. Basu, S.K., J.E. Thomas, and S.N. Acharya. 2007. Prospects for growth in global nutraceutical and functional food markets: A Canadian perspective. *Australian Journal of Basic and Applied Sciences* 1:637–649.

4. Health Canada. 1988. *Nutraceutical and Functional Food and Health Claims on Foods Final Policy*. Ottawa, Ontario, Canada: Therapeutic Products Programme and the Food Directorate from the Health Protection Branch.

5. http://www.fimdefelice.org/p2455.html

6. Health Canada. *Nutraceuticals/Functional Foods and Health Claims on Foods*. Available at http://www.hc-sc.gc.ca/fn-an/label-etiquet/claims-reclam/nutra-funct_foods-nutra-fonct_aliment-eng.php.

7. Pathak, Y.W. 2011. "Definitions, Formulations, and Challenges," in *Handbook of Nutraceuticals: Ingredients, Formulations, and Applications*. Boca Raton, FL: CRC Press.

8. PricewaterhouseCoopers. *Leveraging Growth in the Emerging Functional Foods Industry: Trends and Market Opportunities*. Available at http://www.pwc.com/us/en/transaction-services/publications/functional-foods.jhtml.

9. http://genelinkbio.com/company.shtml

10. Richman, A. 2009. *The Gut-Brain Connection*. Available at http://www.nutraceuticals-world.com/contents/view_online-exclusives/2009-10-01/the-gut-brain-connection.

11. Jacobsen, B., and S. Wagner. 2008. *So Many Vitamins, So Little Time*. Available at http://abcnews.go.com/Health/story?id=5304089&page=3.

12. Bomdashubodha, P. 1970. *Rasaratna Samuchhaya* 1:76. Calcutta, India.

13. Bomdashubodha, P. 1970. *Rasaratna Samuchhaya* 1:78. Calcutta, India.

2

Roles of Metals in Metabolism

Jatin Shah and Jayant Lokhande

Contents

2.1 Metals and minerals in nutritional physiology

Plants harness energy from the sun and transfer it through the food cycle. Although nonmetals such as carbon, hydrogen, nitrogen, and oxygen are key elements synthesized to complex molecules during this process, other metals and nonmetals are also used in varying amounts. As a result, energy is transferred along with a range of metals and nonmetals in foods of plant origin. Animals and humans not only use the trapped energy in food by breaking complex molecules to simpler ones, but they also absorb and make use of the accompanying metallic and nonmetallic elements.

Based on their normal concentration in the human body and the quantity needed to support daily bodily functions (**Table 2.1**), minerals are classified as macrominerals, microminerals, and trace elements:

- Macrominerals are usually present in the human body in quantities greater than 5 g and are needed in quantities greater than 100 mg daily as a source of nutrition. Calcium, iron, magnesium, phosphorus, potassium, and sodium are examples of macrominerals.
- Microminerals are usually present in the human body in quantities less than 5 g and are needed in quantities of 1–100 mg daily as a source of nutrition. Chromium, copper, manganese, selenium, sulfur, and zinc are examples of microminerals.
- Trace elements (as the name suggests) are present and needed in very small amounts. Fluorine, iodine, cobalt, molybdenum, silicon, and others are included in this group.

Table 2.1 Daily Requirements of Elements and Their Sources

Minerals	Daily Requirements		Sources
	Men	Women	
Calcium	700 mg	700 mg	Milk, cheese, and other dairy foods; green leafy vegetables, such as broccoli, cabbage, and okra (but not spinach); soybeans, tofu, and soy drinks with added calcium; nuts, bread, and anything made with fortified flour; fish in which the bones are consumed, such as sardines and pilchards
Iodine	0.14 mg	0.14 mg	Ocean fish and shellfish; cereals and grains
Iron	8.7 mg	14.8 mg	Liver and meat; beans, nuts, and dried fruit, such as dried apricots; whole grains, such as brown rice; fortified breakfast cereals; soybean flour; most dark-green leafy vegetables, such as watercress and curly kale
Beta carotene	7 mg	7 mg	Yellow and green (leafy) vegetables, such as spinach, carrots, and red peppers; yellow fruit, such as mango, melon, and apricots
Boron	<6 mg	<6 mg	Green vegetables, fruit, and nuts
Chromium	0.025 mg	0.025 mg	Meat; whole grains, such as wholemeal bread and whole oats; lentils; spices
Cobalt	0.0015 mg	0.0015 mg	Fish; nuts; green leafy vegetables, such as broccoli and spinach; cereals, such as oats
Copper	1.2 mg	1.2 mg	Nuts, shellfish, and offal
Magnesium	300 mg	270 mg	Nuts, spinach, bread, fish, meat, and dairy foods
Manganese	<0.5 mg	<0.5 mg	Tea, bread, nuts, cereals, and green vegetables, such as peas and runner beans
Phosphorus	550 mg	550 mg	Red meat, dairy foods, fish, poultry, bread, rice, and oats
Potassium	3,500 mg	3,500 mg	Fruit, such as bananas; vegetables; lentils, nuts, and seeds; milk and bread; fish and shellfish; beef, chicken, and turkey
Selenium	0.075 mg	0.06 mg	Brazil nuts, bread, fish, meat, and eggs
Sodium chloride (salt)	<6 g	<6 g	Ready-to-eat meals; meat products, such as bacon; some breakfast cereals; cheese; some tinned vegetables; some bread and savory snacks
Zinc	9 mg	7 mg	Meat and shellfish; milk and dairy foods, such as cheese; bread and cereal products, such as wheat germ

Source: Data from http://www.nhs.uk/Conditions/vitamins-minerals/Pages/vitamins-minerals.aspx.

Although they are present in small quantities (**Figure 2.1**), trace elements have their own significant roles in animal and human nutrition physiology and may be referred to as minerals. Because they are required in small amounts, they are also referred to as micronutrients.

2.1.1 Calcium

Calcium makes up 1.9% of a person's body weight; thus, it is the fifth most abundant element in the body. It constitutes 2% of an adult's fat-free weight—most of which is concentrated in the skeleton, which serves as its chief reservoir. Other locations of calcium include the teeth, soft tissue, and extracellular fluid (ECF).

2.1.1.1 Calcium physiology

Calcium influx and efflux occurs across all cellular membranes, thus influencing vital neuromuscular and cellular functions. It enters the ECF either by absorption from food in the gastrointestinal tract (GIT) or through resorption from bones. It enters bones through the process of bone formation and is excreted from the ECF via kidneys, GIT, or skin. Calcium metabolism is governed by calcium receptors in the parathyroid gland, which secrete parathyroid hormone, and vitamin D3 status.

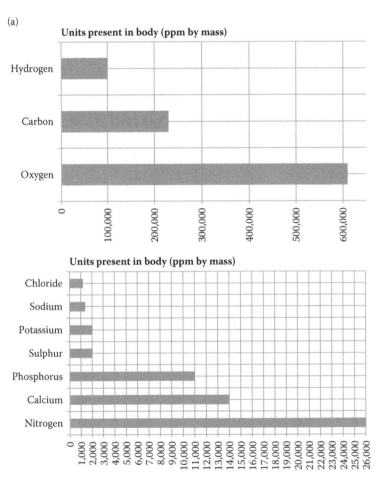

Figure 2.1 *(See color insert.) Elements and their concentration. (Modified from https:// people.ok.ubc.ca/wsmcneil/335/intro-abund.pdf.)*

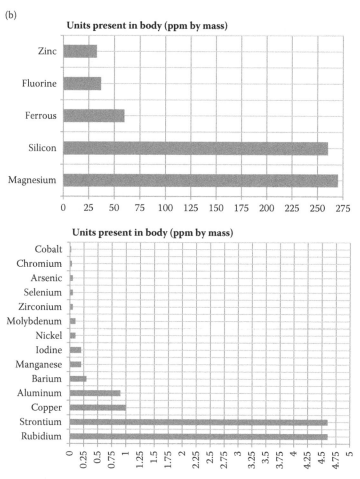

Figure 2.1 (See color insert.) *Elements and their concentration. (Modified from https://people.ok.ubc.ca/wsmcneil/335/intro-abund.pdf.)*

2.1.1.2 Calcium deficiency

Short-term negative calcium balance and deficiency is usually prevented by bone stores, which maintain adequate blood calcium levels. In the long term, bone stores may be depleted, making the bones weak and more prone to fractures and impairing muscle and nerve functions. Low blood calcium levels are often a consequence of a disturbance in the calcium-regulating mechanism (e.g., insufficient parathyroid hormone or vitamin D) rather than dietary deficiency.

2.1.1.3 Calcium toxicity

The GIT limits the amount of absorbed calcium; hence, toxicity is uncommon. Short-term intake of large amounts of calcium does not produce any ill effects except constipation and a higher risk of renal calculi. However, long-term intake of excess calcium or higher absorption in the presence of higher vitamin D levels can lead to abnormal circulating levels and deposition in tissues. Significantly high levels of calcium can lead to loss of appetite, nausea, vomiting, abdominal pain, or seizures, among other symptoms (1).

2.1.2 Iron

Total body iron in a healthy individual amounts to 3.5 g (50 mg/kg) and is distributed in red blood cell hemoglobin (65%), muscle fibers (10%), and other tissues. The remaining iron resides in the liver, macrophages of the reticulo-endothelial system, and bone marrow. On average, a person's daily diet contains 15–20 mg of iron, of which 1–2 mg is absorbed. Intake is balanced by losses occurring from the loss of intestinal mucosal cells, menstruation, and other blood loss.

2.1.2.1 Iron physiology

Iron intake is essential for cellular metabolism and aerobic respiration; it also supports erythropoiesis. Iron serves as a transport medium for oxygen movement from the lungs to each individual cell and for electron movement within cells. It is also an integral part of various enzyme systems in the body. Because iron is not actively excreted from the body and its overload leads to a variety of symptoms, iron regulation and homeostasis are essential to ensure smooth functioning of the body.

Two types of dietary iron exist: heme iron and nonheme iron. The hemoglobin and myoglobin present in meat, poultry, and fish are the chief sources of heme iron. Nonheme iron is absorbed from cereals, lentils, legumes, fruits, and vegetables. Heme iron is degraded to nonheme iron if food is cooked at a high temperature for long duration. Calcium is essential for the absorption of heme iron, whereas a variety of dietary factors influence the absorption of nonheme iron. For example, ascorbic acid significantly enhances nonheme iron absorption.

Iron is regulated in the body by three mechanisms:

1. Continuous iron recycling from catabolized erythrocytes
2. Access to the stored protein, ferritin
3. Regulation of absorption of iron from the intestines

2.1.2.2 Iron deficiency

The populations most at risk for iron deficiency are infants, children, adolescents, and women of childbearing age, especially pregnant women.

2.1.2.3. Iron toxicity

Iron overload can lead to toxicity and cell death via free radical formation and lipid peroxidation.

2.1.3 Magnesium

Much of the body's magnesium is found in muscles and soft tissues (30–40%). A small portion is located in the ECF (1%), with the remaining magnesium (50–60%) residing in the bony skeleton. Magnesium may be present in the ionic form or bound to proteins. In the bones, it is present as a surface constituent of the hydroxyapatite (calcium phosphate) mineral component. Because it is easily exchangeable with serum, hydroxyapatite serves as a moderately accessible magnesium store. However, the proportion of bone magnesium in this exchangeable form declines significantly with age.

2.1.3.1 Magnesium physiology

Magnesium in the soft tissues is a critical cofactor of enzymes that play an important role in energy metabolism, protein synthesis, RNA and DNA synthesis, and maintenance of electric potential in nerve tissues and cellular membranes. Magnesium is also essential in the regulation of potassium fluxes and calcium metabolism. It is needed for both the manufacture and action of insulin. At the same time, magnesium relies on insulin for its transport from blood into the cells, where it is most needed.

2.1.3.2 Magnesium deficiency

Failure to consume adequate magnesium has a significant impact on human physiology. Inadequate magnesium intake depresses cellular and extracellular potassium and results in low serum calcium. Although acute deficiency is rare, it may occur as an adverse effect of chronic alcohol/diuretic use or in individuals who are fed intravenously for long periods of time. Early symptoms include confusion, apathy, fatigue, muscle soreness, muscle twitches and spasms, insomnia, and poor memory. Severe deficiency results in numbness, paresthesia, muscle spasms, tension headache, hallucinations, and delirium. Previous research has established the association of magnesium deficiency with cardiovascular diseases, diabetes, anxiety disorders, migraines, osteoporosis, and cerebral infarction.

2.1.3.3 Magnesium toxicity

High intake of magnesium leads to hypermagnesemia, which makes it difficult for the body to absorb calcium. Although magnesium from dietary sources rarely causes harm, contaminated food or water leads to nausea, hypotension, and diarrhea. Magnesium toxicity can occur in individuals with end-stage renal disease and after ingestion of high amounts of magnesium salts. Muscle paralysis and coma may ensue as a consequence (1).

2.1.4   Phosphorus

Phosphorus ranks second in terms of abundance in the human body. It makes up 1% of an individual's total body weight. Although primarily located in bones and teeth, it is present throughout the cells and tissues of the body. Phosphorus can be found in virtually every bodily fluid. It exists as inorganic phosphorus or phosphate (Pi), lipid phosphorus, and phosphoric ester phosphorus in plasma.

2.1.4.1   Phosphorus physiology

Phosphorus is essential for the growth, maintenance, and repair of the body's cells and tissues. It is needed for a variety of cellular processes, ranging from nucleic acid synthesis and metabolism, energy metabolism, lipid metabolism, cell signaling, membrane integrity, muscle function, enzyme activity, and bone mineralization. It plays a role in the balance and use of vitamins and minerals. It helps filter waste products in the kidney, determines how the body stores and uses energy, maintains the regularity of heartbeat, assists in the contraction of muscles, and facilitates nerve conduction.

2.1.4.2   Phosphorus deficiency

Phosphorus deficiency is rare due to its presence in a variety of foods and efficient renal conservation, even after consumption of low-phosphorus diets. Phosphorus deficiency usually affects alcoholics and individuals with liver disease. The condition results in muscle weakness, impaired leukocyte function, osteomalacia, or rickets. Research indicates that gastrointestinal malabsorption, diabetes mellitus, renal tubular dysfunction, and premature birth are associated with low circulating levels and depleted phosphorus stores (2–4).

2.1.4.3   Phosphorus toxicity

Phosphorus excess is more common than deficiency. It frequently occurs as a result of renal disease, severe dysfunction of calcium regulation, high dietary intake of phosphorus-rich food. or low intake of dietary calcium. Given the delicate balance between phosphorus and calcium, high phosphorus intake demands an equivalent increase in calcium, which may affect bone density. Excessively circulating levels of phosphorus combine with calcium to form deposits in soft tissues, such as muscle.

2.1.5   Potassium

Potassium is the most abundant cation in the intracellular fluid. In terms of mass, it forms 0.2% of the human body. It exists primarily in the intracellular fluid at concentrations of 140–150 mEq/L and in the extracellular fluid at concentrations of 3.5–5 mEq/L (5).

2.1.5.1 Potassium physiology

Potassium cations are necessary for the smooth functioning of the body's cells. They play an important role in neuron function through potassium ion channels, influence osmotic balance among cells and the interstitial fluid through the Na+/K+-ATPase ion pump, and are responsible for fluid and electrolyte balance in the body.

Serum potassium concentration is maintained through a complex bodily mechanism as a consequence of a critical balance between intake, excretion, and distribution between intracellular and extracellular spaces. In the short term (minutes), potassium homeostasis is influenced by insulin, pH, beta adrenergic agonists, and bicarbonate concentration in the body. In the long term, potassium excretion and balance are primarily influenced by the kidney and aldosterone.

2.1.5.2 Potassium deficiency

Potassium deficiency is uncommon because the kidneys optimally regulate the level of potassium in the body. It may result from excessive potassium losses due to vomiting, diarrhea, poor diabetic control, low-calorie diets (less than 800 calories per day), chronic alcoholism, excessive exercise, or the use of some diuretics and laxatives. Potassium deficiency leads to muscle fatigue, decreased reflex responses, and paralytic ileus. A significant reduction of potassium in bodily fluids may lead to a potentially fatal condition called hypokalemia, which can result in cardiac arrhythmia, alkalosis, and respiratory paralysis.

2.1.5.3 Potassium toxicity

An elevation in serum potassium (>5.5 mEq/L) is called hyperkalemia. Hyperkalemia may occur in various disease states, in response to drugs that reduce excretion of potassium from the kidneys, or after sudden transcellular shifts of potassium from the intracellular to the extracellular fluid. Additionally, any condition or drug resulting in adrenal inhibition or decreasing aldosterone levels can lead to potassium retention. Hyperkalemia leads to muscle weakness and, in some cases, paralysis and cardiac arrhythmia.

2.1.6 Sodium

Sodium is an essential macronutrient of the body. Sodium cations are abundantly present in the extracellular fluid compartment. At equilibrium, their distribution in plasma is 5–6 times greater than in the interstitial fluid.

2.1.6.1 Sodium physiology

Sodium cations regulate blood volume and blood pressure, maintain fluid and electrolyte balance, facilitate the transmission of nerve impulses, and influence muscle contraction and relaxation. Sodium cations are responsible for the osmotic equilibrium between cells and the interstitial fluid through the Na+/K+-ATPase pump. They also maintain acid-base balance in the body.

Sodium ions bear an inverse relationship with potassium and hydrogen. Elevated sodium levels in the body lead to depletion of potassium and hydrogen (and vice versa). Aldosterone secreted in the adrenal glands is crucial for sodium ion homeostasis and is secreted in response to low blood volume or hypovolemia. Aldosterone secretion leads to sodium conservation in the renal tubules, which in turn leads to excretion of potassium and hydrogen, thus regulating blood pressure. Most of the sodium filtered in the kidney is reabsorbed, with only 1% excreted in urine. The daily intake of sodium approximately equals the daily excretion through urine.

2.1.6.2 Sodium deficiency

Low dietary intake can lead to sodium deficiency, which is also known as hyponatremia (plasma sodium <135 mEq/L). Extrarenal sodium loss may occur due to profuse sweating, burns, severe vomiting, or diarrhea. Other causes include adrenal cortex insufficiency or excessive secretion of the antidiuretic hormone by the posterior pituitary gland. Notably, the most common cause of hyponatremia is not a deficiency of total body sodium but an excess of total body water. Signs and symptoms, which vary according to the level of hyponatremia, include nausea, vomiting, visual disturbances, a depressed level of consciousness, agitation, confusion, coma, seizure, muscle cramps, weakness, and myoclonus.

2.1.6.3 Sodium toxicity

Hypernatremia is characterized by an increase in extracellular sodium concentration with or without a parallel presence of low, normal, or high total body sodium content. It usually occurs from excessive water loss, inadequate water intake, a lack of antidiuretic hormone (vasopressin), or excessive intake of sodium. It may also result from any systemic or renal disease that impairs tubular function.

2.1.7 Chromium

Chromium is an essential micronutrient that plays an important role in carbohydrate and lipid metabolism.

2.1.7.1 Chromium physiology

Chromium's biological action is closely associated with that of insulin, and most chromium-dependent reactions are also insulin dependent. Adequate chromium nutrition leads to a reduced requirement for insulin, improved sugar metabolism in patients with hypoglycemia or hyperglycemia, and an improved blood lipid profile. Chromium also improves lean body mass. Because inorganic chromium cannot potentiate insulin action, it must be converted to an organic and biologically active form.

2.1.7.2 Chromium deficiency

Chromium deficiency may present as glucose intolerance or cardiovascular disease. It is prominent among older people with type 2 diabetes and in infants with protein-calorie malnutrition.

2.1.7.3 Chromium toxicity

Acute chromium toxicity is uncommon given the low absorption and high excretion rates of chromium.

2.1.8 Copper

Copper as a trace mineral influences numerous biochemical processes. It is distributed in the skeletal muscles, brain, liver, and other parts of the body. It is usually associated with areas of high metabolic activity.

2.1.8.1 Copper physiology

Copper is essential for the smooth functioning of cellular enzymes. Because of their ability to adopt both redox states (oxidized/reduced), copper ions can function as catalytic cofactors in the redox chemistry of many cellular processes. Copper is needed for cardiovascular homeostasis. It helps in the formation of red blood cells and assists in keeping blood vessels, nerves, bones, and the immune system healthy. Dietary copper is also needed for many microvascular control mechanisms that affect inflammation, microhemostasis, and regulation of peripheral blood flow. Because free copper ions can damage cellular components, cellular uptake, distribution, and efflux of copper ions is coordinated by efficient regulatory mechanisms. Copper is also involved in melatonin formation.

2.1.8.2 Copper deficiency

Although uncommon, copper deficiency presents with symptoms such as fatigue, malaise, loss of appetite, anemia, damage to blood vessels, cardiomegaly, skin lesions, and neuropathy. Abnormal bone development, skin pigmentation, and hair pigmentation are other common symptoms.

2.1.8.3 Copper toxicity

When present in excess, copper may be harmful to the body. Wilson's disease is a rare inherited disease in which excess copper is deposited in the liver, brain, and other organs of the body, leading to hepatitis, brain disorders, and kidney disorders.

2.1.9 Manganese

Manganese is distributed throughout the body, with the liver, thyroid, pituitary, pancreas, kidneys, and bone having the highest concentrations. Manganese is located primarily in the mitochondria of these organs. An adult male has 12–20 mg of manganese (6).

2.1.9.1 Manganese physiology

Manganese plays an important role in normal amino acid, lipid, protein, and carbohydrate metabolism. It works as a cofactor or an activator of many enzymes (7). Manganese is responsible for regulating blood glucose, reproduction, digestion, bone growth, and removal of free radicals. In conjunction with vitamin K, it also supports blood clotting and hemostasis (6). Manganese is primarily absorbed from the small intestine.

Approximately 1–5% of ingested manganese is normally absorbed. Stable tissue levels are maintained through a tight homeostatic mechanism involving absorption and excretion. Tissue manganese levels are directly proportional to its concentration in the diet. Although manganese is present in all diets in low concentrations, vegetarian foods have a higher concentration. However, a vegetarian diet does not improve manganese status because meat enhances the bioavailability of manganese from diet.

2.1.9.2 Manganese deficiency

Manganese deficiency is rare. However, when present, it leads to impaired growth, poor bone formation and skeletal defects, reduced fertility and birth defects, abnormal glucose tolerance, and altered lipid and carbohydrate metabolism (8).

2.1.9.3 Manganese toxicity

Excessive exposure to manganese through respiratory and dietary routes, called manganism, leads to accumulation in the central nervous system. Toxicity is classified into three grades:

1. Mild toxicity leads to manganese psychosis and is characterized by asthenia, anorexia, insomnia, muscular pains, mental excitement, hallucinations, unaccountable laughter, impaired memory, and compulsive action.
2. Moderate toxicity is characterized by the presence of symptoms such as speech disorders, clumsy movements, abnormal gait, poor balance, hyperreflexia in the lower limbs, and fine tremors.
3. Severe toxicity resembles Parkinson disease, with rigidity, spasmodic laughter, and a mask-like face.

2.1.10   Selenium

Selenium is a trace mineral that is involved in many life-supporting biochemical processes. Cellular respiration, cellular utilization of oxygen, DNA and RNA reproduction, maintenance of cell membrane integrity, and destruction of free radicals all require selenium (9). Selenium is distributed throughout the body, with the kidneys and liver having the highest concentrations. Selenium concentration in the body varies according to regional soil concentration. An estimated 3–6 mg is present in individuals living in low-selenium areas, whereas approximately 13 mg is present in individuals who live in other regions.

2.1.10.1   Selenium physiology

Selenium exists as a constituent of selenoproteins, and it has a variety of structural and enzymatic roles. Selenium functions as an antioxidant and catalyst for the production of active thyroid hormone, is required for optimal functioning of the immune system, is essential for sperm motility, and may reduce the risk of miscarriage (10).

2.1.10.2   Selenium deficiency

Selenium deficiency is rare, often seen in areas where selenium concentration in soil is low. Research suggests that selenium deficiency does not usually cause illness by itself. Rather, it may make the body more susceptible to illnesses caused by other nutritional, biochemical or infectious stresses. It has been associated with Keshan disease, Kashin-Beck disease, and myxedematous endemic cretinism. There is evidence that selenium deficiency may contribute to development of a form of heart disease, hypothyroidism, and a weakened immune system (11).

2.1.10.3   Selenium toxicity

Selenium toxicity is rare and is usually associated with industrial accidents. High circulating levels of selenium (>100 µg/dL) can lead to a condition called selenosis, which is characterized by gastrointestinal distress, hair loss, white blotchy nails, garlic breath odor, fatigue, irritability, and mild nerve damage. Chronic exposure produces hypochromic anemia, leukopenia, irregular menses, and garlic breath or a metallic taste (12).

2.1.11   Sulfur

After calcium and phosphorus, sulfur is the most abundant macromineral in the body, based on percentage of total body weight. It is the sixth most abundant macromineral in breast milk (11).

2.1.11.1 Sulfur physiology

Sulfur amino acids are required for normal growth and maintenance of nitrogen balance. Sulfur is exclusively derived from proteins, which contain between 3% and 6% of sulfur amino acids (13). A very small percentage of sulfur is derived from inorganic sulfates; other forms of organic sulfur are present in food, such as garlic, onion, and broccoli.

Only 2 of the 20 amino acids normally present in proteins contain sulfur: methionine and cysteine. Methionine cannot be synthesized by the human body, whereas cysteine needs a steady supply of sulfur. Keratin (present in the skin, hair, and nails) derives its strength from sulfur-containing cysteine. Sulfur is essential for collagen synthesis and cellular respiration.

Sulfur is absorbed from the small intestine in the form of the two sulfur-containing amino acids or in the form of sulfites present in water, fruits, or vegetables. It is stored in the body's cells, especially in skin, hair, and nails. Excess sulfur is excreted through urine or feces.

2.1.11.2 Sulfur deficiency

Sulfur deficiency is rare. It occurs when the sulfur concentration in soil is low or when a person's diet is deprived of proteins. It is difficult to differentiate from protein deficiency.

2.1.11.3 Sulfur toxicity

Sulfur toxicity is rare.

2.1.12 Zinc

Zinc is a nutritionally essential trace element and an abundant element in the human body. Most of the body's zinc is concentrated in muscle and bone (85%) and skin and liver (11%), with the remaining amounts distributed in other tissues. The highest concentration of zinc is found in the choroid of the eye and the optic nerve.

2.1.12.1 Zinc physiology

Zinc is essential to the structure and function of a large number of macromolecules and is responsible for over 300 enzymatic reactions. It has both catalytic and structural roles in enzymes. Zinc plays a role in immune function, protein synthesis, wound healing, DNA synthesis, and cell division (14–16). Zinc also retards oxidative processes and is involved in the synthesis of RNA. It supports normal growth and development during pregnancy, childhood, and adolescence and is required for proper sense of taste and smell.

The zinc ion (Zn++) does not participate in redox reactions, thus imparting stability in a biological medium whose potential is in constant flux. Zinc ions are hydrophilic and do not cross cell membranes by passive diffusion. Zinc transport thus has both saturable and nonsaturable components, depending on the prevailing Zn(II) concentrations. Zinc ions exist primarily in the form of complexes with proteins and nucleic acids and participate in all aspects of intermediary metabolism, transmission, and regulation of the expression of genetic information, storage, synthesis, and action of peptide hormones and structural maintenance of chromatin and biomembranes (17).

2.1.12.2 Zinc deficiency

Zinc deficiency is quite common. It is the fifth leading risk factor for disease in the developing world. Approximately a quarter of the world's population is at risk of zinc deficiency (18). It is usually a nutritional deficiency. Severe cases of deficiency affect the epidermal, gastrointestinal, central nervous, immune, skeletal, and reproductive systems. Acrodermatitis enteropathica (a genetic disorder in children) is a serious manifestation of zinc deficiency. Other signs include hair loss, skin lesions, diarrhea, and wasting of body tissues.

2.1.12.3 Zinc toxicity

Patients with acute zinc toxicity can present with nausea, vomiting, loss of appetite, abdominal cramps, diarrhea, and headaches. The long-term intake of 150–450 mg of zinc per day has been associated with chronic effects, such as low copper status, altered iron function, reduced immune function, and reduced levels of high-density lipoproteins (19).

Table 2.2 Possible Molecular Biological Functions of Metals and Minerals

Elements	Function
Calcium	Cell membrane potential maintenance
Iron	Transport medium
Magnesium	Energy metabolism; RNA and DNA synthesis
Phosphorus	Repair of cells and cell signaling
Potassium	Osmotic pressure homeostasis at cell membrane
Sodium	Fluid and electrolyte balance
Chromium	Carbohydrate and lipid metabolism
Copper	Redox reactions in cell processes
Manganese	Activation of enzymes as cofactor
Selenium	Enzymatic reactions through proteins
Sulfur	Protein formation
Zinc	Immune function and enzymatic reactions

2.2 Potential functional modes of metals and minerals

The various biological action pathways of macroelements, microelements, and trace elements can be postulated from their physiological descriptions (**Table 2.2**).

2.3 Conclusion

Metal and mineral nutrients are instrumental in facilitating biological reactions in the body; hence, they are frequently referred to as *biocatalysts*. They can serve as facilitators in the conduction of electrical impulses in the body due to their presence in the nerve synaptic spaces. Metal and mineral nutrients help to maintain the blood pH and transport of other nutrients within the body. Their presence and participation are essential for the proper assimilation of vitamins, food, and drugs. In addition, they are also involved in the cellular transport mechanism, wherein they regulate the entry and detoxification of chemicals from the cells. In this respect, many other unexplored metals and minerals can be evaluated by considering their ethnic usage and manufacturing techniques coupled with modern quality assurance methods. Ancient medical texts, such as *Ayurved* (Science of Life) and *Ras-Shastra* (Indian Iatrochemistry and Alchemy), describe the pharmaceutical engineering and fundamental concepts of elements and how they can be used in nutrition. These fundamental concepts can be further used to devise brand-new nutritional formulations and usages.

References

1. Larsson, S.C., M.J. Virtanen, M. Mars, et al. 2008. Magnesium, calcium, potassium, and sodium intakes and risk of stroke in male smokers. *Archives of Internal Medicine* 168:459–465.
2. Shaikh, A., T. Berndt, and R. Kumar. 2008. Regulation of phosphate homeostasis by the phosphatonins and other novel mediators. *Pediatric Nephrology* 23:1203–1210.
3. Knochel, J.P. 1977. The pathophysiology and clinical characteristics of severe hypophosphatemia. *Archive of Internal Medicine* 137:203–220.
4. Knochel J.P., C. Barcenas, J.R. Cotton, T.J. Fuller, R. Haller, and N.W. Carter. 1978. Hypophosphatemia and rhabdomyolysis. *Transactions of the Association of American Physicians* 91:156–168.
5. Chang, R. 2007. *Chemistry.* New York: McGraw-Hill.
6. Aschner, J.L., and M. Aschner. 2005. Nutritional aspects of manganese homeostasis. *Molecular Aspects of Medicine* 26:353–362.
7. Marschner, H. 1995. *Mineral nutrition of higher plants,* 2nd ed. Cambridge, UK: Academic Press.
8. Freeland-Graves, J., and C. Llanes. 1994. "Models to study manganese deficiency" (pp. 59–86), in Klimis-Tavantzis, D.L., ed., *Manganese in health and disease.* Boca Raton, FL: CRC Press.
9. Chan, S., B. Gerson, and S. Subramaniam. 1998. The role of copper, molybdenum, selenium, and zinc in nutrition and health. *Clinics in Laboratory Medicine* 18:673–685.
10. Rayman, M.P. 2000. The importance of selenium to human health. *Lancet* 356:233–241.
11. Parcell, S. 2002. Sulfur in human nutrition and applications in medicine. *Alternative Medicine Review* 7:22–44.

12. Koller, L.D., and J.H. Exon. 1986. The two faces of selenium-deficiency and toxicity are similar in animals and man. *Canadian Journal of Veterinary Research* 50:297–306.
13. Nimni, M.E., B. Han, and F. Cordoba. 2007. Are we getting enough sulfur in our diet? *Nutrition and Metabolism* 4:24.
14. Solomons N.W. 1998. Mild human zinc deficiency produces an imbalance between cell-mediated and humoral immunity. *Nutrition Reviews* 56:27–28.
15. Prasad, A.S. 1995. Zinc: An overview. *Nutrition* 11 (1 Suppl): 93–99.
16. Heyneman, C.A. 1996. Zinc deficiency and taste disorders. *The Annals of Pharmacotherapy* 30:186–187.
17. Tapiero, H., and K.D. Tew. 2003. Trace elements in human physiology and pathology: zinc and metallothioneins. *Biomedicine and Pharmacotherapy* 57:399–411.
18. Maret, W., and H.H. Sandstead. 2006. Zinc requirements and the risks and benefits of zinc supplementation. *Journal of Trace Elements in Medicine and Biology* 20:3–18.
19. Institute of Medicine, Food and Nutrition Board. 2001. *Dietary reference intakes for vitamin A, vitamin K, arsenic, boron, chromium, copper, iodine, iron, manganese, molybdenum, nickel, silicon, vanadium, and zinc.* Washington, DC: National Academies Press.

Ethnopharmacology and Ethnomedicine of Metals

Jayant Lokhande, Shriniwas Samant,
Ganesh Shinde, and Yashwant Pathak

Contents

3.1　Metals in the ancient era

3.1.1　Definition and classification of metals and their basis

Metals were named and used differently in Vedic era (around 5000 B.C.). *Yajurveda* and some of the most ancient medicinal scriptures, such as *Caraka Samhita* and *Sushruta Samhita* (3000 B.C.) described metals in the form of *rucha* (chanting prayers) and *sutra* (a concise form of expression), such as in the following:

　　Hiranyamchame shyammachame lohamchame
　　Sisamchame trapuchamchame yadnyen kalpantam (1)
　　Swarna rupya tamrani trapusisamayani cha (2)
　　Trapusisatamrarajatkrushnalohsuvarnani lohbhasmachyeti (3)

Each of these four verses mentions metals, such as *hiranya* (gold), *shyam* (silver), *loh* (iron), *sisam* (lead), and *trapu* (tin).

The Sanskrit name for metals is *loh*. The term *loh* is derived from the etymological roots *luh* and *karsane*—meaning materials that are obtained by extraction from their respective ores/minerals. The other synonym for metals in Sanskrit is *dhatu*, which means *deha dharanat* (sustenance of body tissues). *Dhatu* provide nourishment and strength to body tissues; thus, the body is sustained in its natural healthy form.

Six metals were known and employed in ancient internal medicine until the twelfth century A.D.: *suvarna* (gold), *rajata* (silver), *tamra* (copper), *lauha* (iron), *naga* (lead), and *vanga* (tin). These metals were classified into three groups by their applications in medicine:

1. *Sara loha* (excellent, with regeneration as a basic biological action): gold and silver
2. *Suddha loha* (pure, with modulation as a basic biological action): copper and iron
3. *Puti loha* (cleansing, with translation as a basic biological action): lead and tin

In the twelfth century A.D., three more metals were added into the domain and classified into a fourth group—*Misra loha* (mixed, with optimization as a basic biological action): *yashada* (zinc), *pittala* (brass), *kamsya* (bell metal), and *varta* (an alloy made of gold, silver, copper, iron, and tin) (4).

The mythological origin of the six metals were said to be from different Hindu deities (**Table 3.1**). The nine metals were also believed to be related to the nine planets (**Table 3.2**).

3.1.2 Alchemy: the conversion of natural mine metals into bioacceptable metals

3.1.2.1 *Alchemy concepts of* rasa shastra *(Indian iatrochemistry)*

The primordial alchemical concept of *rasa shastra* (Indian iatrochemistry) was *lohavedha* (metallic transformation)—that is, to transform lower or base metals into noble and/or higher metals—and their subsequent use in creating strong and healthy body tissues by which one can eliminate poverty, senility, diseases, and death.

In the medieval period, *rasa shastra* evolved and advanced in the well-known alchemy science to accomplish both *lohavedha* and *dehavedha* (transformation of body tissues). In order to accomplish *dehavedha*, *Rasa shastra* placed emphasis on the *rasayana* (anti-aging) concept of *Ayurved* (science of life).

Table 3.1 Metals, Mythology, and Biological Analogies

Metal	Mythological God of Origin	Biological Function of the God
Gold	*Vishnu*	Anabolism
Silver	*Siva*	Metabolism
Copper	*Sun*	Catabolism
Tin	*Indra*	Nuclear function
Lead	*Vasuki*	Energy provision
Iron	*Yama*	Apoptosis

Table 3.2 Metals, Mythology, and Therapeutic Analogies

Metal	Correlated Planet	Physiological and Medical Analogy of the Planet
Suvarna (gold)	Guru (Jupiter)	Bone disorders, connective tissue disorders, cell regeneration
Rajata (silver)	Candra (moon)	Respiratory, immunology and lymphatic system disorders, protein metabolism, hormonal control
Tamra (copper)	Surya (sun)	Blood and reticuloendothelial system disorders, cardiovascular system disorders, glucose metabolism
Lauha (iron)	Shani (Saturn)	Nervous system disorders, cell growth regulator
Naga (lead)	Budha (Mercury)	Fat metabolism, adipose tissue disorders, enzyme control
Vanga (tin)	Shukra (Venus)	Reproductive system disorders, tissue formation
Pittala (brass)	Mangala (Mars)	Musculoskeletal system disorders, catabolism regulation
Kamsya (bell metal)	Rahu (dragon's head)	Infection disorders, chronic metabolic disorders, psychiatric disorders, genetic disorders, positive feedback system
Varta lauha (alloy of five metals)	Ketu (dragon's tail)	Acute metabolic disorders, acute infectious diseases, negative feedback system

With such practice, one can achieve the best quality of *rasadi dhatus* (body tissues) to prevent diseases and the aging process, as well as to transform the body into a healthy and youthful state.

The origin of *rasayana* therapy can be traced to the B.C. era. Ancient scholars such as Nagarjuna, Nandi, Somadeva, Rasa Vagbhata, and Dhudhuka Natha formulated highly effective and stable *rasayana* drugs and helped this therapy to reach its zenith. In this period, they developed a number of valuable *rasa* (mercury) preparations exhibiting the best *rasayana* properties by employing numerous alchemical and pharmaceutical techniques with the help of metals/minerals. The alchemical and pharmaceutical processes employed are shown in **Figure 3.1**.

3.1.2.1.1 Satvapatana (extraction of metal from the core mineral complex) The term *satvapatana* comes from *satva* (pure essence) and *patana* (application of energy for purity), meaning the extraction and purification of metals from its core mineral complex.

This procedure occurs as follows:

1. Ore minerals are mixed and ground thoroughly with fusion materials, such as *amlas* (acids), *ksharas* (alkalis), and certain organic materials, such as *tankana* (borax).
2. This mineral mixture is altered into small pellets, which are put in *musha* (heat-resistant pots) for drying.
3. The pellets are placed in *kosthi yantras* (furnaces), connected with air blowers. This system facilitates steady and gradually increasing

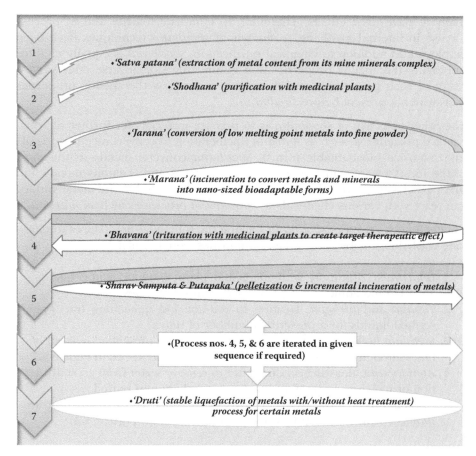

Figure 3.1 *(See color insert.)* *Alchemical steps for metals.*

heat, so that the mineral mixture melts and is maintained in the same state for 1.5–2 hours, or until the metal content of the mineral is separated and accumulates at one focal point in the *musha*. These mineral transformation phases are called *bijavarta* (seed of metal) and *shud-dhavarta* (pure phase of metal), which are the melting and boiling states of molten metal, respectively.

4. After separation, *satva* (the pure metal content of mineral) is achieved. The duration of heat and heating temperature are optimized based on the metal's nature. The maintenance of this optimum temperature is vitally important for extracting the *satva* from its mineral complex.

3.1.2.1.2 The shodhana *(purification) process* The *shodhana* (purification) concept is fundamental to *ayurvidya* (the knowledge of life) and is an essential key for *gunantar adhana* (the alteration of properties and induction of extra properties in the substances). Generally, the metals and/or minerals are likely to have toxic effects. Therefore, to eliminate toxicity and to process for use in internal medicine, a number of *shodhana* techniques have been employed. In addition to chemical purification, *shodhana* is equally vital for improving therapeutic properties and minimizing side/adverse effects. As a result of such procedures, metals acquire many new therapeutic properties that were not present before *shodhana*.

The other objective of *shodhana* is to prepare materials for further pharmaceutical procedures, such as *marana* (incineration to convert metals/minerals into nanosize bioadaptable forms). *Shodhana* converts metals in the most brittle forms by inducing cracks on their surfaces and fractions metals into fine particles to expose the maximum surface area, so that the metal surface can react with various chemicals in subsequent processes such as *marana*.

The techniques used in *shodhana* include the following procedures to eliminate washable, soluble, and volatile impurities from the metals:

1. *Bhavana:* Trituration with medicinal plant extracts, acidic liquids, and alkaline liquids for a prescribed number of times
2. *Tapana* and *nirvapa:* Heating to red-hot and quenching into prescribed liquids for a prescribed number of times
3. *Svedana:* Boiling in acidic or alkaline liquids for a prescribed number of times through *dola yantra* (distillation apparatus), as shown in **Figure 3.2**
4. *Agni tapana:* Roasting on direct fire to remove water from crystallization or moisture content and make metals light and puffed
5. *Urdhvapatana:* Upward and transverse sublimation or distillation in *damaru yantra* (a distillation apparatus, shown in **Figure 3.3**)

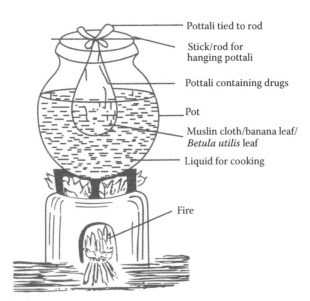

Pottali tied to rod

Stick/rod for hanging pottali

Pottali containing drugs

Pot

Muslin cloth/banana leaf/ *Betula utilis* leaf

Liquid for cooking

Fire

Figure 3.2 Dola yantra.

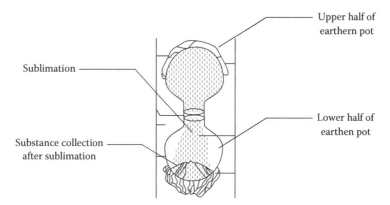

Figure 3.3 *Damaru yantra.*

6. *Kshalana*: Washing with hot water or *kanjika* (fermented rice water)
7. *Bharjana:* Frying in *ghrut* (clarified butter of nonhybrid brown-colored cow's milk)
8. *Sneha nishkasana:* De-oiling or absorption of oily content

There are two types of *shodhana* processes: *samanya shodhana* (common purification) and *vishesha shodhana* (specific purification). *Samanya shodhana* is commonly applicable for certain metal groups. Metals such as silver, copper, and iron are first converted into ultrafine sheets. These sheets are heated red-hot on fire and quenched in each of the following liquids seven times:

1. *Tila taila* (sesame oil)
2. *Takra* (buttermilk)
3. *Gomutra* (cow urine)
4. *Kaanjika/aranala* (fermented rice water)
5. *Kulattha kashaya* (horse gram decoction)

Note that this *samanya shodhana* method is not applicable for gold.

Metals in the *putiloha* group (i.e., lead, tin, and zinc), which have low melting points, are first melted on fire and then poured into the above-mentioned five liquids through a *pithara yantra* (a stone plate with a hole in the middle) seven times for each of liquid; the resulting molten and powdered metals are collected for subsequent processes.

The *vishesha shodhana* purification process is specific for a particular metal and is used to induce a new therapeutic property. In *vishehsa shodhana*, the same procedure as in the *samanya shodhana* method is applied, but the liquids are changed. For example, to improve the color of gold, as well as its shine and immunomodulating property, *panchamrittika* (five clays: brick clay, ash, ant hill earth, red ochre, salt) paste is applied on gold sheets and heated in a *laghu puta* (an incinerator with a maximum temperature of 300°C).

3.1.2.1.3 Jarana *(conversion of molten metal into an ultrafine powder form)*   The Sanskrit word *jarana* means "to make grow" or "long-standing." *Jarana* (the conversion of molten metal into an ultrafine powder form) is an intermediary and essential application in between the *shodhana* (purification) and *marana* (incineration) processes. It is used for metals with low melting points and vapor pressures, such as those in the *puti loha* (cleansing, translator) group, including lead and tin.

In this procedure, these metals are first melted on fire in an iron pan. Then, certain medicinal plants are added, such as *apamarga panchanga* (*Achyranthes aspera* whole plant) or *ashwattha twak* (*Ficus religiosa* bark skin) powder. The plant material is of equal weight to metals and burnt in combination by rubbing until the whole metal mixture converts into black powder. If necessary, more medicinal plant powder is added until complete conversion of metal mixture into powder form occurs. Then, the metal mixture with the medicinal plant powder is aggregated at the center of an iron pan and covered with an earthen lid. Strong heat is applied again for 2–3 hours to complete the redox reaction. After *swang sheetam* (self-cooling), the ash is sieved to separate the metal particles (if any). This powder is used in the *marana* process (see next section).

The objective of this procedure is to create eligibility and sustainability of metals with low melting points in their respective subsequent alchemical process in order to acquire bioadaptability in their nanosize forms.

3.1.2.1.4 Marana *process (incineration to convert metals/minerals into nanosized bioadaptable forms)*   *Marana* is a Sanskrit word meaning "to kill and convert into another form." This is the most important consecutive pharmaceutical process applied to drugs of mineral origin, converting them into a fine-ash *bhasma* (bioadaptable nanosized) form. The *marana* process is shown in **Figure 3.4**.

To achieve this form, the mineral-origin drugs are first mixed with *marana* substances homogeneously. The mixture is subjected to *bhavana* (trituration) with specific medicinal plants extracts and acidic or alkaline liquids. On drying, *putapaka* (transformation by energy) is applied. These two steps—*bhavana* and *putapaka*—are repeated for a specific number of times or until the metal/mineral is "killed" and converted into the desired compound form.

Rasa shastra (Indian iatrochemistry) scholars have categorized *marana* substances into several different categories. The first category of *marana* drugs are *rasa* (purified mercury) compounds, such as *kajjali* (purified mercury + purified sulfur), *hingula* (purified cinnabar), and *rasa sindura* (purified mercury + sulfur + alum + ammonium chloride). These are considered to be the best compounds because they aid in the disintegration of metal/mineral particles very quickly. *Shodhit* (purified mercury) forms *pishti* (amalgam) with any metal, and metal loses its own form while performing *bhavana* (trituration)

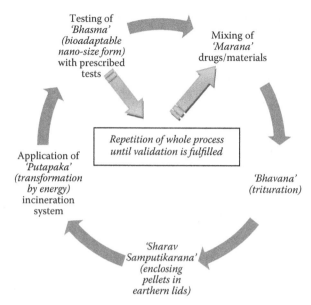

Testing of
'Bhasma'
(bioadaptable
nano-size form)
with prescribed
tests

Mixing of
'Marana'
drugs/materials

Repetition of whole process
until validation is fulfilled

Application of
'Putapaka'
(transformation
by energy)
incineration
system

'Bhavana'
(trituration)

'Sharav
Samputikarana'
(enclosing
pellets in
earthern lids)

Figure 3.4 *Process pathway for making bioadaptable nanometals.*

with certain medicinal plant extracts. On drying, the mixture is made into *chakrika* (pellets) and subjected to the *putapaka* procedure (transformation by energy). After completion of the process, mercury or mercurial compounds may not be traced as present in any form. Various methods of *marana* are used for single metals to create a targeted therapeutic effect with the help of different herbs.

The second-best category of *marana* drugs includes medicinal plants in the form of *amla* (acidic) or *kshar* (alkaline) *kwath* (decoctions). The *shodhit* (purified) metals are *bhavit* (triturated) with medicinal plant decoctions and dry pellets are made. After incineration, inorganic material of media exists in the form of traced elements.

Bhavana (trituration) is used to create desirable effects in metals. It is performed for the prescribed number of times as per the given metal for the following purposes:

- *Bhavana* helps in the proper mixing of the *marana* drugs.
- *Bhavana* helps to make the metal/mineral particles finer and exposes the unreacted metal or mineral particles for continuing the chemical reaction in *putapaka* (transformation by energy).
- *Bhavana* helps in organic coating. When heat is applied, different kinds of physicochemical reactions are formed. It imparts specific color and therapeutic effects to the *bhasma* (bioadaptable nanosized form). By changing the *bhavana* drugs/extracts, the *bhasma* color and properties are changed accordingly.

- Some *bhavana* extracts also help to remove the toxic effects of metals/minerals.
- *Bhavana* also helps to change the metals/minerals into organometallic compounds, which in due course help to make the *bhasma* product more and more acceptable by the body tissues and organs.

The *bhavit* (triturated) semisolid mass mixture is translated into a thin, small, and flat *chakrika* (circular pellet) and dried in sunlight. The purpose behind making pellets and enclosing in lids is to expose the maximum surface of the metal/mineral and to impart gradual incineration. It is then enclosed in *sharav samputa* (an assembly made of two earthen lids, as shown in **Figure 3.5**) and sealed properly. This helps in protecting against material process loss from heat sublimation during *putapaka* (transformation by energy) and keeps it safe after *putapaka* is completed.

Puta is a term used to represent the quantum of heat required in the *marana* material according to heat tolerance capacity. Not every metal or mineral can melt and transform at the same temperature. *Rasa shastra* scholars have described different kinds of *puta* according to temperature generation (**Figure 3.6**).

Important points to determine for the *putapaka* process include the following:

1. Which type of *puta* is suitable for which metal/mineral for its *bhasma* (bioadaptable nanosized form) phase?
2. What is the quality and mode of heat?
3. How many times should it be iterated?

In this process, grinding during *bhavana* also plays a very important role. If grinding is done properly, then fewer repetition cycles of *bhavana* and *putapaka* are needed to accomplish *bhasma*.

For gold, silver, and lead, *kukkuta* (hen-size scale) heat needs to be applied first at 350°C for the first few hours, followed by *kapota* (pigeon-size scale) heat at 300°C for the last half hour. Within this range of temperature, the

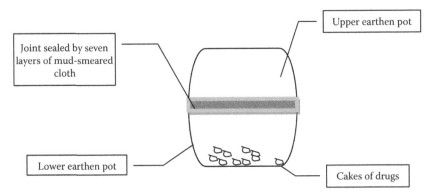

Figure 3.5 Sharav samputa.

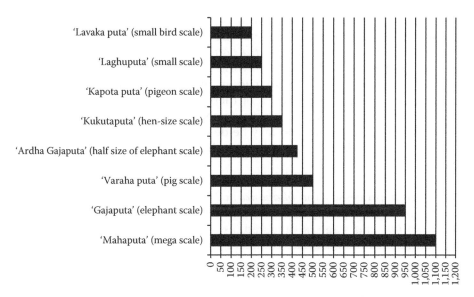

Figure 3.6 *Thermal temperature of ancient furnace systems (in °C).*

bhasma of gold, silver, and lead can be prepared. Other metals need higher temperatures for their conversion into *bhasma*. These metals require *gaja-puta* (elephant-size scale) heat at 950°C for a shorter duration (i.e., approximately 15–20 minutes), and then the heat is tapered down.

Bhasma can be validated using a few different techniques:

1. *Varitara* (water buoyancy): Despite the higher specific gravity of metal, after acquiring true *bhasma,* it forms a uniform colloidal suspension in water. Moreover, if higher gravity particles are added, it still does not sink due to unreeling of the water surface tension.
2. *Rekha purnatva* (microfurrows filling): The particles are microns in size, so that they easily fit in human palm furrows.
3. *Lochananjana sannibha* (application on eyelids, like collyrium): If particles are put on eyelids, they should not generate any kind of hypersensitivity reaction from the ultrasensitive mucus membrane of the eye.
4. *Ishtavarnattwa* (specific respective color): *Bhasma* should develop a specific color because it depends on the formation of a required kind of chemical compound, such as sulfide, oxide, sulfate, or carbonate. Until the required color of *bhasma* is achieved, its preparation process is not considered to be complete.
5. *Gatarasatva* (amicable to the senses): A *bhasma* should not have any taste or odor.
6. *Mrudutwa* (soft): A good *bhasma* should be soft to the touch.

7. *Niruttha* (nonrising) or *apunarbhava* (nonreversible): *Bhasma* should not return to its original metallic form, even if the smallest amount is heated with *mitra pancaka* (group of five friends)— *ghrita* (clarified butter of a nonhybrid cow), *madhu* (honey), *guggulu* (*Commiphora* guggulu gum), *Gunja* (*Abrus precatorius* seeds), and *tankana* (borax). Irrespective of their organic and alkaline nature, these drugs are melted at a lower temperature than usual, which on cooling becomes hard and may be detected in the *bhasma* samples. Therefore, if *bhasma* is melted with them, it should not be found in traces.

The *marana* processes for *bhavana* and *putapaka* are repeated several times, until a good quality of the *rasa abhibhava* (bioadaptable subjugated powering) *bhasma* stage is validated, as described previously. Within one *bhavana* and *putapaka* cycle, complete conversion of metal particles into the desired bioadaptable compound is not possible. Within one *bhavana* and heat treatment, the metal particle starts to convert into the desired compound form; a layer of that compound invades the metal particle and does not allow the reaction to go deep into the unreacted metal particle. Therefore, further exposure of the metal particles is needed. By repetition of *bhavana,* grinding and trituration is performed, which removes the upper layer of the formed compound and exposes the metal particle surface for further reaction by heat treatment. Thus, by repetition of the whole process, the unreacted metal particle surface becomes exposed to have subsequent reactions.

After completion of the *marana* process, *bhasma* develops the following qualities:

1. It becomes curative, such that it can be used in internal medicine to cure even severe diseases.
2. It is suitable for long-term use because it is now nontoxic for humans, even after long-term continuous use.
3. It becomes *rasayana* (anti-aging) to create a healthy, disease-free body.

3.1.2.1.5 Druti (stable liquefied state of metals/minerals) *Druti* is a unique ancient Indian alchemical process in which a metal/mineral is converted into a stable colloidal liquid form. The therapeutic properties of metals and minerals are multifold compared to *bhasma* (bioadaptable nanosized form). *Druti* is different from the melted state of a metal or mineral.

Ancient alchemists performed two types of *druti: garbha druti* (internal digestion) and *bahya druti* (external digestion) in the context of 18 *samskaras* (transformations) in the alchemical purification of mercury. *Garbha druti* requires *bija* (seed of metal) to prepare liquefied metals. For this process, there are two types of *bida* (catalytic agents), general and specific, as described in the following.

General *bida* is prepared by *sauvarcala* (salt), *sunthi* (*Zingiber officinale Rosc.*), *pippali* (*Piper longum Linn.*), *maricha* (*Piper nigrum Linn.*), *sphatika* (alum), *kasisa* (from sulfate), and *gandhaka* (sulfur). All these substances are taken in equal quantities to be impregnated with the leaf and fruit pulp juice of *shobhanjana* (*Moringa oleifera Lam.*) for 100 times. By more impregnation and trituration, the *bida* becomes more and more effective. The *bida* is then added to mercury in one-eighth of its quantity. The *bida* is to be kept below and above mercury for processing.

Specific *bida* are different for every metal. For example, for *garbha druti* (internal digestion) of gold, a specific type of *bida* is needed. This *bida* is prepared by the *kshara* (alkali) of *mulaka* (*Raphanus sativus Linn.*) by adding cow urine. Gold is impregnated and triturated by adding this alkali preparation 100 times. This *bida* aids in the immediate *druti* (digestion) of gold for *jarana* (assimilation) in mercury. Mercury is mixed with equal quantities of sulfur, and *kajjali* (black powder of collyrium) is prepared. Into this, a one-eighth quantity of *bida* is added. Cow urine should be added again and triturated until it becomes a paste. This paste is smeared inside a crucible *musha* (heat-resistant pot) that is specially prepared for this purpose (**Figure 3.7**).

In the bottom of an earthen pot, *haritala* (orpiment), *manah shila* (realgar), or *gandhaka* (sulfur) is kept in powder form in the appropriate quantity. Over it, the *musha*, smeared with the paste of mercury, etc., is inverted (i.e., placed face down). The earthen pot is placed over an oven. Cow dung should be used as fuel in the oven. As the earthen pot is heated up, the fumes of *manha shila* come up. A part of the fume will go into the *musha*. Very mild heat should be applied until all *manha shila* is burnt. It takes about 6–12 hours to obtain stable liquefied metal.

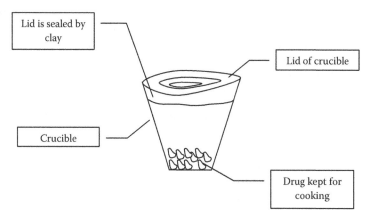

Figure 3.7 Musha.

In *bahya druti* (external digestion), the metal is reduced to liquid form outside and then added to mercury. The fruit of *kapi tinduka* (*Feronia limonia Swingle*) is taken in powder form and has to be impregnated and triturated with goat urine 100 times. After each cycle of impregnating and triturating, the paste should be dried by sunlight. This powder should be added to the *abhraka* (mica) *bhasma*, kept in a *musha*, and heated over fire. After this, the metal should be poured over a pot containing *til tailam* (*Sesamum indicum*) oil. This process has to be repeated three times to get the metal in the appropriate liquid form.

For *bahya druti* (external digestion) of gold, the powder of *deva dali* (*Luffa echinata Roxb.*) should be impregnated with its juice 100 times. After every impregnation and trituration, it should be dried in the shade.

1. Gold should be heated until it melts. In this, the powder of *deva dali* should be added.
2. This liquid metal should be added to mercury, and the powder of *krishna aguru* (*Aquilaria agallocha Roxb.*), *kasturi* (musk), sugar, garlic, the white variety of asafetida, sulfur, and powder of the seed of *palasha* (*Butea monosperma Kuntze*) should be added.
3. This whole mixture should be triturated in *tapta khalva* (hot mortar and pestle) made up of iron.

By this method, the gold remains in liquid form. This process is called *dvandva melapana* (combination of two items).

Some scholars did not follow the rules regarding the quantity of metal to be added to mercury as described in *Grasa Mana Samskara*. They added mercury in the liquefied metal in equal quantity only. However, it was found that it is better to add the liquid metal in gradually increasing quantities (e.g., 16 times, 32 times, or even 64 times of mercury). This makes the mercury progressively more and more potent. Mercury that has digested the liquid metal for 64 times becomes most potent.

In *jarana samskara* (assimilation), the mercury does not change in weight, despite the addition of other metals. This is the actual test for the successful completion of the procedure. *Jarana* is of two types: *bala jarana* (tender) and *vruddha jarana* (mature). *Bala jarana* is performed to make mercury suitable for *deha siddhi* (rejuvenation) in curing diseases. *Vrddha jarana* is for both *deha siddhi* and *lauha siddhi* (alchemy).

Ancient scholars have also identified the following five *lakshanas* (specific signs) of *druti*:

1. *Nirlepatvam* (nonsticking)
2. *Drutatvam* (liquefaction)
3. *Tejastvam* (shining or brightness)
4. *Laghuta* (lightness)
5. *Drutam yogasya sutena* (quick mixing with mercury)

3.1.3 Metals and their respective alchemy processes

3.1.3.1 Suvarna (gold)

The varieties of gold, according to *Ayurved*, are as follows:

Prakrutam Sahajam Vanhisambhutam Khanisambhavam | Rasend-
ravedhsanjatam Swarnam panchavidham smrutam (5)

1. *Prakrita*, which originates from *rajoguna* and covers the whole *brah-
   manda* (universe). It is not available even to *deva* (god).
2. *Sahaja*, which is native gold obtained from Mount Sumeru.
3. *Bahni sambhuta*, which is said to have originated from the god *Agni*
   and is claimed to make the body *ajara* (free from senility) and *amara*
   (free from death) by mere *dharana* (bearing).
4. *Khanija*, which is from mines.
5. *Rasa* or *rasendra vedhaja*, which is obtained by mercurial alchemical
   transformation and is considered to be *rasayana* (anti-aging), *pavitra*
   (auspicious), and *srestha* (best).

In modern metallurgy, gold is obtained from either native gold, minerals/ores
(i.e., mineral gold), or transformation.

The metallurgical properties of gold are as follows:

1. Hardness (Vickers): 25
2. Density: 19.32 g/cc
3. Melting point: 1062°C
4. Color: gray to yellow
5. Solubility: Insoluble in water and dissolves in nitro-hydrochloric acid

Gold is malleable and ductile, and it has a bright metallic luster. In nature, it
is found in its free state as well as in compound form. It is frequently alloyed
with silver and often contains traces of iron and copper. Pyrite is nearly always
associated with gold. Its streak is yellow. The correlation between modern
metallurgical and ancient alchemical processes is described in **Table 3.3**.

According to *rasa shastra*, the qualities of gold should be as follows:

Ghrushtam varnedhusrunsadrusham raktavarnamcha dahe chhede
kinchit sitkapilam nirdalam bhuribharam
Snigdham swarnam ravivirahitam styanraktapramandhyam shreshtham
dwishtamtulitalasachharuvarnamcha swarnam (6)

Regarding the alchemical processes of gold per *rasa shastra*, the gold *shodhana* (purification) procedure is described as follows:

Balmikmruttika dhumam gairikam cheshtikapatuhu | ityetaha mruttika
pancha jambirairarnalakai

Table 3.3 Modern Metallurgy Applications of *Rasa Shastra* (Indian Iatrochemistry)

Metallurgy Parameter	Determination Factor
Hardness	How many *bhavana* (triturations) will be required for metal?
Density	What will be the pattern of *marana* (incineration to convert metals/minerals into nanosized bioadaptable forms) for metal?
Melting point	How many *putapaka* (transformation by energy) by *puta* will be required for metal?
Color	How can the *bhasma* (bioadaptable nanosized form) of metal be differentiated from its mineral origin?

Pishtava kantakvedhyni swarnapatrani lepyet | putetapruthuhasantyam tu nirvate trinshadutpalaihe (7)

1. Prepare the five clay pastes (brick clay, ash, ant hill earth, red ochre, salt) by mixing each of the materials with lemon juice in an iron mortar.
2. This paste is applied on thin gold sheets and these sheets are kept in *sharav samputa* (an assembly made of two earthen lids), sealed with a piece of cloth smeared in mud.
3. On drying, the mixture is put in *laghu puta*, as explained in *putapaka*.
4. This whole process is replicated three times.
5. At the end of *swang sheet* (self-cooling), the *sharav samputa* is opened and the gold sheets are cleaned with a piece of cloth to shine.

For the gold *marana* procedure (incineration to convert metals/minerals into a nanosized bioadaptable form), the *makaradhwaja* preparation is followed. In this method, the following substances are required:

1. *Shodhit* (purified) gold sheets: 1 part
2. Purified *asta samskarita parada* (8× alchemized mercury): 8 parts
3. Purified *gandhaka* (sulfur): 16 parts
4. Lemon juice (as required)

The procedure is described as follows:

1. *Shodhit* (purified) gold sheets are cut into very small pieces, mixed with *ashta samskarita parada* (mercury) in an iron mortar, and triturated with lemon juice until a good-quality *pishti* (amalgam) of gold is formed.
2. *Suddha gandhaka* (purified sulfur) powder, which is twice the purified mercury in weight, is added and grounded carefully until good-quality *kajjali* (purified mercury + purified sulfur) paste is made.
3. *Kajjali* is filled in a *kupi* (a glass bottle covered with seven layers of mud cloth) up to one-third of its volume only, keeping two-thirds of the bottle empty. The *kupi* is then heated in a *baluka yantra* (sand

furnace) in the *kramagni paka* (gradual escalation of heat) mode. The heating starts first by *mrudu agni* (soft fire; 150–200°C) until the *kajjali* melts. Then, the heat is raised gradually to *madhyamagni* (intermediate fire; 350–400°C). The profuse fuming and boiling of *kajjali* takes place at this stage. During this stage, profuse fumes come out of the bottle mouth. The neck of the *kupipakwa* bottle must be cleared from the blockage of purified sulfur fumes by frequently inserting a red-hot iron rod. The melted *kajjali* inside the bottle also starts boiling; hence, care must be taken to prevent it from coming out of the bottle. The temperature of the furnace should not be allowed to rise. This stage may be allowed to continue for a long time (i.e., 6–8 hours) to allow the extra purified sulfur that is present in the compound to burn to its maximum.

4. Now, the temperature of the furnace is raised further to the *tivrargni* (strong fire) stage (550–600°C) to allow the extra purified sulfur that is present in the compound to burn quickly. This stage is known as the flaming stage, in which flames start to appear in the *kupipakwa* and rise as high as 2–3 inches, depending upon the quantity of purified sulfur in the starting mixture. If the sulfur quantity is at least double the purified mercury in weight, then its height and continuance time will be greater (and vice versa). After some time, its height is reduced and disappears slowly from the *kupipakwa* mouth, which indicates that the extra purified sulfur is almost burnt. One should observe the bottle for other signs, such as redness at the bottom of the bottle.

5. In *shitashalaka pariksha* (cold-needle test), a cold iron rod is inserted in the *kupipakwa* to its bottom and removed. If the rod comes out completely clean, there is no extra sulfur in the *kupipakwa* and it can be corked. However, if some burnt and melted purified sulfur adheres to the *shitashalaka*, then one should wait for some time to allow the extra purified sulfur to burn. Allow the same heat temperature to continue until the *shitashalaka* test is positive.

6. A delay in corking is also not desired because the compound formed at the bottom of the *kupipakwa* starts to disassociate and freed purified sulfur starts to evaporate. This can be confirmed by *tamra mudra pariksha* (copper coin test). In this test, a copper coin is put on the mouth of the *kupipakwa* and watched for the presence of purified mercury. If purified mercury is deposited on the copper coin, the *kupipakwa* mouth should be corked immediately. After corking and sealing of the *kupipakwa* mouth, the heating may be continued for 1–2 hours to get the formed compound sublimated at the neck of *kupipakwa*.

7. The *baluka yantra* (sand furnace) is allowed to become *swang sheetam* (self-cooled). On self-cooling, the *kupipakwa* is taken out of the furnace and its external *mruttika lepa* (seven mud cloth layers) are removed. The *kupipakwa* is then broken by tying and burning spirit-soaked cotton thread at the middle of *kupipakwa*. On complete

burning, it is removed and water drops are sprinkled immediately to break open *kupipakwa* in two halves. The upper broken half contains the *makaradhwaja* compound, which has to be removed meticulously. The lower half contains the partially prepared gold *bhasma* (bioadaptable nanosize form), with some amount of purified mercury and purified sulfur (**Figure 3.8**).

The procedure for gold *bhasma* (bioadaptable nanosize form) preparation is described in the following steps:

Kanchanar jataputaihe kukkuta namdhyehe (8)

1. The partially prepared gold *bhasma* powder is put in an iron mortar and subjected to *bhavana* (trituration) with *kanchanara* (*Bauhinia variegata* leaves) decoction and made into small flat pellets, which on drying are closed in a *sharav samputa* (an assembly made of two earthen lids) and sealed.
2. It is then subjected to heating in a *laghuputa*, *kapota*, or *kukutta puta* (300–350°C capacity furnace) initially for 1 hour and then maintained at 400–450°C for another hour. In this way, 10–14 cycles of *puta* with *kanchanara* (*Bauhinia variegata* leaves) decoction are applied to prepare red-color gold *bhasma*.

3.1.3.2 Rajat (silver)

The varieties of silver, according to *Ayurved*, are as follows:

Sahajam khanisanjatam krutrimamcha tridhamtam | rajatam purvapurvamhi swagunairuttamomam (9)

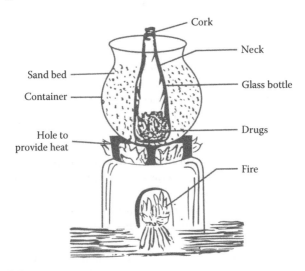

Figure 3.8 Valuka yantra.

1. *Sahaja*: The best variety, available at *Kailasa Parvata* (Kailash Mountain), which can remove diseases by mere touch but is claimed to be *durlabha* (rare)
2. *Khanija*: A better variety, available at *Himalaya Parvata* (Himalaya Mountain), which is commonly available and is said to possess *rasayana* (anti-aging) properties
3. *Krutrima*: A good variety, which occurs as result of a *vedha* (alchemy) process, is made through transformation from lower metals (e.g., tin) by purified mercury

The metallurgical properties of silver are as follows:

1. Hardness (Vickers): 25
2. Density: 10.49 g/cc
3. Melting point: 960°C
4. Color: White in color, but may be grey, black, or bluish black due to the action of atmosphere or of solutions
5. Solubility: Soluble in nitric acid

Silver is malleable, ductile, and an excellent conductor of heat and electricity. In nature, silver is usually found in large irregular masses in flat scales, granules, crystals, cubes, and octahedrons. Sometimes, silver occurs as native silver in large, twisting, branching masses. An important source of silver is sulfide ore (argentite, AgS). The mineral of black silver usually contains varying amounts of gold (up to 28%), as well as copper, arsenic, mercury, iron, antimony, and platinum. It is also associated with lead (galena, PbS). Silver is also found in veins with calcite ($CaCO_3$) and quartz (SiO_2).

According to *rasa shastra*, the qualities of silver should be

Guru snigdham mrudu shwetam dahechhyedaghanakshamam | varnadhyam chandravatswachhyam rupyam navagunam shubham (10)

Regarding the alchemical processes of silver per *rasa shastra*, the silver *shodhana* (purification) procedure is described as follows:

Sukshmapatrikrutam rupyam prataptam jatvedasi | nirvapitamagastyasya rase vartrayamshuchi (11)

The fine sheets of silver are treated by a *samanya shodhana* (common purification) method, in which the sheets are heated to red hot on burning fire and quenched in juice from *agasti patra* (*Sesbania grandiflora* leaves). This procedure is repeated three times to complete purification.

For the silver *marana* procedure (incineration to convert metals/minerals into a nanosized bioadaptable form), the process is as follows:

1. After *shodhana,* silver sheets are cut into small pieces and mixed with two parts of *hingula* (purified cinnabar) and ground in iron mortar with lemon juice until a good-quality *pishti* (amalgam) is formed.
2. This *pishti* is converted into small pellets and dried further. Pellets are kept in the lower pot of a *damaru yantra* (hourglass-shaped furnace). Its joint is sealed tightly, and it is heated at the bottom by wood fire for 6–12 hours continuously. The upper pot of the *damaru yantra* is kept cool by a wet cloth that is changed frequently, so that mercury vapors coming out of the *hingula* are condensed properly.
3. This whole process is repeated three times. In subsequent heatings, equal parts of *hingula* are used. If good-quality, free-shining, black-colored *bhasma* is not prepared within three heatings with *damaru yantra*, then additional *puta* heating is required. In such *puta*, sulfur is used in one-fourth part to silver.
4. The color of the *bhasma* should be black.

3.1.3.3 Tamra (copper)

The varieties of copper, according to *Ayurved*, are as follows:

Mlechham nepalkanchyeti tayornepalkanwaram | Nepaladanyakhanyu thhammlechhammityabhidhiyate (12)

1. *Nepalaja* (country of Nepal): Best quality
2. *Mlecchaja* (country of Sindha): Inferior quality

The colors of copper are as follows:

1. *Rakta* (red): Best variety
2. *Krsna* (black): Inferior variety

Copper that has red color and softness is considered to be the best because it is claimed to be pure and free from impurities. Blackish and hard copper is considered inferior because it may contain iron or lead as impurities, which also make it hard.

The metallurgical properties of copper are as follows:

1. Hardness (Vickers): 100
2. Density: 7.94 g/cc
3. Melting point: 1083°C
4. Color: Red

In nature, copper is found in its native form as well as in the form of sulfides, oxides, and carbonates. Native copper is rare. Generally, copper is obtained in the form of its ores. As many as 300 copper ores are found in nature, in which the sulfide group of minerals are common. The sulfides include chalcocite, chalcopyrite, covellite, and cuprite, whereas the carbonates include malachite and azurite. Native copper is found in the form of crystals and cubes, but scales, grains, plates, and masses are also not uncommon. It is an excellent conductor of heat and electricity. Native copper is generally almost pure, but sometimes it contains small amounts of silver, arsenic, bismuth, and antimony. It occurs principally in volcanic rocks, veins, and dark-colored igneous rocks. It is hard to mine. The sulfides and carbonates are easier to handle. The sulfides are black, purple, and yellow in color; the oxides and native copper are dull red; and the carbonates are blue and green. It is commonly associated with calcite, quartz, and silver.

According to *rasa shastra*, the qualities of silver should be

Susnigdham mrudulam shonam ghanaghatkshamam guru | Nirvikaram gunaihe shreshtham nepalam tamrammuchyate (13)

Regarding the alchemical processes of copper per *rasa shastra*, the copper *shodhana* (purification) procedure is described as follows:

Snuhyarkakshirsindhutthaistamrapatrani lepayet | Agnou pratapya nirgundirase sansechayettrishaha (14)

Samanya sodhita (common purified) copper sheets are pasted. The paste is prepared with *saindhava lavana* (rock salt), either with *snuhi kshira* (*Euphorbia nerifolia* oozing fluid) or *arka kshira* (*Calotropis procera* oozing fluid). These sheets are heated to red hot on fire and quenched in the juice of *Nirgundi* (*Vitex negundo*) leaves three times.

For the copper *marana* (incineration to convert metals/minerals into nanosized bioadaptable forms) process, two methods are employed. The first method is described as follows:

1. *Sodhita tamra* (purified copper) is cut into small pieces, mixed with two parts of purified mercury, and triturated with lemon juice until a good amalgam is formed.
2. This mixture is added to *kajjali* (purified mercury + purified sulfur), which is a mixture of two parts of purified sulfur and one part of purified mercury.
3. *Bhavana* (trituration) by *changeri* (*Oxalis corniculata*) juice is applied on this mixture and small flat pellets are prepared. Pellets are subjected to *gajaputa* (900°C) heat. This whole procedure is performed three times. Sulfur in equal quantity is used in all successive *puta* (transformation by energy).
4. After three such heatings, black-colored copper *bhasma* (bioadaptable nanosized form) is prepared. The copper needs to be converted to the sulfide form only (not sulfate form), for which the end product

of *bhasma* can be tested with a piece of sour curd. In this test, after putting the curd on the *bhasma*, there should not be a bluish tinge. Only then is the process said to be completed.

The second method of the copper *marana* process is described as follows:

1. *Shodhit* (purified) copper sheets are mixed with one part purified mercury, one part purified sulfur, a half-part of *talaka* (purified realgar), and a quarter-part of *manahsila* (red arsenic). In this, copper sheets are mixed with purified mercury and triturated with lemon juice until an amalgam is made. Then, purified sulfur is added to prepare *kajjali* (purified mercury + purified sulfur). Thereafter, *talaka* and *manshila* are added and mixed properly.
2. This *kajjali* is then filled in a *kupi* (glass bottle covered with seven layers of mud cloth), up to a third of the bottle, and heated by the *valuka yantra* (sand furnace) method using *kramagnipaka* (gradual escalation of heat) for at least 18 hours.
3. On getting the red compound, *kupi* is corked and sealed to sublime the compound at the neck, leaving the copper *bhasma* (bioadaptable nanosize form) at the bottom.
4. In this procedure, the addition of *talaka* and *manshila* is important to form a copper-sulfide compound and to reduce the toxic effects of copper as per *vishasya vishamaushadham* (poison cures poison). Therefore, *talaka* and *manshila* are added in the mixture to neutralize the toxic effects of copper. The copper *bhasma* is black in color.

3.1.3.4 Loha (iron)

The varieties of iron, according to *Ayurved*, are as follows:

Kharsaramcha hrinnalam taravahamcha vajiram | kallohabhidhanamcha shadavidhamtikshnamuchyate (15)

Kantloham chaturdhoktam romakam bhramakamtatha | Chumbakam dravakchyeti teshushreshtham paramparam (16)

Mrudumundamkadaramcha trividham mundauchyate (17)

1. *Tiksna* (six subtypes), which is known as *gajavalli*
2. *Kanta* (five subtypes), which is known as *chumbaka*
3. *Munda* (three subtypes), which is *vartula* (round-shaped) and found in earth or hills

Tiksna loha (iron) is recommended for *marana* (incineration to convert metals/minerals into nanosized bioadaptable forms).

The metallurgical properties of copper are as follows:

1. Hardness (Vickers): 150
2. Density: 7.8 g/cc
3. Melting point: 1535°C
4. Color: Grayish black

Iron is obtained from its oxide minerals, of which hematite, limonite, magnatite, and siderite are the most important. Sometimes, sulfide minerals are also used for obtaining iron, in which minerals have to be first converted into an oxide form.

Regarding the qualities of *kant loh* (iron), if it is pasted with *kasis* (iron sulfate) and *amalaki* (*Emblica officinalis* fruit) and washed with water, it should produce a clear reflection, like a mirror.

Regarding the alchemical processes of iron per *rasa shastra*, the *shodhana* (purification) procedure is described as follows:

Tapnani sarvalohani kadalimulavarini | Saptadhatvabhishiktani shudh-himayatyanuttamam (18)

Loh (iron) should be heated up and quenched in the water of *kadali mulam* (*Musa paradisiacal* roots) seven times.

For the iron *marana* (incineration to convert metals/minerals into nanosized bioadaptable forms) process, *shodhit* (purified) *loh churna* (iron powder) is treated consecutively by the following procedures: *bhanupaka* (transformation by sun), *sthalipaka* (transformation in pot), and *putapaka* (transformation by energy).

The *bhanupaka* (transformation by sun) method consists of the following steps:

1. In this procedure, a mixture of *triphala kvatha* (equal parts of *Terminalia chebula* fruit, *Terminalia belerica* fruit, and *Emblica officinalis* fruit decoction) is taken in the weight of the iron powder. Two parts of water are added to the *triphala* and boiled to reduce it to one-quarter of its weight.
2. This *triphala kvatha* is mixed with iron powder and dried in sunlight continuously for 3 days. This whole procedure is repeated seven times.

The *sthalipaka* (transformation by pot) method consists of the following steps:

1. *Triphala kvatha* is made by taking *triphala* at three times the weight of iron powder, adding 16 times water, and reducing it to one-eighth by boiling.
2. This *kvatha* (decoction) is mixed with *bhanupaka loha* (sun-transformed iron) in a *sthali* (wide-mouth iron pot) and heated until dry.
3. The *sthalipaka loha* is washed with clean water and dried. Iron powder is then taken for *putapaka*.

The *putapaka* (transformation by energy) method consists of the following steps:

1. *Sthalipaka loha* iron powder is impregnated and triturated with cow urine for 3 days. This paste is converted into small cakes and kept in a *sharav samputa* (an assembly made of two earthen lids) and subjected to *gajputa* (incineration at 950°C), which ends with *swangsheetam* (self-cooling). This procedure is repeated three times.
2. Subsequently, it is incinerated three times by impregnating with *triphala kvatha* (*Terminalia chebula* fruit, *Terminalia belerica* fruit, *Emblica officinalis* fruit decoction).
3. *Gajputa* incinerations 7–9 are impregnated with the juice of *kumari ras* (aloe vera).
4. *Gajputa* incinerations 10–12 are impregnated with the juice of *punarnava ras* (*Boerhhavia diffusa*).
5. In the 13th *ardha gajputa* (425°C) incineration, 1/12th of purified *hingula* (cinnabar) is added, triturating with *arka ras* (*Calotropis procera* juice).

3.1.3.5 Naga (lead)

The varieties of lead, according to *Ayurved*, are as follows:

Drutdravam mahabharam chhede krushnasamujjwalam | putigandham bahikrushnam shuddhasismatoanyatha (19)

Lead melts at low temperatures, is heavy in weight, looks shiny black when cut, has a foul smell, and is black on the outer surface. Lead that has such properties is considered to be pure or of good quality.

The metallurgical properties of lead are as follows:

1. Hardness (Vickers): 5
2. Density: 10.22 at 800°C
3. Melting point: 327°C
4. Color: Black

Lead is of two types: native and mineral. Native lead is rare. Mineral lead is generally obtained from galena, a sulfide mineral (PbS). The other important mineral of lead is cerussite (lead carbonate). Anglecite is another mineral of lead, from which it can be obtained or extracted. Over a dozen lead minerals exist, but only two of them (the secondary minerals derived from galena) are important. Lead occurs very rarely in crystals, thin plates, or small modular masses. Its minerals usually contain small quantities of silver and antimony.

Regarding the alchemical processes of lead per *rasa shastra*, the *shodhana* (purification) procedure is described as follows:

Phaltrikkashayeva Kumarirasevakarivar Salilevagayetsaptavaram (20)

In an iron pot, lead is melted and immersed in the juice of three myrobalans and aloe vera juice seven times.

The lead *marana* (incineration to convert metals/minerals into nanosized bio-adaptable forms) process consists of the following steps:

1. As mentioned earlier, *naga* (lead) is from the *putiloha* group, so *jarana* (conversion of molten metal into ultrafine powder form) is first required before *marana* is performed (see section 3.1.2.1.3).
2. After *jarana*, equal parts of purified *manahsila* (red arsenic) powder are added with *jarita* (lead powder), triturated with *vasa* (*Adhatoda vasica* juice), and enclosed in *sharav samputa* (an assembly made of two earthen lids).
3. After drying, *laghu (kukkuta) puta* (450°C heat) is applied.
4. Steps 1–3 are repeated seven times.
5. From the second step onwards, Manahsila (red arsenic) is added in one-quarter part of the lead.

3.1.3.6 Vanga (tin)

The varieties of tin, according to *Ayurved*, are as follows:

Khurakammishrakamchyeti dwividham vangamuchhyate | Khurakam shresh zthamuddishtam mishrakam chawaram smrutam (21)

Dhawalamcha mrudusnigadham drutdravam sagauravam | Nishabdam khurvangsyat mishrakam shyamshubhrakam (22)

Khura vanga is considered to be the best variety. It is white in color and is obtained in the Bengal part of India; hence, it was named as *vanga*.

The metallurgical properties of tin are as follows:

1. Hardness (Brinell): 3.9
2. Density: 5.76 g/cc
3. Melting point: 232°C
4. Color: Gray-white

Regarding the alchemical processes of tin per *rasa shastra*, the *shodhana* (purification) procedure is described as follows:

Khurabhidhanam khalu shukraloham nidhay davyambhishajam vare-nyaha | Chulligatam chapyatha galyitva vishudhhachurnodakpurna-game | Gadham tu sachhidrayidhanken kshipetsamachhatramukhe tu kumme | ethham nishikte bhishajashwashwavaram vangam vishudhhi samupaiti nunam (23)

1. It is purified using the *samanya shodhana* method prescribed for metals.
2. In *visesa sodhana*, it is melted and poured either in *churnodaka* (lime water) or in *nisayukta nirgundi drava* (*Vitex negundo* juice with *Curcuma longa* powder) through *pithara yantra* seven times.
3. It is also purified by applying *svedana* (steam) in *dola yantra* (distillation apparatus) with *churnodaka* for 1.5 hours. It should be poured in liquids very carefully to avoid being jolted.

By these methods, a good portion of tin turns into powder form, which should not be discarded. This is due to the oxidation process, with the effects of heat and cold treatment performed several times. By heat treatment, some of the volatile impurities are also removed. The metal that is not turned into powder form through this treatment is then subjected to the *jarana* process, as described in 3.1.2.A.c).

The tin *marana* (incineration to convert metals/minerals into nanosized bioadaptable forms) process consists of the following steps:

1. First complete the *jarana* treatment (see section 3.1.2.1.3).
2. The metal ash is subjected to *kumari swarasa bhavana* (trituration with aloe vera juice). Then, the pellets are dried and are enclosed in a *sarava samputa* (an assembly made of two earthen lids). They are then subjected to *putapaka* (transformation by energy) through *gajaputa* (700°C heat) for 1 hour.
3. On self-cooling, the *sharav samputa* is taken out. The *kumari swarasa bhavana* and *putapaka* processes are repeated several times until a whitish-brown *bhasma* (bioadaptable nanosized form) is prepared.
4. Generally, within seven repetitions of the *putapaka* process, good-quality *bhasma* is prepared.

3.1.3.7 Yasada (zinc)

According to *Ayurved*, zinc was known to ancient Indians since the 14th century. Madanpala was the first scholar to mention zinc as the seventh *dhatu* (metal), as stated in his text *Nighantu* (Medicinal Plants Compendium). Bhavamisra subsequently included zinc in the in *dhatu* (*loha*) group.

The metallurgical properties of zinc are as follows:

1. Hardness (Vickers): 30
2. Density: 7.1 g/cc
3. Melting point: 419°C
4. Color: White with a bluish tinge

Zinc minerals are generally associated with lead and copper minerals in the form of veins and replacement deposits in igneous or carbonate rocks. Zinc is not found in its native form; rather, it is obtained mostly from minerals. Zinc sulfide (sphelerite, SnS) is yellowish, brick red, gray, or blackish in color and has a resinous luster. Zinc oxide (zincite, ZnO) is yellowish red (orange) in color. Zinc carbonate (smithsonite, $ZnCO_3$) is usually dull, like soil. Zinc silicate (hemimorphite, $ZnSiO_3$) is a crystalline mineral that contains some amount of water.

The *shodhana* (purification) procedure for zinc is described as follows:

Yashadam girijam tasya doshaha shodhanmarane | vangasyeva hi bodhawya gananstu pravadamyaham (24)

The *shodhana* for zinc is performed just like the procedure for tin (see section 3.1.3.5).

The *marana* (incineration to convert metals/minerals into nanosized bioadaptable forms) process for zinc consists of the following steps:

1. Complete the *jarana* treatment (see section 3.1.2.1.3.).
2. The metal ash is subjected to *kumari swarasa bhavana* (tirturation with aloe vera juice). The pellets are prepared and enclosed in *sarava samputa* (an assembly made of two earthen lids). This *sarava samputa* is subjected to *putapaka* (transformation by energy) through *gajaputa* (700°C heat) mode for 1 hour.
3. On self-cooling, the *sharav samputa* is taken out, and the *kumari swarasa bhavana* and *putapaka* processes repeated several times until a whitish-brown *bhasma* (bioadaptable nanosize form) is prepared.
4. Generally, within seven repetitions of the *putapaka* process, good-quality *bhasma* is prepared.

3.1.3.8 Pittala (brass)

Pittala (brass) is one of the *mishra loha*, which has been known since ancient times. Initially, it was included in the *loha* (metal) group, but later was considered to be a *mishra loha* (alloy metal).

Ancient Indian scholars described two varieties of *pittala*. After being heated and put in *kanjika* (fermented rice liquid), it is considered to be *ritika* if it becomes red like copper. If it becomes black, then it is considered to be *kakatundi*.

The metallurgical properties of brass are as follows:

1. Hardness (Rockwell B): 25
2. Density: 7.6 g/cc
3. Melting point: 900°C
4. Color: Yellow

The *samanya shodhana* (purification) procedure for brass is described as follows:

1. First, fine sheets of *pittala* are prepared and heated strongly on fire.
2. These are quenched in *tila taila* (sesame oil), *takra* (buttermilk), *gomutra* (cow urine), *kaanjika/aranala* (fermented rice water), and *kulattha kashaya* (horse-gram decoction) three times in each liquid.

The *vishesha shodhana* (purification) procedure for brass is described as follows:

Taptakshipta cha nirgundi rase shyamarajoanvinte | Panchavaren sanshuddhiam ritirayati nichhitam (25)

Pittala sheets are heated and then quenched in *Nirgundi rasa* (*Vitex negundo leaves juice*) mixed with *Trivrit* (*Operculiana turpethum*) powder. The process is repeated 5 times.

The *marana* (incineration to convert metals/minerals into nanosized bioadaptable forms) process for brass consists of the following steps:

1. A paste of purified sulfur with *arka kshira* (*Calotropis procera* exudate) is made and applied over the *pittala* sheets.
2. These sheets are put in *sharav samputa* (pellets enclosed in earthen lids) and subjected to *gajaputa* (950°C) heat. In two or three repetitions, black-colored *bhasma* (bioadaptable nanosized form) is prepared.

3.1.3.9 Kamsya (bell metal)

Like *pittala* (brass), *kamsya* (bell metal) is also included in the group of *mishra loha* (alloy metals). This alloy is made of 4 parts of copper and 1 part of tin and is melted together. Today, *kamsya* is made by mixing 78 parts of copper with 22 parts of tin.

The varieties of bell metal, according to *Ayurved*, are as follows:

Kamsyam tu dwividham proktam pushpatailakabhedata (26)

The first variety is *pushpaka*, which is white in color and used for medicinal purposes. *Taila kamsya* is *kapisha* (blackish red) in color; this variety is not recommended for use in medicine.

The metallurgical properties of bell metal are as follows:

1. Hardness (Brinell): 3.9
2. Density: 5.76 g/cc

3. Melting point: 422°C
4. Color: Gray-yellow

In the *shodhana* (purification) procedure for bell metal, *kamsya* sheets are heated to red hot and quenched in *tila taila* (sesame oil), *takra* (buttermilk), *gomutra* (cow urine), *kaanjika/aranala* (fermented rice water), and *kulattha kashaya* (horse-gram decoction) for three to seven times.

The *marana* (incineration to convert metals/minerals into nanosized bio-adaptable forms) process for bell metal consists of the following steps:

1. A paste of purified sulfur and purified orpiment with lemon juice is prepared. It is applied on the thin sheets of *kamsya*.
2. These sheets are put in the *sharav samputa* (pellets enclosed in earthen lids) and subjected to *gajaputa* (950°C) heating. In five repetitions, black-colored *pittala bhasma* (bioadaptable nanosized form) is prepared.

3.1.3.10   Varta loha (alloy of five metals)

Varta loha is included in the *mishra loha* (alloy metal) group. Because it is made of five metals (*kamsya, tamra, pittala, loha, naga*: bell metal, copper, brass, iron, lead), it is also named *pancha loha*. These metals arc mixed in equal proportion and melted to make an alloy.

The metallurgical properties of *varta loha* are as follows:

1. Hardness (Vickers): 50
2. Density: 3.4 g/cc
3. Melting point: 640°C
4. Color: Gray-black

In the *shodhana* (purification) procedure, *varta loha* sheets are heated to red hot then are quenched in *tila taila* (sesame oil), *takra* (buttermilk), *gomutra* (cow urine), *kaanjika/aranala* (fermented rice water), and *kulattha kashaya* (horse-gram decoction) for three to seven times.

The *marana* (incineration to convert metals/minerals into nanosized bio-adaptable forms) process for *varta loha* consists of the following steps:

1. A paste of purified sulfur and purified arsenic with lemon juice is prepared. It is applied on the *varta loha* thin sheets.
2. These sheets are kept in the *sharav samputa* (pellets enclosed in earthen lids) and subjected to *gajaputa* (950°C) heating. In five repetitions, black-colored *bhasma* (bioadaptable nanosize form) is prepared.

3.1.4   Properties of bioacceptable metals

The metals and their respective *bhasma* colors after alchemy are shown in **Table 3.4**.

Table 3.4 Metals and Their Respective *Bhasma* (Bioacceptable Nanosized Form) Color after Alchemy

Metal	Before Alchemy	After Alchemy
Suvarna (gold)	Bright yellow	Violet red
Rajata (silver)	Bright white	Black
Tamra (copper)	Red	Black
Lauha (iron)	Black	Red/dark brown
Naga (lead)	Black	Light red
Vanga (tin)	Gray	Grayish white
Jasada (zinc)	Gray	Reddish white
Kamsya (bell metal)	Gray	Black
Pittala (brass)	Yellow	Black
Varta lauha (alloy of five metals)	Gray	Black

Ancient scholars of *Ayurved* described techniques to make stable liquids from these metals so that they can be absorbed directly through the oral mucosa when put in the mouth:

> Tikshnachurnamtusaptaham pakwadhatriphaladravai:
> Lolitambhavyeddharme kshirkandadravai: puna:
> Saptahambhavitamsamyakstravasamputaketatah:
> Dhamitamdravatayamtichiramtishtantisutavata (27)

Any metal *bhasma* should be triturated and immersed in the fruit juice of *Emblica officinalis* for 7 days. Later, it should be put in a *musha* (crucible) and undergo *putapaka* (transformation by energy) at 500°C. These procedures will convert all metals into liquid forms.

The ancient Vedic tradition in India also sang the praises of a mysterious sacred substance called *soma*—the nectar of immortality. It was prepared and consumed by the Vedic priests in order to attain health, longevity, spiritual illumination, and entry into the abode of the immortal gods. *Rig Veda 8:48* says:

> I have tasted, as one who knows its secret, the honeyed (*Soma*) drink that inspires and grants freedom, the drink that all, both Gods and mortals, seek to obtain, calling it nectar. We have drunk the *Soma*, we have become immortal; we have gone to the light; we have found the gods.... These glorious *Soma* drops are my health and salvation: they strengthen my joints as thongs do a cart. May these droplets guard my foot lest it stumble and chase from my body all manner of ills? Far-famed *Soma*, stretch out our life-spans so that we may live... Make me shine brightly like fire produced by friction. Illumine us.... Enter within us for our well-being. With hearts inspired may

we relish the Juice like treasure inherited from our Fathers! Lengthen our days, King *Soma*, as the sun causes the shining days to grow longer.... It is you, O *Soma*, who guard our bodies; in each of our limbs you have made your abode. Our weariness and pains are now far removed; the forces of darkness have fled in fear. *Soma* has surged within us mightily. We have reached our goal! Life is prolonged! The drop that we have drunk has entered our hearts, an immortal inside mortals.

In this tradition, *soma* is also closely related to gold. Although modern scholars have traditionally assumed that *soma* was a hallucinogen produced from plants, some scholars have now begun to suggest that the concept of *soma* involves a metallurgical or alchemical allegory. In the ancient *Rig Veda*, *soma* is said to have been fetched from heaven by a divine eagle who extracted it from a bronze (or metal) fortress (28). The eagle who fetched the *soma* from heaven may very well be a Vedic reference to the divine phoenix. The fact that the *soma* was kept in a metal fortress provides a clue that the *soma* had its true source in the metals.

However, the Vedic texts generally state that *soma* has mountains (*girau*) and stones (*adrau*). *Soma* is also described as dwelling in the mountains (*giristha*) or growing on the mountains (29). Terrestrial mountains are the abode of *soma* (30). The texts state that the *soma* was plucked from the rock (*adri*) by the mountain dwellers and then bought by the priests (*kavis*) who prepared the *soma*. The Sanskrit term *kavi* not only means priest-poet, but also smith, metallurgist, or alchemist.

Although the *soma* is said to grow on or in the rocks in the form of shoots or stalks (*amshu*), this term can easily be interpreted as the veins or stalk-like protrusions that are often found associated with certain ores, such as quartzite gold. The texts state that the *soma* was specifically plucked from two rocks (*adrau*), which are either reddish brown (*aruna*) or yellow (*hari*) in color—colors that are typical of gold-bearing ores.

Once the ores were collected by mountain dwellers and bought by the priest-metallurgists, the ores would be crushed and washed free of impurities. The crushing or pressing of the *soma* is abundantly described in the *Rig Veda*. It involved the use of grinding stones, which were said to make a loud noise. The use of grinding stones instead of a mortar and pestle was a common practice in ancient metallurgy.

It appears that several stages of grinding, washing, cooking, and filtering were involved. A woolen fleece (*avi*) was used as the filter. The use of woolen filters was also very common in ancient metallurgy; it was often used as a means of collecting gold particles from running streams. The grease or oil on the fleece would capture the tiny flecks of gold, but it would not capture particles of sand or grit, which would be wetted by the water and washed away.

The use of a woolen filter during the washing of the crushed ores would thus allow the priest-metallurgists to collect the gold particles.

After crushing, washing, and filtering, the soma would be cooked in water. As the juice began to mature, it was said to become clothed in robes of milk. In other words, it assumed a white appearance, resembling milk or the moon. Due to its white color, it was often referred to as the "milk of heaven" or the "milk of the gods" and was compared to divine semen, indicating its creative spiritual potency. It is interesting to note that when the monatomic elements are placed in water, they form a permanent suspension that looks exactly like pure white milk.

In the ancient era, the therapeutic properties of alchemized bioacceptable metals were mentioned as tacit concise forms. There are various such scriptures, including *Rasendrasarsangraha*, *Rasaratnakara*, *Rasendrachudamani*, *Bruhattantrasar*, and *Rasratnasamuchhaya*. This section provides a few basic Sanskrit terminologies used by ancient alchemy scholars to describe the physiological and therapeutic properties of metals.

Rasa describes the basic elemental constitutional properties before metabolic transformation:

Kshambhognikshambuteja: Khavayvagnyanilgonile:
Dwayolbane: kramdhutairmadhuradirasodbhava (31)

- *Madhur rasa*: Constitution made up of *pruthvi*, *jala*, and their bio-activity promotion
- *Amla rasa*: Constitution made up of *pruthvi*, *tejas*, and their bio-activity promotion
- *Lavana rasa*: Constitution made up of *tejas*, *jala*, and their bioactivity promotion
- *Katu rasa*: Constitution made up of *tejas*, *vayu*, and their bioactivity promotion
- *Tikta rasa*: Constitution made up of *akash*, *vayu*, and their bioactivity promotion
- *Kashay rasa*: Constitution made up of *pruthvi*, *vayu*, and their bio-activity promotion

Note that *pruthvi*, *jala*, *tejas*, *vayu*, and *akash* should not be taken literally as earth, water, fire, wind, and space, respectively. The Sanskrit definitions are far broader, denoting the basic constitutional elements' behavior in every aspect of the world.

- *Akash bhuta*: Fundamental reason for existence of any living or non-living object
- *Vayu bhuta*: Fundamental movement required for the creation of any existence

- *Teja bhuta*: Fundamental energy required for modification in order to create existence
- *Jala bhuta*: Fundamental aggregation force of all elements to create existence
- *Pruthvi bhuta*: Fundamental occupied space by any existence
- *Akash mahabhuta*: Combined reason for existence of living and non-living objects
- *Vayu mahabhuta*: Movement required to create combined existence
- *Teja mahabhuta*: Modification required to create combined existence
- *Jala mahabhuta*: Aggregation of all elements to create combined existence
- *Pruthvi mahabhuta*: Space occupied by combined existence

Virya describes anabolic and/or catabolic properties and cellular respiration:

Ushnam Sheetam Dwidhaivanye Verrmachakshateapicha (32)

- *Ushna virya*: A drug will be predominant in the body's catabolism activation during its own metabolism.
- *Sheet virya*: A drug will be predominant in the body's anabolism activation during its own metabolism.

Vipaka describes the properties after the first and second metabolic transformations:

Jatharenagnina Yodyadudeti Rasantaram
Rasanam Parinamante Sa Vipak iti Smruta:
Swaduhu Patuchya Madhurmamloamlam Pachyate Rasa:
Tiktoshanakashayanam Vipak: Prayasha: Katu (33)

- *Madhur vipak*: A drug will promote the basic elements *pruthvi* and *jala* after metabolic transformation.
- *Amla vipak*: A drug will promote the basic elements *pruthvi* and *tejas* after metabolic transformation.
- *Katu vipak*: A drug will promote the basic elements *pruthvi* and *jala* after metabolic transformation.

Guna describes the tissue and cellular-level properties:

Gurumandahimasnigdhashlashnasandramrudusthira:
Guna: Sasukshmavishada Vinshati: Saviparyaya (34)

- *Guru guna: Bruhane* (increase in cell energy reservoir)
- *Laghu guna: Langhane* (increase in cell energy utility)
- *Sheet guna: Stambhane* (negative feedback to cells for any reaction)
- *Ushna guna: Swedane* (increase in tissue permeability)

- *Snigdha guna: Kledane* (increase in cellular fluid)
- *Ruksha guna: Shoshane* (decrease in cellular fluid)
- *Manda guna: Shamane* (cellular homeostasis)
- *Tikshna guna: Shodhane* (cell break)
- *Sthir guna: Dharane* (cell stability)
- *Chal guna: Prerane* (cell stimulus)
- *Mrudu guna: Shlathane* (increase in cellular osmosis)
- *Kathin guna: Drudhikarane* (increase in cell rigidity)
- *Vishad guna: Kshalane* (increase in tissue breakdown)
- *Picchila guna: Lepane* (tissue defense)
- *Shlashna guna: Ropane* (tissue regeneration)
- *Khar guna: Lekhane* (cell molecular weight reduction)
- *Sthul guna: Sanwarane* (increase in tissue molecular weight)
- *Sukshma guna: Vivarane* (increase in cell permeability)
- *Sandra guna: Prasadane* (tissue rearrangement)
- *Drava guna: Vilodane* (increase in extracellular fluid)

Karma describes nutrient properties. According to *Ayurved*, these follow the fundamental elements in the body. If these elements are working normally at both the macrolevel and microlevel, then the body is healthy:

Trayodosha dhatavachya purisham mutramevacha
Deham sandharayantyete hyavyapanna rasairhitai (35)

Three *dosha* and seven *dhatu*, when in normal limits, sustain the health of every individual:

- *Dosha*: *vata, pitta,* and *kapha*
- *Dhatu*: *rasa, rakta, mansa, meda, asthi, majja,* and *shukra*

All nutrients and medicines therefore influence the *dosha* and *dhatu*:

- *Dosha prabhava*: Physiological properties
- *Vyadhi prabhava*: Therapeutic properties

All such properties of metals are a result of their respective *rasa, guna, veerya, vipaka,* and *prabhava*.

Kinchitrasenakurute karma paken chaparam
Gunantaren veeryen prabhaven kinchan (36)

Sometimes, metals exhibit altogether different nutrients and therapeutic properties because of their *prabhav* (invisible) serendipitous core elements. According to *Ayurved*, the doses of various metal *bhasma* (bioadaptable nano-sized form) for therapeutic purposes are as follows:

Sewanasyapramanamtukathayishyamisampratam
Ballardhakanakamhisuprakathitamrupyamchashulbamtatha

Tikshnavangabhujangamarnichayo ballardhaballonmita:
Tatulyashudhhapippalinigadi takshodramchakarshonmitam
Sewyamsamparihrityagrishmasharadotamramsusevyamnarai (37)

- Gold, silver, and copper = 200 mg/day
- Iron, tin, lead, and brass = 500 mg/day

The *anupana* (drug and food vehicle) is a very unique concept of *Ayurved*. For every metal administration, a specific drug vehicle is given in the texts to enhance the metal's nutraceutical and therapeutic properties in the body. It is said that all metal *bhasma* should be taken with equal doses of *Piper longum* fruit and honey.

3.1.4.1 Nutraceutical and therapeutic properties of gold according to ayurved

Snigdham medhyam garvishaharam bruhanam vrushyamgrayam | yakshmonmadprashamanaparam deharogpramathi
Medha buddhi smruti sukhakaram sarvadoshamayaghnam ruchyam dipi prashamitjaram swadupakam suvarnam (38)

Nihishehsarogvidwhansi bhutpretbhayapaham | bandhanam bhaviroganam vishatrayagadapaham (39)

I. *Rasa—Kasaya, tikta, madhura,* and *katu*
II. *Guna—Sita, guru, snigdha,* and *picchila*
III. *Virya—Sita*
IV. *Vipaka—Madhura*
V. *Karma—*Nutraceutical properties
   i. *Vrusya* (increases libido)
   ii. *Balya* (provides energy to tissues)
   iii. *Brimhana* (anabolic)
   iv. *Rasayana, pustipradayi, ojovivardhana, vayasthairyakara* (anti-aging)
   v. *Medhya* (improves cognition)
   vi. *Matismritiprada* (improves short- and long-term memory)
   vii. *Kantikara* (improves skin complexion)
   viii. *Vak visuddhikara* (improves vocal cord function)
   ix. *Pavitra, papaghna* (generates creative thoughts)
   x. *Lekhana* (weight reduction)
   xi. *Visagarahara* (removes toxins)
   xii. *Bhutavesa-prasantikara* (antipsychosis)
   xiii. *Rucya* (improves anorexia)
   xiv. *Varnya* (anti-aging for skin)
   xv. *Pathya* (adjuvant to diet)

VI. *Dosa prabhava—tridosaghna* (pacify all three dosha): Due to this property, gold is a very important metal for the body because it can restore physiological balance in any given condition.

VII. *Vyadhi prabhava* (therapeutic properties)
    i. *Ksaya* (cachexia)
    ii. *Unmada* (psychosis)
    iii. *Jwara* (pyrexia of unknown origin)
    iv. *Sosa* (muscular dystrophy)
    v. *Swasa* (asthma)
    vi. *Kasa* (chronic bronchitis)
    vii. *Agnimandya* (anorexia)
    viii. *Aruchi* (dyspepsia)
    ix. *Arsa* (rectoanal diseases)
    x. *Prameha* (diabetes mellitus)
    xi. *Karsya* (chronic malnourishment)
    xii. *Pandu* (all kinds of anemia)
    xiii. *Dusta grahani* (chronic colitis)
    xiv. *Visa garodbhava roga* (acute and chronic toxemia)
    xv. *Agantuka roga* (infectious diseases)
    xvi. *Papajaroga* (psychiatric disorders)
    xvii. *Tridosaja roga* (chronic metabolic disorders)
    xviii. *Yaksma* (pulmonary and extrapulmonary tuberculosis)

VIII. *Anupana* (drug vehicle)—*Navnit* (butter made from Indian cow milk) and *madhu* (honey)

IX. To remove the toxic effects of gold if it is overdosed or not prepared scientifically:

Abhayasitayayuktamtridinamnrubhirangane
Hemdoshaharikhyata satyamsatyamnasanshaya (40)

One should take *Terminalia chebula* fruit with rock sugar twice a day for 3 days.

3.1.4.2 Nutraceutical and therapeutic properties of silver according to ayurved

Rupyam Vipakmadhuram Tuvaramlasaram Sheetsaram Paramlekhannakam cha Ruchyam
Snigdham cha vatakaphajijjatharagnidipi Balyapradam Sthiravayaskaranam cha Vrushyam (41)

Raupyam Sheetam Kashayamlam Snigdham Vataharagurum | Rasayanam Vidhanena Sarvarogapaharam (42)

Rupyamsheetam   Kashayamlam   Swadupakarasam   Saram |
Vayasansthapanam Snigdham Lekhanam Vatapittajit

Pramehadika Rogancha Nashayatyachiraddhruvam | Gutikasya Dhruta Vakye Trushnashoshavinashini (43)

I. *Rasa—Kasaya, amla, madhura*
II. *Guna—Snigdha, sita, guru, sara*
III. *Virya—Sita*
IV. *Vipaka—Madhura*
V. *Karma—*nutraceutical properties
    i. *Lekhana* (lipid metabolism)
    ii. *Vayasthapana, vayah-sthairyakara, ayuprada, viryavardhaka* (anti-aging)
    iii. *Varnya* (melatonin balance)
    iv. *Vrisya* (increases libido)
    v. *Paramabalya* (improves athletic performance)
    vi. *Rucya* (improves taste centers)
    vii. *Medhya* (improves memory)
    viii. *Gadahara* (maintains disease-free condition)
    ix. *Jatharagnidipi, pustikari* (improves anabolism)
    x. *Ajaramarakara, rasayana* (anti-aging)
VI. *Dosa prabhava—Vata, prakopajit, vatapittajit*: Influences the combination function of *vata* and *pitta*, as well as *vata* alone
VII. *Vyadhi prabhava* (therapeutic properties)
    i. *Pittaja roga* (reticuloendothelial disorders)
    ii. *Netra roga, timira roga* (ophthalmic diseases)
    iii. *Pandu roga* (anemia)
    iv. *Sarva dosaja roga* (chronic metabolic disorders)
    v. *Dirgha roga* (chronic infectious diseases)
    vi. *Yaksma swasa* (active tuberculosis)
    vii. *Gulma* (neoplastic disorders)
    viii. *Udara* (hepatic disorders)
    ix. *Arsa* (anorectal disorders)
VIII. *Anupana* (drug vehicle)—*Navnit* (butter made from Indian cow milk) and *madhu* (honey)
IX. To remove the toxic effects of silver if it is overdosed or not prepared scientifically:

Sharkaramadhusanyuktam sevayedyodintrayam
Apakwaraupyadoshen vimuktasukhamashnute (44)

One should take rock sugar and honey twice a day for 3 days.

3.1.4.3 Nutraceutical and therapeutic properties of copper according to ayurved

Tamram Tiktakashayakchya Madhuram Pake cha Viryoshnakam Samlam Pittakapham Jatharukam Kushthamajurtyankruta

Udhvardha Parishodhanam Vishayakrutsthaulyapaham Shutkaram Durnamkshayapandurogshamanam Netryam Param Lekhanam (45)

Tattarogharanupanasahitam Tamram Dwivallonmitam Samlidham Parinamshulamudaram Shulam cha Pandu Jwaram Gulmamplihamyakrutkshayagnisadanam Meham cha Mulamayam Dustam cha Grahani Hared Dhruvamidam Tatsomanathabhidham (46)

I. *Rasa—Kasaya, tikta, madhura,* and *amla*
II. *Guna—Sita, laghu, sara, snigdha*
III. *Virya—Usna*
IV. *Vipaka—Katu*
V. *Karma* (nutraceutical properties)
    i. *Lekhana* (lipid metabolism)
    ii. *Alpa brimhana* (controls delivery of nutrition)
    iii. *Ksutkara* (water metabolism)
    iv. *Hridvi* (heart function)
    v. *Krimighna* (worm infestation)
    vi. *Rasayana* (anti-aging)
    vii. *Ropana* (internal and external wound healing)
    viii. *Netrya* (ophthalmic disorders)
    ix. *Shodhana* (antioxidant)
VI. *Dosa prabhava—Vata kaphahara, pitta kaphahara*: Influences the combined action of *vata-kapha* and *pitta-kapha*
VII. *Vyadhi prabhava* (therapeutic properties)
    i. *Pandu* (anemia)
    ii. *Udara roga* (ascites)
    iii. *Arsa* (hemorrhoids)
    iv. *Gulma* (neoplasm)
    v. *Pliha roga* (spleen disorders)
    vi. *Yakridroga* (hepatic disorders)
    vii. *Krimi* (worms)
    viii. *Sotha* (hepatic edema)
    ix. *Sula* (visceral pain)
    x. *Udarasula* (gastric ulcer)
    xi. *Parinamasula* (duodenal ulcer)
    xii. *Ksaya* (tuberculosis)
    xiii. *Kustha* (leprosy)
    xiv. *Pinasa* (sinusitis)
    xv. *Agnimandya* (dyspepsia)
    xvi. *Ksayaja-prameha* (diabetes with weight loss)
    xvii. *Grahaniroga* (ulcerative colitis)
    xviii. *Mutrakricchha* (urinary incontinence)
    xix. *Jirna jwara* (chronic pyrexia)
    xx. *Murccha* (syncope)

xxi. *Amadosa* (collagen disorders)

xxii. *Amavata* (rheumatoid arthritis)

xxiii. *Vridhhi* (hernia)

xxiv. *Sthaulya* (obesity)

xxv. *Jara* (geriatric debility)

xxvi. *Visadosa* (antioxidant)

xxvii. *Mrityu* (lifesaving drug in an emergency)

VIII. *Anupana* (drug vehicle)—*Madhu* (honey) and *ghrit* (clarified butter from Indian cow milk)

IX. To remove the toxic effects of copper if it is overdosed or not prepared scientifically:

Munibrihisitapanam dhanyakamvasitasaha:
Tamradoshamsheshamvai pibnhanyadinatrayai: (47)

One should take coriander seeds with rock sugar followed by room-temperature water twice a day for 3 days.

3.1.4.4 Nutraceutical and therapeutic properties of iron according to ayurved

Mundam   Param   Mrudulakam   Kaphavatashulamulammehanada Kamalapanduhari
Gulmabhavata   Jatharatiharam   Pradipi   Shopapahamrudhirakrut Khalukoshthashodhi (48)

Kshayam   pandugadam   Gulmam   Shulamulamayam   Tatha |
Mehamedoagnimandyam cha Yakrutplihamkamalam (49)

Shwasam kasam cha Kushtham cha Jwaram Shulanwitam Tatha
Krushnayaha Shothashularsha Krumipandutvashoshanuta
Vayasyam guru Chakshushyam Sarvamedoanilapaham (50)

I. *Rasa—Tikta*

II. *Guna—Usna/sita, ruksa*

III. *Veerya—Ushna*

IV. *Vipaka—Madhura*

V. *Dosa prabhava—Kaphapitta kara*

VI. *Karma* (nutraceutical properties)

   i. *Dipana* (increase in appetite)

   ii. *Rasayana* (anti-aging)

   iii. *Balakara* (increase in anabolism)

VII. *Vyadhi Prabhava* (therapeutic properties)

   i. *Yakrid vikara* (liver disorders)

   ii. *Pliharoga* (spleen disorders)

iii. *Amasula* (pain from connective tissue disorders)
iv. *Pandu* (anemia)
v. *Ksaya* (hemoptysis in tuberculosis)
vi. *Kustha* (leprosy)
vii. *Prameha* (diabetes)
viii. *Amavat* (rheumatoid arthritis)
ix. *Medoroga* (lipid disorders)
x. *Chardi* (hyperemesis)
xi. *Atisara* (chronic diarrhea)
xii. *Ardhanga vata* (hemiplegia)
xiii. *Sarvanga vata* (paraplegia)
VIII. *Anupana* (drug vehicle)—*Madhu* (honey), *ghrit* (clarified butter from Indian cow milk), three myrobalans decoction, *Embelia ribes* fruit powder
IX. To remove the toxic effects of iron if it is overdosed or not prepared scientifically:

Munirasapishtavidangam munirasalidhamchirsthitamghame
Dravayatilohdoshan vanhinavanitapindamiva (51)

One should take leaf juice from *Sesbania grandiflora* and the fruits of *Embellia ribes*, followed by a sun bath, once per day for 7 days.

3.1.4.5 Nutraceutical and therapeutic properties of lead according to ayurved

Nagah Samirkaphapittavikarhanta Sarvapramehavanarajikrupityonihi Ushnah Saro Rajatranjankrut Vranarsho Gulmagrahanyatisrutikshanado mshumali
Nagastu Nagshatatulyabalam Dadati Vyadhim Vinashayati Jivanmatanoti Vanhim Pradipyati kambalamkaroti Mrutyuchanashayati Satatam Sevitah Sah (52)

I. *Rasa—Tikta, katu, lavana,* and *ksara*
II. *Guna—Usna, guru/laghu, snigdha,sara*
III. *Virya—Usna*
IV. *Vipaka—Katu*
V. *Dosa prabhava—Vatahara, kapha vatahara, vatapitta kaphapaha*
VI. *Karma* (nutraceutical properties)
i. *Balya, satanaga tulya balaprada* (energy liberation through catabolism)
ii. *Virya vardhana* (increases bioavailability of other drugs)
iii. *Ayuvardhana* (increases cell longevity)
iv. *Kanti vardhana* (increases skin complexion)
v. *Caksusya* (improves eye vision)

vi. *Vilekbana* (target tissue oxidant removal)

VII. *Vyadbi prabhava* (therapeutic properties)
   i. *Prameha* (chronic diabetes)
   ii. *Dhanurvatadi* (tetanus)
   iii. *Vatajaroga* (nervous system disorders)
   iv. *Gulma* (neoplasm)
   v. *Kustha* (leprosy)
   vi. *Vrana* (tissue repair)
   vii. *Sosa* (muscular dystrophy)
   viii. *Visa* (antivenom)
   ix. *Trisna* (kidney disorders)
   x. *Amashotha* (edema)
   xi. *Amasula* (connective tissue disorders pain)
   xii. *Amavata* (rheumatism)
   xiii. *Medoroga* (lipid disorders)
   xiv. *Pandu* (anemia)

VIII. *Anupana* (drug vehicle)—*Madhu* (honey) and *ghrit* (clarified butter from Indian cow milk)

IX. To remove the toxic effects of lead if it is overdosed or not prepared scientifically:

Hemamharitakisevet sitayuktamdinatrayam
Apakwanagdoshen vimuktasukhamedhate (53)

One should take gold *bhasma* and fruits of *Terminalia chebula* twice a day for 3 days after food.

3.1.4.6 Nutraceutical and therapeutic properties of tin according to ayurved

Vanga Tikshnoshnaruksham Kaphakrumivamijinmehamedoanilghnam Kashshwaskshayaghnam Prashamitahutbhunga mandyamadhmanahari Balyam Vrushyam Prabhakrunmanasijajanakam Sarvamehapramathi Pradnyakrudvaranamuchaiiralaghuratirasasyapadam Bruhanamcha (54)

Balyam   Dipanpachanam   Ruchikaram   Pradnyapadam   Shitalm Saundaryekvivardhanam Hitkaram Nirogtakarakam Dhatusthoulyakaram Kshashikshayaharam Sarvapramehapaham vanga Bhakshayato Narasya Nabhaved Swapreapi Shukrakshayah (55)

I. *Rasa*—*Tikta, amla, katu, ksara*
II. *Guna*—*Ruksa, sita/usna, sara,* and *laghu*
III. *Virya*—*Usna/sita*
IV. *Vipaka*—*Katu*
V. *Dosa prabhava*—*Kaphapitta hanta*

VI. *Karma* (nutraceutical properties)
    i. *Caksusya* (eye disorders)
    ii. *Medodhara* (obesity)
    iii. *Prajnakar* (improves memory)
    iv. *Varnya, saundaryakara* (improves skin complexion)
    v. *Ayusya, vrisya, saktidayi, pustikara, rasayana* (age longevity)
    vi. *Madanajanana (vajikara,* aphrodisiac)
    vii. *Dahahara* (anodyne)
    viii. *Visahara* (antioxidant)
    ix. *Saukhyakara* (antipsychosis)
    x. *Vilekhana* (local oxidant removal)
    xi. *Matiprada* (improves memory)
    xii. *Ratirasa* (increases libido)
VII. *Vyadhi prabhava* (therapeutic properties)
    i. *Prameha* (diabetes)
    ii. *Medoroga* (obesity)
    iii. *Kaphaja roga* (collagen and connective tissue disorders)
    iv. *Asiti vataja roga* (bone marrow disorders)
    v. *Gulma* (neoplasm)
    vi. *Sukraksaya* (oligospermia and azoospermia)
    vii. *Jararoga* (anti-aging)
    viii. *Mutrakrichhra* (kidney disorders)
    ix. *Svetapradara* (leukorrhea)
    x. *Amavata* (rheumatism)
VIII. *Anupana* (drug vehicle)—*Madhu* (honey) and *ghrit* (clarified butter from Indian cow milk)
  IX. To remove the toxic effects of iron if it is overdosed or not prepared scientifically:

Meshashrungisitayuktam ya: sevatedinatrayam
Vangadoshavimuktaso sukhamjivatimanava (56)

One should take leaf juice from *Gymnema sylvestrae* and rock sugar twice a day for 3 days.

3.1.4.7 Nutraceutical and therapeutic properties of zinc according to ayurved

Yashadam Tuvaram Tiktam Shitalam Kaphapittahrita
Chakshushya Param Mehan Pandum Shwasamcha Nashayet (57)

    I. *Rasa—Tikta, kasaya,* and *katu*
    II. *Guna—Sita*
    III. *Virya—Ushna*
    IV. *Vipaka—Katu*

V. *Dosa prabhava—Kapha pittanut*

VI. *Karma* (nutraceutical properties)

    i. *Parama caksusya* (extremely useful in eye disorders)

    ii. *Bala-virya-buddhi-vardhaka* (increases immunity)

    iii. *Slesmakala sankocakara* (removes cellulite, extra adipose tissues, and lipids)

VII. *Vyadhi prabhava* (therapeutic properties)

    i. *Prameha* (diabetes)

    ii. *Pandu* (anemia)

    iii. *Swasa* (asthma)

    iv. *Nisa sveda* (diaphoresis)

    v. *Vrana srava* (wound healing)

    vi. *Kampavata* (motor neuron diseases)

VIII. *Anupana* (drug vehicle)—*Madhu* (honey) and *ghrit* (clarified butter from Indian cow milk)

IX. To remove the toxic effects of zinc if it is overdosed or not prepared scientifically:

Balabhayasitayuktam sevayetdyodinatrayam
Jasadvikarosya nashamayatinanyatha (58)

One should take the fruits of *Terminalia chebula* and rock sugar twice a day for 3 days after food.

3.1.4.8 Nutraceutical and therapeutic properties of brass according to ayurved

Ritistiktarasa Ruksha Jantukhghni Sastrapittanuta
Krumikushthahara Yogat Soshanaviryacha Shitala (59)

Ritikayugalam Rukshatiktamcha Lavanrase
Shodhanam Panduroghnam Krumighanam Natilekhanam (60)

    I. *Rasa—Tikta, katu, kasaya,* and *lavana*

    II. *Guna—Ruksa, sita,* and *sara (usna)*

    III. *Virya—Usna*

    IV. *Vipaka—Katu*

    V. *Dosa prabhava—Vatapittajit, slesmapittaghna*

VI. *Karma* (nutraceutical properties)

    i. *Rasayana, vrisya* (anti-aging)

    ii. *Krimighna* (worm infestation)

    iii. *Sodhana, visahara, jantughna* (infectious disorders)

    iv. *Lekhana* (adipolysis)

    v. *Kustha* (skin disorders)

    vi. *Pliharoga* (reticuloendothelial function)

VII. *Vyadhi prabhava* (therapeutic properties)
  i. *Raktapitta* (hemolysis disorders)
  ii. *Kustha* (leprosy)
  iii. *Pandu* (anemia)
  iv. *Pliharoga* (reticuloendothelial system disorders)
  v. *Visajanya roga* (oxidizing diseases)
VIII. *Anupana* (drug vehicle)—*Madhu* (honey) and *ghrit* (clarified butter from Indian cow milk)

3.1.4.9 Nutraceutical and therapeutic properties of bell metal according to ayurved

Kansyakashaymtiktoshnam Lekhanamvishadamsaram
Gurunetrahimamrukshamkaphapittaharam (61)

  I. *Rasa—Tikta* and *kasaya*
  II. *Guna—Usna, laghu, visada, sara, ruksa, guru*
  III. *Virya—Usna*
  IV. *Vipaka—Katu*
  V. *Dosa prabhava—Kaphavatajit, vatahara, vatapittaghna, kaphapittahara*
  VI. *Karma* (nutraceutical properties)
    i. *Dipana* (increases digestion)
    ii. *Lekhana* (obesity)
    iii. *Drik-prasadna* (improves vision)
    iv. *Satmyakara* (body homogeneity)
    v. *Sukhakara* (improves mood)
    vi. *Arogyakara* (age longevity)
    vii. *Dridha-dehadayi* (improves body mass index)
    viii. *Swarya* (throat diseases)
  VII. *Vyadhi prabhava* (therapeutic properties)
    i. *Prameha* (diabetes)
    ii. *Kustha* (skin disorders)
  VIII. *Anupana* (drug vehicle)—*Madhu* (honey) and *ghrit* (clarified butter from Indian cow milk)

3.1.4.10 Nutraceutical and therapeutic properties of varta loh (alloy of five metals) according to ayurved

Himalam katukam rukhsham kaphapittavinashanam
Ruchyam twachyam krumighnam netram malavishodhanam (62)

  I. *Rasa—Amla, katu*
  II. *Guna—Hima/sita, ruksa*
  III. *Virya—Sita*
  IV. *Vipaka—Katu*
  V. *Dosa prabhava—Kaphapittanasana*

VI. *Karma* (nutraceutical properties)
   i. *Rucya* (dyspepsia)
   ii. *Twacya* (skin complexion)
   iii. *Krimighna* (infectious diseases)
   iv. *Malavisodhana* (improves intestinal absorption)
   v. *Yogvahi* (augments properties of other nutrient)
VII. *Vyadhi prabhava* (therapeutic properties)
   i. *Netraroga* (eye disorders)
   ii. *Galaroga* (throat disorders)
   iii. *Sarvaroga* (can be used in every disease and acts as a catalyst to enhance certain therapeutic properties of other medicines)
VIII. *Anupana* (drug vehicle)—*Madhu* (honey) and *ghrit* (clarified butter from Indian cow milk)

3.2 Bioacceptable metals as metallonutraceutics

3.2.1 Metals as metallopharmaceutics and metallonutraceutics

All ancient references show that all bioacceptable metals can act as nutrients and/or pharmaceuticals. They can also be new hybrid products with properties of both nutrients and pharmaceuticals. Currently, no official definition exists for the term *metallonutraceuticals*, although they can be considered to be metals that enrich foods for the health and well-being of the individual.

The scope of metallonutraceuticals is significantly different from functional foods for several reasons, including the following:

I. Prevention and treatment of disease (i.e., medical claims) are relevant to metallonutraceuticals. However, only the reduction of disease, not the prevention and treatment of disease, is involved with functional foods.
II. Metallonutraceuticals can be sold in forms that are similar to drugs (e.g., pills, extracts, tablets), as well as other types of foods. Functional foods are in the form of ordinary food (63, 64).

However, the boundary between metallonutraceuticals and functional foods is not clear due to a nondistinct regulatory framework for functional foods and metallonutraceuticals (65). Metallonutraceuticals can be categorized in two ways:

I. Potential metallonutraceuticals (e.g., gold, silver, platinum) hold the promise of a particular health or medical benefit. A potential nutraceutical can become an established metallonutraceutical only after there is sufficient clinical data to demonstrate such a benefit.
II. Established metallonutraceuticals (e.g., zinc) have sufficient clinical data to demonstrate the promised or labeled benefit (66).

Table 3.5 Metallopharmaceuticals and Metallonutraceuticals: Differentiating Factors

Parameter	Metallopharmaceutical	Metallonutraceutical
Regulations	To be recognized as a drug in respective countries	To be recognized as a functional nutrient in respective countries
Claim	Disease claim	Prevention and mitigation claim
Dose	Therapeutic dose as per claim	Nutrient dose as per nutrient claim
Safety and toxicity	To be safe and nontoxic as per claim and dose	To be safe as an added nutrient
Standardization	As per claim and regulations in respective countries	As per claim and regulations in respective countries as a food

The major differential points for the use of any bioacceptable metal as a metallopharmaceutical or metallonutraceutical are shown in **Table 3.5**.

3.2.2 Strategy for the conversion of inorganic metals into bioacceptable forms

Metallonutraceuticals (*bhasmas*) are herbominerals in constitution. Nanoparticle-sized *bhasmas* were the focus of one study (67), in which it was proposed that nanoparticles are responsible for the fast and targeted action of the *bhasma*. Subsequent actions upon DNA/RNA molecules and protein synthesis within the cell are further hypothesized as possible mechanisms for the rapid onset of therapeutic actions with *bhasma* preparations (68).

Jasada (zinc) *bhasma* prepared by such traditional alchemical processes has been physicochemically characterized using modern state-of-the-art techniques, such as x-ray photoelectron spectroscopy, inductively coupled plasma, elemental analysis with energy dispersive x-ray analysis, dynamic light scattering, and transmission electron microscopy (TEM). The analysis shows that the *jasada bhasma* particles are in an oxygen-deficient state and a clearly identifiable fraction of particles are in the nanometer size range. These properties may contribute to the therapeutic property of this particular type of medicine (69).

Ancient alchemical processes (as explained in **Figure 3.1**) are nothing but the nanofabrication of inorganic metals to convert them into bioacceptable nanometals. The process flow of alchemy was based on nanofabrication preparation methods, such as *putapaka* (top-down nanofabrication for single metallic nanoparticles) and *kupipakwa* (bottom-up nanofabrication for nanopolymetal composites). In the *putapaka* method (performed without mixing metal in purified mercury and sulfur), *bhasma* are prepared by subjecting metals or minerals to a three-step procedures: *shodhana* (purification), *bhavana* (trituration), and *marana* (incineration to convert metals/minerals into nanosized bioadaptable forms).

In the *kupipakwa* method (which includes mixing in purified mercury and sulfur), *bhasma* are prepared by subjecting metals (e.g., gold, silver, copper)

to a four-step procedure: *shodhana, kajjali* (purified mercury amalgamation), *bhavana,* and *marana.* After *shodhana,* the metals are subjected to amalgamation with purified mercury, and then purified sulfur is mixed and triturated until a black, lusterless, fine, and smooth mass is prepared. This procedure is called the *kajjali* preparation. Prepared *kajjali* is levigated by particular liquid media for a certain period. It is allowed to completely dry and then is transferred to a *kachkupi* (glass bottle) covered by seven layers of mud-smeared cloth. This bottle is then subjected to *valukayantra* (a sand bath) for indirect and homogeneous heating for a certain period (**Figure 3.9**). After self-cooling, the glass bottle is broken, sublimed product is collected from its neck, and the *bhasma* is collected from the bottom of the bottle and ground to powder.

All four procedures are repeated for a prescribed number of times for every metal to achieve the final product. The repetitions of the processing cycles are based on the following parameters:

1. What is the target end therapeutic effect?
2. How potent should it be?
3. What is the threshold point at which it will become bioacceptable and/or bioadaptable?

Certain physicochemical changes occur during alchemical processing. During *shodhana,* tension is increased in metals by the application of heat, causing linear expansion. After heating, immediate cooling in liquid media leads to a

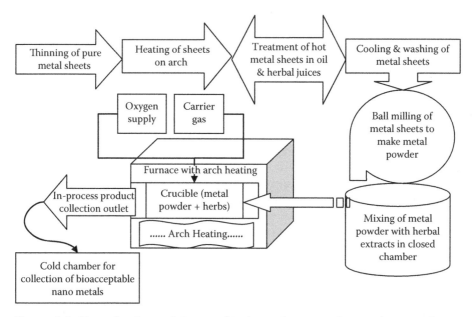

Figure 3.9 (See color insert.) Proposed industrial process ideation for manufacturing bioacceptable nanometals.

decrease in tension and increase in compression force. Repetition in heating and cooling causes disruption in the compression-tension equilibrium, which leads to increased brittleness, reduction in hardness, and finally reduction in particle size. During the red-hot state, some metals and minerals react with atmospheric oxygen or steam and form a chemical compound. Iron, when heated to red hot, reacts with atmospheric oxygen or steam to form ferroso-ferric oxide (Fe_3O_4). Copper in moist air is converted to basic copper sulfate, which in the red-hot state is completely decomposed to cupric oxide (70).

In the *bhavana* process, materials with liquid media are rubbed between the surface of the pestle and mortar. This process involves breaking down material by a rubbing action between two surfaces when stress in the form of attrition is applied; particle surfaces chip and produce small particles. Wet grinding eliminates the hazards of dust. A finer size can be achieved by wet grinding than by dry grinding (71).

Oxidation of metals occurs during heating in open air in the *jarana* procedure. The melting point of metals also increases due to oxidation. The inorganic part of plant material supplies trace elements as capping agents to materials. During incineration, compounds are generally formed on the metal surface. Repetition of this process leads to a reduction in particle size (72).

After *marana*, metals generally convert to their compound forms, which are biologically favorable to the body. Organic molecules derived from the *marana* herbs remain present in *bhasma* and probably act as the coating material on the metal compound used in the *bhasma* (73).

3.2.3 Industrial scale-up strategies

The ancient *putapaka* and *kupipakwa* manufacturing methods for *bhasma* formed the principles of industrial pharmaceutical manufacturing for bioacceptable nanometals (BNMs) in the form of metallonutraceuticals. Nanoparticles are nanosized structures in which at least one of the phases has one or more dimensions (length, width, or thickness) in the nanometer size range (1–100 nm). Nanocrystalline materials are solids composed of crystallites that are less than 100 nm in at least one dimension (74).

I. *Zero-dimensional nanostructures*: All dimensions are nanosized, such as uniform particle arrays (quantum dots), heterogeneous particle arrays, core-shell quantum dots, hollow spheres, and nanolenses

II. *One-dimensional nanostructures*: Any one of the dimensions is outside of the nanometric size (e.g., nanowires, nanorods, nanotubes, nanobelts, nanoribbons)

III. *Two-dimensional nanostructures*: Any two dimensions are outside of the nanometric size range (e.g., thin film junctions [continuous islands], branched structures, nanoprisms, nanoplates, nanosheets, nanowalls, nanodisks)

IV. *Three-dimensional nanostructures*: None of the dimensions are of nanometric size (e.g., quantum dots, hollow spheres, nanoballs [dendritic structures], nanocoils, nanocones, nanopillers, nanoflowers) (75)

The following issues have to be considered for the future development of metallonutraceuticals:

I. Development of concepts for bioacceptable nanometals, particularly their elaboration
II. Development of synthesis and/or fabrication methods for raw materials (metal powders, herbs) as well as for targeted bioacceptable nanometals
III. Understanding of the influence of the size of building blocks in bioacceptable nanometals, as well as the influence on the microstructure, physical, chemical, and mechanical properties of this material
IV. Understanding of the influence of interfaces on the properties of bioacceptable nanometals
V. Investigation of catalytic applications of monometallic and plurimetallic bioacceptable nanometals
VI. Transfer of developed technologies into industrial applications, including the development of the industrial scale of synthesis methods of bioacceptable nanometals (76)

There are two primary approaches for manufacturing bioacceptable nanometals: synthesis methods and physical methods. Synthesis methods include the following:

I. *Chemical reduction of metal salts:* Metal oxide nanocrystals are generally prepared by the decomposition of precursor compounds such as metal acetates, acetylacetonates, and cupferronates in appropriate solvents, often under solvothermal conditions. Metal chalcogenide or pnictide nanocrystals are obtained by the reaction of metal salts with a chalcogen or pnicogen source or the decomposition of single-source precursors under solvothermal or thermolysis conditions. The addition of suitable capping agents, such as long-chain alkane thiols, alkyl amines, and trioctylphosphine oxide, during the synthesis of nanocrystals enables the control of size and shape. Monodisperse nanocrystals are obtained by postsynthesis size-selective precipitation. Solvothermal and other reaction conditions are employed for synthesis to exercise control over the size and shape of the nanocrystals (77).
II. *Sol-gel technique:* The sol-gel process, also known as chemical solution deposition, is a wet chemical synthesis approach that can be used to generate nanoparticles by gelation, precipitation, and hydrothermal treatment. The sol-gel method is widely used in the fields of materials and chemical science. Such methods are used primarily

for the fabrication of materials starting from a chemical solution (or sol) that acts as the precursor for an integrated network (or gel) of either discrete particles or network polymers. By changing certain experimental parameters (including dopant introduction and heat treatment) and properly choosing some other surfactants (including inverted micelles, polymer matrix architecture based on block copolymers or polymer blends, porous glasses, and ex situ particle capping), it is possible to control the size distribution and stability control of quantum-confined semiconductors, metal, and metal oxide nanoparticles. However, the basic chemistry of the sol–gel process is complex due to the different reactivities of the network-forming and the network-modifying components and the wide variety of reaction parameters (78).

III. *Polyol:* This process uses the precipitation of a solid while heating sufficient precursors in multivalent and high boiling alcohol. The alcohol itself acts as a stabilizer, limiting particle growth and prohibiting particle agglomeration. Due to the high temperature that can be applied, highly crystallized oxides are often yielded. The synthesis is also relatively easy to perform (79).

IV. *Microemulsions:* The microemulsion has been demonstrated as a very versatile and reproducible method that allows control over nanoparticle size and yields nanoparticles with a narrow size distribution. Microemulsions are homogeneous in macroscale and microheterogeneous in nanoscale dispersion, with two immiscible liquids consisting of nanosized domains of one or both liquids stabilized by an interfacial film of surface active molecules. The essential distinction between normal emulsion and microemulsion is their particle size and stability. Normal emulsions age by coalescence of droplets and Ostwald ripening. Microemulsions are thermodynamically stable, single optically isotropic, and usually form spontaneously. Microemulsions have ultralow interfacial tension, large interfacial area, and a capacity to solubilize both aqueous and oil-soluble compounds (80).

V. *Electrochemical synthesis:* Electrodeposition is a process that uses electrical current to deposit a composite layer nanostructure-containing material onto a desired substrate. Basically, this method involves the use of a two-electrode or three-electrode electrochemical system. Recently, the electrochemical deposition techniques have been used by many researchers for zero-, one-, two-, and three-dimensional nanostructures using pulse electrodeposition in a three-electrode cell system. In pulse electrodeposition, there are four operation parameters influencing the depositing of nanoparticles on the substrate: the higher potential, the lower potential, the potential on time, and the potential off time. By applying specific potential pulses with time intervals for the total experimental time, zero-, one-, two-, and three-dimensional nanostructures were deposited on the working electrode (81).

Physical methods include the following:

I. *Exploding wire technique:* This technique is one of the newest and simplest methods for producing metal nanoparticles. The explosion is achieved when a very high current density is applied to a thin metal wire, causing the wire to explode to very small fragments. This process involves wire heating and melting followed by wire evaporation, formation of a high-density core surrounded by low-density ionized corona, coronal compression, and fast expansion of the explosion products (8). The exploding wire experiments were mostly employed for the generation and investigation of plasma and shock waves (82, 83).

II. *Plasma spraying:* Plasma spraying, a branch of thermal spray processes, has attracted attention for synthesizing bulk nanostructured coatings and near net shape components. During plasma spraying, powder is injected into the high-temperature plasma jet, where powder is melted/heated and accelerated towards the substrate. Molten, semi-molten, and heated particles exiting from the plasma jet impact the substrate, flatten, and rapidly solidify on the substrate. Further, plasma spray allows the multilayered structures to be processed in a successive spray process using only one piece of equipment, thus allowing the deposition of several materials with controlled porosity and microstructure. This is attractive in synthesizing the solid oxide fuel cells (SOFCs) via plasma spraying compared to tape casting and screen printing, which include multiple instruments and processes that also lead to interreaction between adjacent cell layers (84).

III. *Chemical vapor deposition:* In general, the chemical vapor deposition process involves three key steps: the generation of active gaseous reactant species, the transport of the gaseous species into the reaction chamber, and the formation of intermediate species when the gaseous reactants undergo gas phase reactions. At a high temperature (greater than the decomposition temperatures of intermediate species inside the reactor), homogeneous gas phase reactions can occur where the intermediate species undergo subsequent decomposition and/or chemical reaction, forming powders and volatile byproducts in the gas phase. The powder will be collected on the substrate surface and may act as crystallization centers, and the byproducts are transported away from the deposition chamber. The deposited film may have poor adhesion due to temperatures below the dissociation of the intermediate phase and diffusion/convection of the intermediate species across the boundary layer (a thin layer close to the substrate surface). These intermediate species subsequently undergo four steps: the absorption of gaseous reactants onto the heated substrate and the heterogeneous reaction at the gas–solid interface (i.e., heated substrate), which produces the deposit and byproduct species. The deposits will diffuse along the heated substrate surface, forming the crystallization center and growth of the film. Gaseous byproducts are removed from the boundary layer through diffusion or convection. The unreacted gaseous precursors and byproducts will be transported away from the deposition chamber (85).

IV. *Microwave irradiation:* The high penetration depth of microwave radiation and the effects of specific heating mechanisms resulted in enhanced reaction kinetics in the manufacturing of nanomaterials. There is an assumption that both nonthermal and thermal effects could be responsible for the fast kinetics and effects on morphology. The effect of microwave irradiation on chemical processes is still not clear. This effect could be due to the existence of so-called nonthermal microwave effects, which cannot be easily estimated by temperature measurements. It might lead to a sudden acceleration of reaction rates. Microwave technology has its advantages, but those advantages have not yet been fully grasped or properly applied to chemical processes (86).

V. *Pulsed laser ablation:* The generation of nanoparticles using pulsed laser ablation has inherent advantages compared to conventional methods, such as the purity and stability of the fabricated nanoparticle aerosols and colloids. The nanoparticle generation rate is 100 times higher in air compared to water. Higher-pulse energies lead to higher productivity; however, due to shielding effects, the optimum was determined at moderate pulse energies around 200 µJ, measured by the highest absorption intensity of the plasmon resonance of the colloids. Pulsed laser ablation is an alternative method addressing the deficits of the conventional methods and offering the access to an unlimited nanomaterial spectrum because the nanoparticles may be generated from almost any solid material. The advantages of this method are the high purity of the nanomaterial, the material variety, and the in situ dispersion of the nanoparticles in a variety of liquids, thus allowing safe and stable handling of the colloids (87).

VI. *Supercritical fluids:* Supercritical fluids have also been proposed as media to produce nanomaterials. The properties that make supercritical fluids particularly attractive, as a rule, are gas-like diffusivities, the continuously tunable solvent power/selectivity, and the possibility of complete elimination at the end of the process. Particularly, the mix of gas-like and liquid-like properties can be useful in many applications. The most widely used supercritical fluid is carbon dioxide (CO_2), which is cheap, nonpolluting, and has critical parameters that are simply obtained in an industrial apparatus. However, ammonia, alcohols, light hydrocarbons, and water also have been proposed, among others, for nanomaterial production at supercritical conditions (88).

VII. *Sonochemical reduction:* Sonochemical processes have been actively researched in the field of synthesis of metal nanoparticles and nanostructured materials. Controlling the reduction of metal ions is important because the reduction processes can be often applied to the synthesis of various functional materials. There are a number of reports using sonochemical processes to synthesize various types of materials, such as Fe nanoparticles, CdS nanoparticles, TiO_2 nanoparticles, Pd-Sn nanoparticles, self-assembled $SrMoO_4$, superstructures, hollow $PbWO_4$ nanospindles, Pt-Ru bimetallic nanoparticles, meso- and macro-porous TiO_2, Pt nanoparticles on multiwalled carbon nanotubes, Ag_2Se nanocrystals, and hollow PbS nanospheres. The process of sonochemistry triggers and accelerates chemical reactions with acoustic energy, which effectively shortens reaction time, improves rates of production, reduces reaction conditions, and achieves some reactions that cannot be accomplished by conventional methods (89).

Bioacceptable nanometals in industrial-scale manufacturing method should have the following characteristics:

- Have control of the shape of particle
- Be able to liberate monodisperse particles
- Be scalable, economic, easy, and cheap
- Have less toxic precursors, in water or more environmentally benign solvents (e.g., ethanol)
- Involve the least number of reagents
- Have reaction temperatures close to room temperature
- Have as few synthetic steps as possible (one-pot reaction), thus minimizing the quantities of generated byproducts and waste

Ayurved alchemical manufacturing methods for BNM are a mix of synthesis and physical methods of nanofabrication. A modern method of nanofabrication of BNM on an industrial scale is described in **Table 3.6**.

Table 3.6 Analogy of Nanofabrication Principles of *Ayurved* (Science of Life) with Proposed Modern Equipment Design

Principle	Proposed Modern Manufacturing Method	Ideation
Shodhana	Metal thinning into sheets; heating and immersing into various liquids	To make metal brittle
Vishesha shodhana	Treatment of thin sheets of metal with respective herb juices, decoctions, or other substances	To minimize further toxic effects of metals
Bhavana	Ball milling of metal sheets for mechanical attrition	To transduce metal for further nanofabrication
Jarana samskara	Mixing of ball-milled metal with various herbs and substances	To induce therapeutic properties
Marana	Thermal decomposition in a furnace	To make one-dimensional bioacceptable nanometals

3.2.3.1 Pure metal from ore

Extraction of a metal refers to all of the processes involved in the conversion of a raw material, such as a metallic ore, to a final form in which the metal can be used. In some instances, metal production involves relatively few steps because the metal already occurs in an elemental form in nature, such as with gold, silver, platinum, and other so-called noble metals. These metals normally occur in nature uncombined with other elements and can therefore be put to use with comparatively little additional treatment. In the majority of cases, however, metals occur in nature as compounds, such as oxides or sulfides, and must be converted to their elemental states (90).

The first step in metal production always involves some form of mining. Mining involves removing the metal in its free or combined state from the earth's surface. The two most common forms of mining are surface and sub-surface mining. In the former case, the metal or its ore can be removed from the upper few meters of the earth's surface. For example, much of the world's copper is obtained from huge open-pit mines ranging in depth to as much as 0.6 miles (1 km) and in width to as much as 2.25 miles (3.5 km). Subsurface mining is used to collect metallic ores that are at greater depths below the earth's surface.

A few metals can be obtained from seawater rather than (or in addition to) being taken from the earth's crust. For example, magnesium is obtained from seawater; every cubic mile of seawater contains about 6 million tons of magnesium, primarily in the form of magnesium chloride.

In most cases, metals and their ores occur as complex mixtures containing rocks, sand, clay, silt, and other impurities (91). In the first step, the ore is ground into a powder and separated from the waste materials with which it occurs. Nowadays, ore can be purified by the froth flotation method, which is used for ores of copper and zinc. In this method, impure ore is mixed with water and a frothing agent, such as pine oil, is added. A stream of air is blown through the mixture, causing it to bubble and froth. Impurities such as sand and rock are wetted by the water and sink to the bottom of the container, while metal ore particles adsorb the pine oil. The oil-coated ore particles float and can be skimmed off.

Metals always occur in their oxidized state in ores, often as the oxide or sulfide of the metal. In order to obtain the metal, it must be reduced. With ores of iron, reduction can be carried out by reacting oxides of iron with carbon and carbon monoxide in the blast furnace. Some metal oxides cannot be reduced easily, such as aluminum oxide; therefore, one has to resort to the electrolysis method.

In some cases, an ore is treated to change its chemical state before being reduced. The most common ores of zinc, for example, are the sulfides. These compounds are first roasted in an excess of air and converted to zinc oxide, which is reduced either by reacting it with coke or by electrolysis.

Mixtures that contain two or more metals are known as alloys. Pure metals themselves sometimes are not suitable for a given application. For example, pure gold is too soft for most uses and is combined with other metals (often copper) to form harder, more resistant mixtures. Steel is the most widely used alloy of iron. An alloy of a metal with mercury is commonly called an amalgam (92).

Purification of an impure metal is called refining. One ancient process for extracting the silver from lead was cupellation, where the unwanted metal undergoes oxidation faster than the required metal. Lead was melted in a bone ash test or cupel and air was blown across the surface. Lead is oxidized to litharge (oxide of lead), leaving behind silver. In the 18th century, the process was carried out using a kind of reverberatory furnace, in which air was blown over the surface of the molten lead from bellows or blowing cylinders.

The Pattinson process (1833) is applicable when two metals have different melting points. Lead and silver melt at different temperatures. The equipment consisted of a row of iron pots, which could be heated from below. Lead containing silver was charged to the central pot and melted. This was allowed to cool; as the lead solidified, it was skimmed off and moved to the next pot in one direction, and the remaining metal was then transferred to the next pot in the opposite direction. The process was repeated in the pots successively, resulting in lead accumulating in the pot at one end and silver at the other.

The Parkes process (1850) uses zinc to form a material that the silver enters. This floats on the lead and can be skimmed off. The silver is then recovered by volatilizing the zinc. The impure copper was black, which was repeatedly melted to purify it, alternately oxidizing and reducing it. In one of the melting stages, lead was added. Gold and silver preferentially dissolved in this, thus providing a means of recovering these precious metals. To produce purer copper suitable for making copper plates or hollow ware, further melting processes were undertaken, using charcoal as fuel. The repeated application of such fire-refining processes was capable of producing copper that was 99.25% pure.

Electrolytic refining, first used in 1869, is the best method to purify a metal to its purest form. The purest copper is obtained by an electrolytic process, undertaken using a slab of impure copper as the anode and a thin sheet of pure copper as the cathode. The electrolyte is an acidic solution of copper sulfate. By passing electricity through the cell, copper is dissolved from the anode and deposited on the cathode. However, impurities either remain in the solution or collect as an insoluble sludge (93).

In order to isolate noble metals, pyrolysis and/or hydrolysis procedures are used. In pyrolysis, the noble metals are released from the other materials by solidifying in a melt to become cinder and then poured off or oxidized. In hydrolysis, the noble-metalliferous products are dissolved either in aqua regia (consisting of hydrochloric acid and nitric acid) or in hydrochloric acid and chlorine gas in solution. Subsequently, certain metals can be precipitated or reduced directly with a salt, gas, organic, and/or nitrohydrate connection. Afterwards, they go through cleaning stages or are recrystallized. The precious metals are separated from the metal salt by calcination. The noble-metalliferous materials are hydrolyzed first and thermally prepared (pyrolyzed) thereafter (94).

3.2.3.2 Ball milling of metals

The fundamental principle of size reduction in mechanical attrition devices lies in the energy imparted to the sample during impacts between the milling media. The nanoparticles from mechanical attrition are produced by a "top-down" process. Such nanoparticles are formed in a mechanical device, generically referred to as a mill, in which energy is imparted to a course-grained material to effect a reduction in particle size. Under certain conditions, the resulting particulate powders can exhibit nanostructural characteristics on at least two levels. First, the particles themselves, which normally possess a distribution of sizes, can be considered nanoparticles if their average characteristic dimension (diameter for spherical particles) is less than 100 nm (95). Second, many of the materials milled in mechanical attrition devices are highly crystalline, such that the crystallite (grain) size after milling is often between 1 and 10 nm in diameter. Such materials are termed *nanocrystalline* (96).

Formation of nanocrystalline material during mechanical alloying and milling was validated by Fecht et al. (97). Similar crystalline sizes may be obtained through conventional ball mills and other techniques, suggesting that it is total strain, rather than milling energy, that decides the minimum attainable grain size by mechanical milling. Various milling parameters (milling temperatures, the nature of products, and the number of phases present during mechanical milling and alloying) have a pronounced influence on limiting attainable grain size and product phases.

According to Harris (98), the major factors contributing to grind limit are as follows:

- Increasing resistance to fracture
- Increasing cohesion between particles, with decreasing particle size causing agglomeration
- Excessive clearance between impacting surfaces
- Coating of the grinding medium by fine particles that cushion the microbed of particles from impact
- Surface roughness of the grinding medium
- Bridging of large particles to protect smaller particles in the microbed
- Increasing apparent viscosity as particle size decreases

3.2.3.3 Mixing of herbs with metal powder in a closed chamber

An increase in the ratio of surface area to volume translates to an exponential increase in the portion of constituent atoms at or near the surface, creating more sites for bonding or reaction with surrounding materials. This results in improved properties, such as increased strength and greater chemical or heat resistance. Due to the attractive forces between the individual nanoparticles, "wetting out" or combining them with the liquid vehicle only really disperses agglomerates of the nanoparticles. In addition to using specialty surfactants, high shear forces are necessary to break up groups of these agglomerates. Therefore, the premix can be fed into an ultra-high shear mixer for creating the final dispersion and to transfer the succeeding phase of the process (99).

3.2.3.4 Thermal decomposition of metals in a furnace

The simplest method to produce metallonutraceuticals could be by heating the desired dispersion mixture of ball-milled metal and respective herbs kept in a heat-resistant crucible. This method will be appropriate only for materials that have a high vapor pressure at the heated temperatures, which can be as high as 2,000°C (e.g., gold, silver, iron). To carry out reactive synthesis, metals with very low vapor pressure have to be fed into the furnace in the form of a suitable precursor, such as organometallics, which decompose in the furnace to produce a condensable material. Therefore, for low vapor-pressure metals, ancient alchemists has suggested *jarana* (conversion of metals with low melting points into fine powder).

The energy is introduced into the crucible by arc heating. The metal atoms are evaporated into the crucible atmosphere, which is either inert (e.g., helium) or reactive (so as to form a compound), as supplied by gas inlets. The nanometal clusters would continue to grow if they remain in the supersaturated region. To control their size, they need to be rapidly removed from the supersaturated environment by a carrier gas. The nanometal cluster size and its distribution are controlled by only three parameters:

1. The rate of evaporation (energy input)
2. The rate of condensation (energy removal)
3. The rate of gas flow (cluster removal)

The hot atoms of the nanometals lose energy by collision with the atoms of the cold gas and undergo condensation into small clusters via homogeneous nucleation. When a compound is being synthesized, these precursors react in the gas phase and form a compound with the material, which is separately injected in the reaction chamber (100).

To stabilize metallic nanoparticles, the following steps are required:

1. Place them in an inert environment, such as an inorganic matrix or polymer.
2. Add surface-protecting reagents.

3. Make organic ligands.
4. React inorganic capping materials.

Therefore, in the collection chamber for the final product, nanometals can be treated with various substrates such as polymers, plant extracts, and emulsifying agents, to give final stability to metallonutraceuticals.

3.3 Conclusion

The ancient scholars of India developed excellent pharmaceutical techniques approximately 6,000 years ago for the fabrication of metallonutraceuticals. These techniques included processes for producing metallonutraceuticals in nanoforms, as well as dosages, the therapeutic properties of the metal compounds, and even preventive measures for hypersensitivity reactions. By taking these basic principles, modern manufacturing techniques can be devised and nutraceutical and therapeutic properties can be established by using reverse pharmacology, evidence-based medicine, and translational research. These ancient nanotechniques can also be used on new metals, such as platinum and manganese, to develop novel metallonutraceutical molecules.

References

1. Shastri, A. 1957. *Yajurveda 18/13, Krishna-Yajurvediya-Taittiriya-Samhita,* 2nd ed. Hyderabad, India: Vijayaraghavan Bashyam.
2. Samhita, C. Sutrasthan 5/70, Ayurved Deepika Commentary, Chaukhamba Sanskrit Sansthan, Varanasi.
3. Sushrut Samhita, Uttarsthan 38/60, Shastri Ambikadatta, Chaukhamba Sanskrit Sansthan, Varanasi.
4. Mishra, S.N., ed. Rasendra Chudamani 14/1, Sharma Somdev, Chaukhamba Orientalia, Varanasi.
5. Mishra, S.N., ed. Rasendra Chudamani 4/2, Sharma Somdev, Chaukhamba Orientalia, Varanasi.
6. Mishra, S.N., ed. Rasendra Chudamani 14/10, Sharma Somdev, Chaukhamba Orientalia, Varanasi.
7. Sharma, G., and S.N. Mishra, eds. *Ayurved Prakash 3/16-17, Upadhyay Madhav,* 2nd ed. Aligarh, India: Agrawal Press.
8. Mishra, S.N., ed. Rasendra Chudamani 14/18, Sharma Somdev, Chaukhamba Orientalia, Varanasi.
9. Mishra, S.N., ed. Rasendra Chudamani 14/26, Sharma Somdev, Chaukhamba Orientalia, Varanasi.
10. Sharma, G., and S.N. Mishra, eds. *Ayurved Prakash, 3/85, Upadhyay Madhav,* 2nd ed. Aligarh, India: Agrawal Press.
11. Sharma, G., and S.N. Mishra, eds. *Ayurved Prakash 3/94, Upadhyay Madhav,* 2nd ed. Aligarh, India: Agrawal Press.
12. Mishra, S.N., ed. Rasendra Chudamani, 14/40, Sharma Somdev, Chaukhamba Orientalia, Varanasi.
13. Mishra, S.N., ed. Rasendra Chudamani 14/42, Sharma Somdev, Chaukhamba Orientalia, Varanasi.
14. Sharma, G., and S.N. Mishra, eds. *Ayurved Prakash, 3/117, Upadhyay Madhav,* 2nd ed. Aligarh, India: Agrawal Press.

15. Mishra, S.N., ed. Rasendra Chudamani, 14/80, Sharma Somdev, Chaukhamba Orientalia, Varanasi.
16. Mishra, S.N., ed. Rasendra Chudamani 14/88, Sharma Somdev, Chaukhamba Orientalia, Varanasi.
17. Mishra, S.N., ed. Rasendra Chudamani, 14/79, Sharma Somdev, Chaukhamba Orientalia, Varanasi.
18. Mishra, K.N.N. *Rasendrasar Sangraha 1/305, Bhatta Shri Gopalkrishna, Sanshodhan*, 4th ed.
19. Kulkarni, D.A. Rasratna Samucchaya. New Delhi, India: ML Publication.
20. Bruhatrasarajsundaram—Nagshodhanam, Dattaram Chaube, Motilal banarasidas.
21. Sharma, G., and S.N. Mishra, eds. *Ayurved Prakash 3/148, Upadhyay Madhav,* 2nd ed. Aligarh, India: Agrawal Press.
22. Mishra, S.N., ed. Rasendra Chudamani, 14/132-33, Sharma Somdev, Chaukhamba Orientalia, Varanasi.
23. Ras Tarangini 18/9, Sharma Sadananda, Motilal Banarasidas Publication, Varanasi.
24. Sharma, G., and S.N. Mishra, eds. *Ayurved Prakash, 3/182, Upadhyay Madhav,* 2nd ed. Aligarh, India: Agrawal Press.
25. Mishra, S.N., ed. Rasendra Chudamani, 14/167, Sharma Somdev, Chaukhamba Orientalia, Varanasi.
26. Sharma, G., and S.N. Mishra, eds. *Ayurved Prakash, 4/65, Upadhyay Madhav,* 2nd ed. Aligarh, India: Agrawal Press.
27. Bruhatrasarajsundaram – Saptadhatudravan, Dattaram Chaube, Motilal banarasidas.
28. Macdonell, A.A. 1922. *Rig Veda 8.100: Hymns from the Rigveda.* Calcutta, India.
29. Macdonell, A.A. 1922. *Parvata_vrdh: Rig Ved. 9:46. Hymns from the Rigveda.* Calcutta, India.
30. Macdonell, A.A. 1922. *Rig Veda. 9,2. Hymns from the Rigveda.* Calcutta, India.
31. Sartha Vagbhatta—Sutrasthan 10/1, Chaukhamba Sanskrit Sansthan, Varanasi, Reprint.
32. Sartha Vagbhatta—Sutrasthan 9/16, Chaukhamba Sanskrit Sansthan, Varanasi, Reprint.
33. Sartha Vagbhatta—Sutrasthana—9/20-21. Chaukhamba Sanskrit Sansthan, Varanasi, Reprint.
34. Sartha Vagbhatta—Sutrasthan 1/18, Chaukhamba Sanskrit Sansthan, Varanasi, Reprint.
35. Sushrut Uttarasthana 6/6, Shastri Ambikadatta, Chaukhamba Sanskrit Sansthan, Varanasi.
36. Sartha Vagbhatta—Sutrasthan 9/22, Chaukhamba Sanskrit Sansthan, Varanasi, Reprint.
37. Bruhatrasarajsundaram—Bhasma Praman, Dattaram Chaube, Motilal banarasidas.
38. Mishra, S.N., ed. Rasendra Chudamani, 14/22, Sharma Somdev. Chaukhamba Orientalia, Varanas.
39. Mishra, S.N., ed. Rasendra Chudamani, 14/24, Sharma Somdev, Chaukhamba Orientalia, Varanasi.
40. Bruhatrasarajsundaram—Suvarnadosha shanti, Dattaram Chaube, Motilal banarasidas.
41. Mishra, S.N., ed. Rasendra Chudamani, 14/38, Sharma Somdev, Chaukhamba Orientalia, Varanas.
42. Mishra, S.N., ed. Rasendra Chudamani, 5/31, Sharma Somdev, Chaukhamba Orientalia, Varanas.
43. Mishra, S.N., ed. Rasendra Chudamani, 14/39, Sharma Somdev, Chaukhamba Orientalia, Varanasi.
44. Bruhatrasarajsundaram—Raupyadosha shanti, Dattaram Chaube, Motilal banarasidas.
45. Mishra, S.N., ed. Rasendra Chudamani 14/69, Sharma Somdev, Chaukhamba Orientalia, Varanasi.
46. Mishra, S.N., ed. Rasendra Chudamani 14/70, Sharma Somdev, Chaukhamba Orientalia, Varanasi.
47. Chaube, D. Bruhatrasarajsundaram—Tamradosha shanti, Motilal banarasidas
48. Mishra, S.N., ed. Rasendra Chudamani 14/79, Sharma Somdev, Chaukhamba Orientalia, Varanasi.
49. Mishra, S.N., ed. Rasendra Chudamani 14/20-21, Sharma Somdev, Chaukhamba Orientalia, Varanasi.

50. Mishra, K.N.N. 1967. Rasendra Sar Sangraha 1/3.6.2, Bhatta Shri Gopalkrishna, Sanshodhan, 4th ed.
51. Chaube, D. Bruhatrasarajsundaram—Lohadosha shanti, Motilal banarasidas.
52. Sharma, G., and S.N. Mishra, eds. *Ayurved Prakash 3/186-187, Upadhyay Madhav,* 2nd ed. Aligarh, India: Agrawal Press.
53. Chaube, D. Bruhatrasarajsundaram—Nagdosha shanti. Motilal banarasidas.
54. Sharma, G., and S.N. Mishra, eds. *Ayurved Prakash, 3/152, Upadhyay Madhav,* 2nd ed. Aligarh, India: Agrawal Press.
55. Sharma, G., and S.N. Mishra, eds. *Ayurved Prakash, 3/153, Upadhyay Madhav,* 2nd ed. Aligarh, India: Agrawal Press.
56. Chaube, D. Bruhatrasarajsundaram—Vangadosha shanti, Motilal banarasidas.
57. Sharma, G., and S.N. Mishra, eds. *Ayurved Prakasha, 3/183, Upadhyay Madhav,* 2nd ed. Aligarh, India: Agrawal Press.
58. Chaube, D. Bruhatrasarajsundaram—Tamradosha shanti, Motilal banarasidas.
59. Mishra, S.N., ed. Rasendra Chudamani, 14/165, Sharma Somdev, Chaukhamba Orientalia, Varanasi.
60. Sharma, G., and S.N. Mishra, eds. Ayurved Prakash, 4/73, Upadhyay Madhav, 2nd ed. Aligarh, India: Agrawal Press.
61. Chaube, D. Bruhatrasarajsundaram—Kamsyaloh, Motilal banarasidas
62. Rasajalanidhi—3/Vartaloham/1.
63. Kwak, N.S., and D.J. Jukes. 2001. Functional foods. Part 2: The impact on current regulatory terminology. *Food Control* 12:109–117.
64. Espin, J.C., M.T. Garcia-Conesa, and F.A. Tomas-Barberan. 2007. Nutraceuticals: Facts and fiction. *Phytochemistry* 68:2986–3008.
65. Kruger, C.L., and S.W. Mann. 2003. Safety evaluation of functional ingredients. *Food and Chemical Toxicology* 41:793–805.
66. DeFelice, S.L. 1995. The nutraceutical revolution: Its impact on food industry R&D. *Trends in Food Science and Technology* 6:59–61.
67. Rastogi, S. 2010. Building bridges between Ayurveda and Modern Science. *International Journal of Ayurveda Research* 1:41–46.
68. Mukherjee, P., R. Bhattacharya, N. Bone, et al. 2007. Potential therapeutic application of gold nanoparticles in B- chronic lymphocytic leukemia (BCLL): Enhancing apoptosis. *Journal of Nanobiotechnology* 5:4.
69. Bhowmick, T.K., A.K. Suresh, S.G. Kane, et al. 2009. Physicochemical characterization of an Indian traditional medicine, Jasada Bhasma: Detection of nanoparticles containing non-stoichiometric zinc oxide. *Journal of Nanoparticle Research* 11:655–664.
70. Dutta, P.K. 1996. *General and Inorganic Chemistry,* 11th ed. Kolkata, India: Sarat Book House.
71. Subrahmanyam, C.V.S., J.T. Setty, S. Suresh, and V.K. Devi. 2002. *Size Reduction: Pharmaceutical Engineering.* Dehli, India: Vallabh Prakashan.
72. Sarkar, P.K., A.K. Chaudhary, B. Ravishankar, et al. 2007. Physico-chemical evaluation of Lauha Bhasma and Mandura Bhasma. *Indian Drugs* 44:21–26.
73. Kumar, A., A.G. Nair, A.V. Reddy, and A.N. Garg. 2006. Availability of essential elements in bhasma: Analysis of Ayurvedic metallic preparations by INAA. *Journal of Radioanalytical and Nuclear Chemistry* 270:173–180.
74. Suryanarayana, C., and F.H. Froes 1992. The structure and mechanical properties of metallic nanocrystals. *Metallurgical Transactions A* 23:1071–1081.
75. Pokropivny, V.V., and V.V. Skorokhod. 2007. Classification of nanostructures by dimensionality and concept of surface forms engineering in nanomaterial science. *Materials Science and Engineering C* 27:990–993.
76. http://ltp.epfl.ch/files/content/sites/ltp/files/shared/Teaching/Master/03-IntroductionToNanomaterials/LectureSupportAll.pdf
77. Rao, C.N.R., S.R.C. Vivekchand, K. Biswasa, and A. Govindaraja. 2007. Synthesis of inorganic nanomaterials. *Dalton Transactions* 3728–3749.

78. Jitendra, N., N. Tiwari, N. Rajanish, et al. 2012. Zero-dimensional, one-dimensional, two-dimensional and three-dimensional nanostructured materials for advanced electrochemical energy devices. *Progress in Materials Science* 57:724–803.
79. Feldmann, C. 2003. Polyol mediated synthesis of nanoscale functional materials. *Advanced Functional Materials* 13:868–873.
80. Zielińska-Jurek, A., J. Reszczyńska, E. Grabowska, and A. Zaleska. 2012. Nanoparticles preparation using microemulsion systems. *In Microemulsions: An Introduction to Properties and Applications*, ed. R. Najjar. Rijeka, Croatia: InTech.
81. Tiwari, J.N., N. Rajanish, and K.S. Kim. 2012. Zero-dimensional, one-dimensional, two-dimensional and three-dimensional nanostructured materials for advanced electrochemical energy devices. *Progress in Materials Science* 57:724–803.
82. Sen, P., J. Ghosh, A. Abdullah, et al. 2004. *Process and Apparatus for Producing Metal Nanoparticles.* Available at http://www.google.com/patents/US20070101823. Accessed November 11, 2013.
83. Sen, P., J. Ghosh, A. Abdullah, et al. 2003. Preparation of Cu, Ag, Fe and Al nanoparticles by the exploding wire technique. *Journal of Chemical Sciences–Indian Academy of Sciences* 115:499–508.
84. Hui, R., Z. Wang, O. Kesler, et al. 2007. Thermal plasma spraying for SOFCs:applications, potential advantages, and challenges. *Power Sources* 172:493.
85. Choy, K.L. 2000. In *Handbook of Nanostructured Materials and Nanotechnology, Vol.1: Synthesis and Processing*, ed. H.S. Nalwa, 533. San Diego, CA: Academic Press.
86. Gizdavic-Nikolaidis, M.R., D.R. Stanisavljev, A.J. Easteal, et al. 2010. Rapid and facile synthesis of nanofibrillar polyaniline using microwave radiation. *Macromolecular Rapid Communications* 31:657–661.
87. Hann, A., S. Barcikowski, and B.N. Chichkov. 2008. Influences on nanoparticle production during pulsed laser ablation. *Journal of Laser Micro/Nanoengineering* 3:73–77.
88. Reverchon, E., and R. Adamia. 2006. Nanomaterials and supercritical fluids. *Journal of Supercritical Fluids* 37:1–22.
89. Zhang, K., B.-J. Park, F.-F. Fang, and H.J. Choi. 2009. Sonochemical preparation of polymer nanocomposites. *Molecules* 14:2095–2110.
90. R.F. Tylecote. 1992. *A History of Metallurgy.* London, UK: Institute of Materials, 157–158.
91. Day, J., and R.F. Tylecote. 1991. *The Industrial Revolution in Metals.* London, UK: Institute of Metals.
92. Engh, T.A. 1992. *Principles of Metal Refining.* New York, NY: Oxford University Press.
93. Klein, C. 2002. *The Manual of Mineral Science*, 22nd ed. New York, NY: Wiley. 94. Patterson, J.W., and R. Passino, eds. 1990. *Metals Speciation, Separation and Recovery*, vol. 2. Boca Raton, FL: Lewis Publications.
95. M.K. Datta, S.K. Pabi, and B.S. Murty. 2000. Thermal stability of nanocrystalline Ni silicides synthesized by mechanical alloying. *Materials Science and Engineering* 219:A284.
96. Gleiter, H. 1989. Nanocrystalline materials. *Progress in Materials Science* 33:223.
97. Fecht, H.J., E. Hellstern, Z. Fu, and W.L. Johnson. 1990. Nanocrystalline metals prepared by high-energy ball milling. *Metallurgical and Materials Transactions* 21:2333–2337.
98. Harris, C.C. 1967. Minerals beneficiation—size reduction-time relationships of batch grinding. *Transactions of the Society of Mining Engineers* 17.
99. Ross. *A Summary of Nanomaterial Mixing Technology.* Available at http://www.mixers.com/whitepapers/nmt.pdf.
100. Alagarasi, A. *Introduction to Nanomaterials.* Available at http://www.nccr.iitm.ac.in/2011.pdf0.

Characterization of Metallonutraceuticals

Jayant Lokhande and Yashwant Pathak

Contents

4.1 Characterization and possible properties of metallonutraceuticals

Metallonutraceuticals are nothing but biological metallic nanocrystals (1). The physical and chemical properties of metals are mostly determined by the motion of electrons. When the motion of electrons is confined to a nanometer-length scale (1–100 nm), which happens in nanomaterials, unusual effects are observed. The physicochemical properties of the metal nanoparticles are completely different than those of their respective bulk materials due to their very small size and high surface-to-volume ratio. The properties of bulk materials are solely composition dependent. However, as the size of the particles decreases to a few nanometers, the electronic structure is altered from the continuous band to discrete electronic levels. Thus, the properties of the nanomaterial become size dependent.

In small nanoparticles, a large fraction of the atoms will be on the surface. The atoms on the surface are chemically more active compared to the bulk atoms because they usually have fewer adjacent coordinate atoms and unsaturated sites. The striking property, which is derived from the increasing surface energy of nanomaterials, is the decrease in the melting point. Furthermore, the overlapping of different grain sizes affects the physical strength of the material. Also, when the crystallites of a material are reduced to the nanometer scale, there is an increase in the role of interfacial defects, such as grain boundaries, triple junctions, and elastically distorted layers (2).

4.1.1 Main characteristics of metallic nanocrystals

The main characteristics of metallic nanocrystals described in this section are as follows:

1. Surface effects and less coordination sites
2. Large surface energy
3. Specific electronic structure
4. Plasmon excitation

5. Quantum confinement
6. Short-range order

4.1.1.1   Surface effects and less coordination sites

This property can be illustrated by the following thought experiment: Consider a cube with an edge width of 1 m. Cut it into two pieces, thereby exposing additional faces of the material. In this way, new visible or usable areas are added while the total volume remains the same. Repeat this exercise until all particles reach a size of approximately 1 nm. The result is a group of small particles with enormous surface area that occupy the same volume you started with.

Some nanomaterials exhibit a surface area that is equivalent to a football field, but they weigh just a few grams. The large surface area makes the nanoparticles highly soluble in liquids. This property is useful in applications such as paints, pigments, medicine pills, and cosmetics. A 30-nm iron particle has 5% of its atoms on the surface, with the remaining atoms residing inside the particle. However, a 3-nm particle has 50% of its atoms on the surface. The atoms on the surface, which are not bonded on one side, are far more active than the atoms residing inside (which are bonded all around; thus, they are "satisfied"). An increase in surface area therefore leads to an increase in reactivity (3).

The atoms on the surface have fewer neighbors than atoms in the bulk. Because of this lower coordination and unsatisfied bonds, the surface atoms are less stabilized than bulk atoms. Smaller particles have a larger fraction of atoms at the surface and a higher average binding energy per atom. The surface-to-volume ratio scales with the inverse size; therefore, numerous properties obey the same scaling law, including the melting point and other phase transition temperatures. Edge and corner atoms have an even lower coordination; thus, they bind foreign atoms and molecules more tightly.

The coordination number is also limited in narrow pores; therefore, the melting point and even the critical point of a fluid are greatly reduced. Phase transitions in metallonutraceuticals could be collective phenomena. With fewer atoms, a phase transition is less well defined; therefore, it is no longer sharp. In such cases, the Gibbs phase rule can lose its meaning because phases and components are no longer properly distinguishable. Small clusters behave more like molecules than bulk matter. It is therefore useful to think of different isomers that coexist over a temperature range rather than of different phases. There are numerous other concepts of thermodynamics that can break down, particularly when the system of interest consists of a single isolated cluster with a small number of atoms (4).

4.1.1.2   Large surface energy

Surface energy is defined as an excess quantity of energy because the atomic structure of the surface differs from the equilibrium bulk structure. For example, when creating a surface by cutting the material, atomic bonds are cut, which leads to an increase of energy and relaxations of atomic positions close to the surface.

At the nanoscale, the thermodynamic behavior of a particle deviates appreciably from that of respective bulk material. First of all, nanoparticles contain surfaces in which atomic sites differing by their distance to the surface are necessarily inequivalent. At the continuous scale, the thermodynamic modeling of surfaces is based on surface quantities that are in excess with respect to the bulk counterparts. In a nanoparticle where the total number of atoms are fixed, the thermodynamic behavior of the surface atoms (and in particular, the surface segregation) has direct and dramatic consequences on the structure and composition of the interior of the particles (5).

4.1.1.3   Specific electronic structure

The electronic structure of a metal particle critically depends on its size. For nanoparticles, the electronic states are not continuous; rather, they are discrete due to confinement of the electron wavefunction. The average spacing of successive quantum levels, d (known as the Kubo gap), is given by $d = 4Ef/3n$, where Ef is the Fermi energy of the bulk metal and n is the number of valence electrons in the nanoparticle (usually taken as its nuclearity). Thus, for example, for a silver nanoparticle of 3-nm diameter containing ~103 atoms, the value of d would be 5–10 meV. Because at room temperature the thermal energy (kT) equals 25 meV, the 3-nm particle would be metallic ($kT > d$). At low temperatures, the level spacings may become comparable to kT (especially in small particles), rendering them nonmetallic. Because of the presence of the Kubo gap, properties such as electrical conductivity and magnetic susceptibility exhibit quantum size effects. Discreteness of energy levels also brings about changes in the spectral features, especially those related to the valence band (6).

4.1.1.4   Plasmon excitation

A collective excitation of the dense electron gas in a metal is called plasmon. In the most general case, the term *plasmon* describes a longitudinally collective oscillation of the electron plasma relative to the crystal lattice. This excitation consists of a coherent motion of a high number of electrons in contrast to those excitations in which the external perturbation acts on a single electron. In general, one has to distinguish between bulk surface and particle (Mie) plasmons. The bulk plasmon denotes a collective excitation of the electron gas in the bulk of the metal, which propagates as a longitudinal charge density fluctuation at a resonance frequency (7).

The position and intensity of plasmon absorption and emission peaks are affected by molecular adsorption, which can be used in molecular sensors. For example, a fully operational prototype device detecting casein in milk has been fabricated. The device is based on detecting a change in absorption of a gold layer (8).

4.1.1.5 Quantum confinement

In any material, substantial variation of fundamental electrical and optical properties with reduced size will be observed when the energy spacing between the electronic levels exceeds the thermal energy (kT). In small nanocrystals, the electronic energy levels are not continuous as in the bulk; rather, they are discrete (finite density of states) because of the confinement of the electronic wavefunction to the physical dimensions of the particles. This phenomenon is called quantum confinement; therefore, nanocrystals are also referred to as quantum dots.

Moreover, nanocrystals possess a high surface area, and a large fraction of the atoms in a nanocrystal are on its surface. Because this fraction depends largely on the size of the particle (30% for a 1-nm crystal, 15% for a 10-nm crystal), it can give rise to size effects in the chemical and physical properties of the nanocrystal. According to the quantum confinement theory, electrons in the conduction band and holes in the valence band are confined spatially by the potential barrier of the surface or trapped by the potential well of the quantum box. Because of the confinement of both the electrons and the holes, the lowest energy optical transition from the valence to the conduction band increases in energy, effectively increasing the EG (confinement effect on the band gap). The sum of kinetic and potential energy of the freely moving carriers is responsible for the EG expansion. Therefore, the width of the confined EG grows as the characteristic dimensions of the crystallite decrease (9).

4.1.1.6 Short-range order

The short-range order of nanostructures determines multiparticle correlations in the arrangement of atoms in their elements. All macroscopic properties of the material are determined by the size and/or mutual arrangement of structural elements. Nanomaterials represent materials from a structural perspective somewhere in between the following two types (10).

Crystalline materials are composed of an orderly repeating array of atoms, molecules, or ions. They come in two general forms: single crystalline, which has single atoms arranged periodically on a three-dimensional lattice, and polycrystalline, which is a consolidated assembly of small single crystals. This material has short- and long-range orders, meaning that the manner in which atoms are arranged at any one location within a crystal is identical to the arrangement at any other location. Most metals are polycrystalline; when they break, the grains or crystallites can be observed as randomly orientated within the material. The surface between grains is called the grain boundary.

Amorphous materials are noncrystalline. They have short-range order but not long-range order. These materials are not composed of grains. A broken edge is observed at a smooth surface because they are considered to be a cooled liquid rather than a true solid. For example, polymers and ordinary glass are amorphous materials.

4.1.2 Important properties of metallic nanoparticles

4.1.2.1 Optical properties

In the linear optical properties, the surface plasma absorption peak reveals a blue shift as the particle size decreases. This can be fully elucidated by the hard spherical model in an inert environment, which appraises the enlargement of energy splitting as the particle size decreases. The surface plasmon in resonance to the discrete energy level due to the splitting of quantum size confinement will be absorbed. The shift of absorbed surface plasmon frequency ω_s is inversely proportional to the particle size and thus recounted as the blue shift (11).

The abnormal red shift of metallic nanoparticles embedded in a reactive matrix is retrieved due to the electron diffusion outside the surface. The electromagnetic field can be enhanced at the shallow surface of the nanoparticles; therefore, it greatly pronounces the nonlinear optical susceptibility. To dispersed metallic nanoparticles, the random distribution of normal surface direction of the particles implies the excitation of surface plasmon, requiring phase-matching conditions.

Charles et al. observed a remarkable phenomenon containing both the blue and red shifts when the size of small silver particles is below 50 Å. Their results are similar to that of sodium particles on NaN_3, as observed by Smithard and Tran for sodium on NaCl (12, 13). Some authors interposed two surface response functions, d_θ and d_r, to explain the occurrence of both the blue and red shifts in the same matrix. The diffusion of electrons becomes severe as the particle size squeezes further or the particles are embedded in an active matrix and the dipole transition matrices are reduced to bring out a red shift. In general, the potential near the surface is relevant to the surrounding dielectric medium, which can suppress or diffuse electrons outside the surface. The diffuseness around the spherical boundary of free metallic particles (e.g., Ag/vacuum) is different from that of a metallic surface adsorbed with other molecules (14).

Nonlinear properties exist in metal nanomaterials, which have particle sizes or film thicknesses that are much smaller than the coherent length. The phase-matching condition is usually neglected and the surface nonlinearity makes a clear contribution due to the enhanced surface-to-volume ratio. Surface second harmonic generation from metals was established on the existence of the nonlinear $E(\Delta E)$ source term that has a large contribution at the boundary due to the discontinuity of the lattice structure and the presence of the bulk magnetic dipole term $E \times \delta H/\delta t$ arising from the Lorentz force of electrons. For metal particles with structures of inversion symmetry, the electric quadruple field within the selvedge region is the dominant source for the generation of second harmonic light. The excitation of surface plasmon (SP), which couples the incident field to propagate along the surface, is thus a main strategy for the enhancement of second harmonic generation. The efficiency of generating surface plasmon depends on the momentum conservation of the electromagnetic waves, which has a rather narrow bandwidth of wave vectors. The random orientation of the scattered light of the innumerable nanoparticles pursues the phase-matching condition (15).

4.1.2.2 Electrical properties

The electrical conductance for the agglomerated metallic particles or the deposited nanosized thin films depends on the filling factor of the connected and empty spaces. At a low filling factor, the particles may be separated individually. The mechanism of conductivity on the thin film can be extensively addressed by a simple tunneling between localized insulating states that imply a high resistivity at low temperatures. When the filling factor is large enough for the particles to aggregate into clusters, and at least a conducting path is connected from one side of the sample to the other, the sample begins to exhibit current conduction. The resistivity will be dependent on the particle size R, the filling factor f, and the aggregation factor A, which represent the extent to which the particles are connected with each other to those that are discrete. Potential barriers usually occur between the boundaries of particles to particles; therefore, the conduction electrons are subjected to grain boundary scattering (16).

4.1.2.3 Magnetic properties

A larger surface-to-volume ratio of the small nanoparticles implies a stronger surface anisotropic field to frustrate and disorder the inner spins, causing quantum tunneling at higher temperatures. Unusual electronic and magnetic characteristics are prevalent at nonzero temperatures, such as the metal-insulator transition in transition metal oxides, non-Fermi-liquid behavior of highly correlated f-electron compounds, abnormal symmetry states of high-temperature superconducting cuprates, and novel bistability of semiconductor heterostructures. The low-temperature magnetic viscosity of these systems shows a constant value below a finite temperature, reflecting the independence of overbarrier thermal transitions and is the signature of quantum tunneling of the magnetization. Many exotic physical properties are associated with quantum size effects, such as the splitting of the continuum conduction band into discrete levels, electromagnetic field enhancement on the surface, and magnetic property changes from diamagnetic into paramagnetic and from ferromagnetic into superparamagnetic (17).

4.1.2.4 Physicochemical properties

In a wide variety of metal materials, a decrease in solid-to-liquid transition temperatures has been observed with decreasing nanocrystal size. Because surface atoms tend to be coordinatively unsaturated, there is a large energy associated with this surface. The key to understanding this melting point depression is the fact that the surface energy is always lower in the liquid phase compared to the solid phase. In the dynamic fluid phase, surface atoms move to minimize surface area and unfavorable surface interactions. In the solid phase, rigid bonding geometries cause stepped surfaces with high-energy edge and corner atoms. By melting, the total surface energy is thus reduced. This stabilizes the liquid phase over the solid phase. The smaller the nanocrystal, the larger the contribution made by the surface energy to the overall energy of the system and, thus, the more dramatic the melting temperature depression. Because melting is believed to start on the surface of a nanocrystal, this surface stabilization is an intrinsic and immediate part of the melting process. It is clear that the melting point decreases as the size of the particles decreases, and there is a dramatic decrease in the melting points for particles that are smaller than 3–4 nm (18).

Once the metal nanoparticles have been created, it is of vital importance to be able to accurately characterize them, both in terms of their structural properties (size and shape) and in terms of their optical properties. These properties are intimately connected. Various analytical instruments can be used to characterize these properties of metallonutraceuticals, as described in the following sections.

4.1.3 Determining surface area and particle size of metallonutraceuticals

4.1.3.1 Gas adsorption technique

Adsorption is defined as the adhesion of atoms or molecules of gas to a surface. The amount of gas adsorbed depends on the exposed surface and also on the temperature, gas pressure, and strength of interaction between the gas and solid. In Brunauer-Emmett-Teller surface area analysis, nitrogen is usually used because of its availability in high purity and its strong interaction with most solids. Because the interaction between gaseous and solid phases is usually weak, the surface is cooled using liquid N_2 to obtain detectable amounts of adsorption. A known amount of nitrogen gas is then released stepwise into the sample cell. Relative pressure that is less than atmospheric pressure is achieved by creating conditions of partial vacuum. After the saturation pressure, no more adsorption occurs, regardless of any further increase in pressure. Highly precise and accurate pressure transducers monitor the pressure changes due to the adsorption process. After the adsorption layers are formed, the sample is removed from the nitrogen atmosphere and heated to cause the adsorbed nitrogen to be released from the material and quantified. The data collected is displayed in the form of a Brunauer-Emmett-Teller isotherm, which plots the amount of gas adsorbed as a function of the relative pressure (19).

4.1.3.2 Electron microscopy technique

In this technique, an electron beam is used to illuminate a specimen and produce a magnified image. Because an electron microscope has greater resolving power than a light microscope, it can reveal the structure of smaller objects because electrons have wavelengths about 100,000 times shorter than visible light photons. The electron microscope uses electrostatic and electromagnetic lenses to control the electron beam and focus it to form an image. Modern electron microscopes produce electron micrographs, using specialized digital cameras or frame grabbers to capture the image (20). This characterization includes ascertaining the morphology of crystalline phases, number of phases, structure of phases, identification of crystallographic defects, and composition of crystalline phases (21).

4.1.3.3 X-ray diffraction technique

X-ray powder diffraction (XRD) is a rapid analytical technique primarily used for identifying the phases of a crystalline material, determining atomic spacing, and obtaining information on unit cell dimensions. The analyzed material is finely ground and homogenized, and the average bulk composition is determined.

In 1912, Max von Laue discovered that crystalline substances act as three-dimensional diffraction gratings for x-ray wavelengths, similar to the spacing of planes in a crystal lattice. XRD is based on constructive interference of monochromatic x-rays and a crystalline sample. These x-rays are generated by a cathode ray tube, filtered to produce monochromatic radiation, collimated to concentrate, and directed toward the sample. A key component of all diffraction is the angle between the incident and diffracted rays. The interaction of the incident rays with the sample produces constructive interference (and a diffracted ray) when conditions satisfy Bragg's law ($n\lambda = 2d \sin\theta$). This law relates the wavelength of electromagnetic radiation to the diffraction angle and the lattice spacing in a crystalline sample. These diffracted x-rays are then detected, processed, and counted.

By scanning the sample through a range of 2θ angles, all possible diffraction directions of the lattice should be attained due to the random orientation of the powdered material. Conversion of the diffraction peaks to d-spacings allows identification of the mineral because each mineral has a set of unique d-spacings. Typically, this is achieved by comparison of d-spacings with standard reference patterns (22).

4.1.4 Determining the surface state and surface complexes of metallonutraceuticals

4.1.4.1 Infrared spectroscopy

Infrared spectroscopy is based on molecular vibrations caused by the oscillation of molecular dipoles. Bonds have characteristic vibrations depending on the atoms in the bond, the number of bonds, and the orientation of those bonds with respect to the rest of the molecule. Thus, different molecules have specific spectra that can be collected for use in distinguishing products or identifying an unknown substance (to an extent).

Collecting spectra through this method occurs in one of three general ways. Nujol mulls (Schering-Plough Corporation, Kenilworth, NJ) and pressed pellets are typically used for collecting spectra of solids, whereas thin-film cells are used for solution-phase infrared (IR) spectroscopy (23). The main goal of IR spectroscopic analysis is to determine the chemical functional groups in the sample. Different functional groups absorb characteristic frequencies of IR radiation. Using various sampling accessories, IR spectrometers can accept a wide range of sample types, such as gases, liquids, and solids. Thus, IR spectroscopy is an important and popular tool for structural elucidation and compound identification (24).

4.1.4.2 Ultraviolet spectroscopy

Ultraviolet-visible (UV-vis) spectroscopy is used to obtain the absorbance spectra of a compound in solution or as a solid. What is actually being observed spectroscopically is the absorbance of light energy or electromagnetic radiation, which excites electrons from the ground state to the first singlet excited state of the compound or material. The UV-vis region of energy for the electromagnetic spectrum covers 1.5–6.2 eV, which relates to a wavelength range of 800–2,000 nm. The Beer-Lambert law is the principle behind absorbance spectroscopy: $A = \varepsilon bc$. For a single wavelength, A is the absorbance (it is unitless, usually shown as arbitrary [arb.] units), ε is the molar absorptivity of the compound or molecule in solution (M^{-1} cm^{-1}), b is the path length of the cuvette or sample holder (usually 1 cm), and c is the concentration of the solution (25).

4.1.4.3 Nuclear magnetic resonance spectroscopy

Nuclear magnetic resonance spectroscopy (NMR) is a powerful method that takes advantage of the magnetic properties of certain nuclei. The basic principle behind NMR is that some nuclei exist in specific nuclear spin states when exposed to an external magnetic field. NMR observes transitions between these spin states that are specific to the particular nuclei in question, as well as that nuclei's chemical environment. However, this only applies to nuclei whose spin, I, is not equal to 0, so nuclei with $I = 0$ are "invisible" to NMR spectroscopy. These properties have led to NMR being used to identify molecular structures, monitor reactions, and study metabolism in cells.

The chemical theory that underlies NMR spectroscopy depends on the intrinsic spin of the nucleus involved, described by the quantum number S. Nuclei with a nonzero spin are always associated with a nonzero magnetic moment, as described by $\mu = \lambda S$, where μ is the magnetic moment, S is the spin, and λ is always nonzero. It is this magnetic moment that allows for NMR to be used; therefore, nuclei whose quantum spin is zero cannot be measured using NMR. Almost all isotopes that have both an even number of protons and neutrons have no magnetic moment and cannot be measured using NMR.

In the presence of an external magnetic field (B) for a nuclei with a spin $I = \frac{1}{2}$, there are two spin states present: $+\frac{1}{2}$ and $-\frac{1}{2}$. The difference in energy between these two states at a specific external magnetic field (B_x) are given by $E = \mu B_x / I$, where E is energy, I is the spin of the nuclei, and μ is the magnetic moment of the specific nuclei being analyzed. The difference in energy shown is always extremely small; therefore, for NMR, strong magnetic fields are required to further separate the two energy states. At the applied magnetic fields used for NMR, most magnetic resonance frequencies tend to fall in the radio frequency range. The reason NMR can differentiate between different elements and isotopes is due to the fact that each specific nuclide will only absorb at a very specific frequency. This specificity means that NMR can generally detect one isotope at a time, which results in different types of NMR, such as ^1H NMR, ^{13}C NMR, and ^{31}P NMR, to name only a few (26).

4.1.4.4 Electron paramagnetic resonance

Electron paramagnetic resonance spectroscopy (EPR) is a powerful tool for investigating paramagnetic species, including organic radicals, inorganic radicals, and triplet states. The basic principles behind EPR are very similar to the more ubiquitous NMR, except that EPR focuses on the interaction of an external magnetic field with the unpaired electron(s) in a molecule, rather than the nuclei of individual atoms. EPR has been used to investigate kinetics, mechanisms, and structures of paramagnetic species.

The degeneracy of the electron spin states is lifted when an unpaired electron is placed in a magnetic field, creating two spin states of $m_s = \pm\frac{1}{2}$, where $m_s = -\frac{1}{2}$, which is the lower energy state, is aligned with the magnetic field. The spin state on the electron can flip when electromagnetic radiation is applied. In the case of electron spin transitions, this corresponds to radiation in the microwave range.

The energy difference between the two spin states is given by the equation $\Delta E = E^+ - E^- = h\nu = g\beta B$, where h is the Planck constant (6.626×10^{-34} J s^{-1}), ν is the frequency of radiation, β is the Bohr magneton (9.274×10^{-24} J T^{-1}), B is the strength of the magnetic field in Tesla, and g is known as the g-factor. The g-factor is a unitless measurement of the intrinsic magnetic moment of the electron, and its value for a free electron is 2.0023. The value of g can vary, however, and can be calculated by rearrangement of the above equation—that is, $g = hv/\beta B$, using the magnetic field and the frequency of the spectrometer. Because h, v, and β should not change during an experiment, g values decrease as B increases. The concept of g can be roughly equated to that of a chemical shift in NMR (27).

4.1.4.5 Raman spectroscopy

Raman activity depends on the polarizability of a bond, which is a measure of the deformability of a bond in an electric field. This factor essentially depends on how easy it is for the electrons in the bond to be displaced, inducing a temporary dipole. When there is a large concentration of loosely held electrons in a bond, the polarizability is also large, and the group or molecule will have an intense Raman signal. Because of this, Raman is typically more sensitive to the molecular framework of a molecule rather than a specific functional group (as in IR). This should not be confused with the polarity of a molecule, which is a measure of the separation of electric charge within a molecule. Polar molecules often have very weak Raman signals due to the fact that electronegative atoms hold electrons so closely.

Raman spectroscopy can provide information about both inorganic and organic chemical species. Many electron atoms, such as metals in coordination compounds, tend to have many loosely bound electrons and therefore tend to be Raman active. Raman can provide information on the metal ligand bond, leading to knowledge of the composition, structure, and stability of these complexes. This can be particularly useful in metal compounds that have low vibrational absorption frequencies in the IR. Raman is also very useful for determining functional groups and fingerprints of organic molecules.

Often, Raman vibrations are highly characteristic to a specific molecule, due to vibrations of a molecule as a whole, not in localized groups. The groups that do appear in Raman spectra have vibrations that are largely localized within the group and often have multiple bonds involved. Raman measurements provide useful characterizations of many materials. However, the Raman signal is inherently weak (less than 0.001% of the source intensity), restricting the usefulness of this analytical tool. Placing the molecule of interest near a metal surface can dramatically increase the Raman signal. This is the basis of surface-enhanced Raman spectroscopy. Several factors lead to the increase in Raman signal intensity near a metal surface (28):

1. The distance to the metal surface:
   - Signal enhancement drops off with distance from the surface.
   - The molecule of interest must be close to the surface for signal enhancement to occur.
2. Details about the metal surface, including morphology and roughness:
   - This determines how close and how many molecules can be near a particular surface area.

3. The properties of the metal:
   - The greatest enhancement occurs when the excitation wavelength is near the plasma frequency of the metal.
4. The relative orientation of the molecule to the normal of the surface:
   - The polarizability of the bonds within the molecule can be affected by the electrons on the surface of the metal.

4.1.5 Determining the surface composition of metallonutraceuticals

4.1.5.1 Auger electron spectroscopy

Auger electron spectroscopy (AES) was developed in the late 1960s, deriving its name from the effect first observed by Pierre Auger, a French physicist, in the mid-1920s. It is a surface-specific technique using the emission of low-energy electrons in the Auger process. AES is one of the most commonly employed surface analytical techniques for determining the composition of the surface layers of a sample. Auger spectroscopy involves three basic steps (29):

1. Atomic ionization (by removal of a core electron)
2. Electron emission (the Auger process)
3. Analysis of the emitted Auger electrons

The last stage is simply a technical problem of detecting charged particles with high sensitivity, with the additional requirement that the kinetic energies of the emitted electrons must be determined.

AES is a surface-sensitive spectroscopic technique used for elemental analysis of surfaces. It offers the following advantages:

1. High sensitivity (typically ~1% monolayer) for all elements except H and He
2. A means of monitoring surface cleanliness of samples
3. Quantitative compositional analysis of the surface region of specimens, by comparison with standard samples of known composition

4.1.5.2 Mossbauer spectroscopy

In 1957, Rudolf Mossbauer achieved the first experimental observation of the resonant absorption and recoil-free emission of nuclear γ-rays in solids during his graduate work at the Institute for Physics of the Max Planck Institute for Medical Research in Heidelberg, Germany. The Mossbauer Effect is the basis of Mossbauer spectroscopy. The Mossbauer Effect can be described very simply by looking at the energy involved in the absorption or emission of a γ-ray from a nucleus. When a free nucleus absorbs or emits a γ-ray to conserve momentum, the nucleus must recoil; therefore, in terms of energy:

$E_{\gamma\text{-ray}} = E_{\text{nuclear transition}} - E_{\text{recoil}}$. In a solid matrix, the recoil energy goes to zero because the effective mass of the nucleus is very large and momentum can be conserved with negligible movement of the nucleus. So, for nuclei in a solid matrix: $E_{\gamma\text{-ray}} = E_{\text{nuclear transition}}$. This is the Mossbauer Effect, which results in the resonant absorption/emission of γ-rays and provides a means to probe the hyperfine interactions of an atom's nucleus and its surroundings. A Mossbauer spectrometer system consists of a γ-ray source that is oscillated toward and away from the sample by a Mossbauer drive, which is a collimator to filter the γ-rays, the sample, and a detector (30).

4.1.5.3 Secondary ion mass spectrometry

Secondary ion mass spectrometry (SIMS) instruments (also known as ion microprobes) use an internally generated beam of either positive (e.g., Cs) or negative (e.g., O) ions (primary beam) focused on a sample surface to generate ions that are then transferred into a mass spectrometer across a high electrostatic potential; they are referred to as secondary ions. In a similar technique, a beam of high-speed neutral atoms (e.g., Ar) can substitute for the primary ion beam; this approach is used primarily for surface analysis of organic compounds. The interaction of the primary ion beam with the sample (under vacuum) provides sufficient energy to ionize many elements (31).

The physical and chemical properties of a solid surface are determined by its uppermost monolayers. The chemical composition of these uppermost monolayers can be investigated by the static method of SIMS. In this method, a relatively large target area (0.1 cm²) is bombarded with a small primary ion current density (10^{-9} A cm^{-2}). Thus, a sputtering time of several hours is achieved for an individual monolayer. A mass analysis of the emitted positive and negative secondary ions gives information about the chemical composition of the uppermost monomolecular layer of the bombarded surface.

Important features of SIMS include the detection of chemical compounds, isotope sensitivity, detection of hydrogen and its compounds, depth resolution in the range of a single monolayer, and low detection limits ($<10^{-6}$ monolayer or $<10^{-14}$ g) for many elements and compounds. The additional information on the chemical composition of the uppermost monolayer can be obtained by the electron-induced ion emission from the surface (32).

4.1.5.4 Ultraviolet and x-ray photoelectron spectroscopy

Photoelectron spectroscopy utilizes photoionization and analysis of the kinetic energy distribution of the emitted photoelectrons to study the composition and electronic state of the surface region of a sample. Photoelectron spectroscopy is based upon a single-photon-in/electron-out process. From many viewpoints, this underlying process is a much simpler phenomenon than the Auger process.

The energy of a photon of all types of electromagnetic radiation is given by the Einstein relationship $E = h\nu$, where h is the Planck constant (6.62×10^{-34} J s) and ν is the frequency (Hz) of the radiation. Photoelectron spectroscopy uses monochromatic sources of radiation (i.e., photons of fixed energy). In x-ray photoelectron spectroscopy (XPS), the photon is absorbed by an atom in a molecule or solid, leading to ionization and the emission of a core (inner shell) electron. By contrast, in ultraviolet photoelectron spectroscopy (UPS), the photon interacts with valence levels of the molecule or solid, leading to ionization by removal of one of these valence electrons. The kinetic energy distribution of the emitted photoelectrons (i.e., the number of emitted photoelectrons as a function of their kinetic energy) can be measured using any appropriate electron energy analyzer and a photoelectron spectrum can thus be recorded. This technique has been used for surface studies and has been subdivided according to the source of exciting radiation (i.e., XPS and UPS) (33).

UPS uses vacuum ultraviolet (UV) radiation (with a photon energy of 10–45 eV) to examine valence levels. Such radiation is only capable of ionizing electrons from the outermost levels of atoms (the valence levels). The advantage of using such UV radiation over x-rays is the very narrow line width of the radiation and the high flux of photons available from simple discharge sources. The main emphasis of work using UPS has been in studying the following:

1. The electronic structure of solids: Detailed angle-resolved studies permit the complete band structure to be mapped out in k-space.
2. The adsorption of relatively simple molecules on metals: The molecular orbitals of the adsorbed species can be compared with those of both the isolated molecule and with calculations.

XPS uses soft x-rays (with a photon energy of 200–2,000 eV) to examine core levels. For each and every element, there will be a characteristic binding energy associated with each core atomic orbital. That is, each element will give rise to a characteristic set of peaks in the photoelectron spectrum at kinetic energies determined by the photon energy and the respective binding energies. The presence of peaks at particular energies therefore indicates the presence of a specific element in the sample under study. Furthermore, the intensity of the peaks is related to the concentration of the element within the sampled region. Thus, the technique provides a quantitative analysis of the surface composition and is sometimes known as *electron spectroscopy for chemical analysis*. Atoms of a higher positive oxidation state exhibit a higher binding energy due to the extra coulombic interaction between the photo-emitted electron and the ion core. This ability to discriminate between different oxidation states and chemical environments is one of the major strengths of the XPS technique.

4.1.5.5 Work function electron tunneling

The work function of solid surfaces can be determined experimentally using absolute or relative approaches. Absolute methods allow one to measure the work function value directly. Here, the electrons in the metal are supplied with sufficient kinetic energy to overcome the barrier at the metal/vacuum interface and can thus escape the metal; the work function can be obtained from the resulting electric current. Absolute methods include measurements based on the following:

1. Thermionic emission
2. Field emission
3. Photoelectric effect

In thermionic emission, electrons are ejected from the material after receiving sufficient thermal energy to overcome the energy barrier at the metal/vacuum interface. The appropriate thermal energy is supplied by incremental heating of the sample to temperatures at which the Fermi-Dirac distribution of the electrons in the metal allows for a substantial population of electrons at energies higher than the interface energy barrier. The resulting electric current is measured as a function of temperature; this allows one to extract the work function of the surface. The temperature range used in the thermionic emission method is often very high (thousands of Kelvins), making this method of limited value for studying materials and surfaces that are unstable at high temperatures (34).

Field emission uses an electric field to accelerate the electrons inside of the metal to kinetic energies that are sufficiently high to overcome the interface barrier. The resulting electric current is analyzed as a function of the applied field and the work function is calculated (35).

In the photoelectric effect method, the light used as the source of energy for the electrons is typically in the UV range. As in other absolute methods, the resulting electric current (here called photocurrent) is analyzed to show the energetics of a photoelectric-effect-based measurement of the work function. In this method, the electrons are provided with a known energy, $h\nu$. Electrons with sufficient kinetic energy to overcome the barrier at the interface are able to escape the metal. These photoelectrons then travel away from the metal surface, experiencing the potential depicted. As the electrons move further away from the surface, their kinetic energy increases. The generated photocurrent is then measured as a function of the photoelectron kinetic energy (36).

4.1.6   Determining the surface structure of metallonutraceuticals

4.1.6.1   Low-energy electron diffraction

Low-energy electron diffraction (LEED) is a powerful method for determining the geometric structure of solid surfaces. It is similar to XRD in the type of information that it provides. The most obvious difference is that a beam of electrons, rather than x-rays, is used. Because electrons have a mean free path (or attenuation length) measured in angstroms (Å) as opposed to microns (μ) for x-rays, LEED is particularly sensitive to surface geometry.

LEED is a structural technique that can provide essentially two levels of information:

1. The positions of the bright spots in the diffraction pattern give information on the symmetry of the crystal surface (i.e., on the Bravais lattice of the surface net) and on the size of the unit cell.
2. The intensities of the spots can be analyzed using methods similar to those used in x-ray crystallography to determine the complete surface structure (i.e., the positions and identities of all the atoms within the unit cell) (37).

In both cases, LEED is most commonly used to determine the structure of a solid surface when the bulk structure of the material is already known by other means (e.g., XRD). It is possible, and indeed straightforwardly simple, to use LEED to determine both the absolute dimensions of the surface unit cell and the unit cell symmetry. One can easily deduce that, for example, a surface has a two-dimensional (2D) unit cell twice as large as that of the bulk; this often allows a reasonable guess at the true structure. Sometimes it is sufficient to know that a particular phase is present without knowing the details of the structure. Determination of the exact atomic positions is more difficult, but quantitative experiments can elucidate this second level of information. Sophisticated calculations, generally run on a workstation, can provide atomic coordinates with a typical precision of ±0.05 Å, which is generally more than adequate to determine the adsorption site of a molecule or the atomic positions in a reconstructed surface (38).

4.1.6.2 Scanning electron microscopy

Accelerated electrons in scanning electron microscopy (SEM) carry significant amounts of kinetic energy, and this energy is dissipated as a variety of signals produced by electron-sample interactions when the incident electrons are decelerated in the solid sample. These signals include secondary electrons (that produce SEM images), backscattered electrons, diffracted backscattered electrons (which are used to determine crystal structures and orientations of minerals), photons (characteristic x-rays that are used for elemental analysis and continuum x-rays), visible light (cathodoluminescence), and heat. Secondary electrons and backscattered electrons are commonly used for imaging samples: secondary electrons are most valuable for showing morphology and topography on samples, and backscattered electrons are most valuable for illustrating contrasts in composition in multiphase samples (i.e., for rapid phase discrimination).

X-ray generation is produced by inelastic collisions of the incident electrons with electrons in discrete orbitals (shells) of atoms in the sample. As the excited electrons return to lower energy states, they yield x-rays that are of a fixed wavelength (i.e., related to the difference in energy levels of electrons in different shells for a given element). Thus, characteristic x-rays are produced for each element in a mineral that is "excited" by the electron beam. SEM analysis is considered to be nondestructive; that is, x-rays generated by electron interactions do not lead to volume loss of the sample, so it is possible to analyze the same materials repeatedly.

SEM uses a focused beam of high-energy electrons to generate a variety of signals at the surface of solid specimens. The signals that derive from electron–sample interactions reveal information about the sample, including external morphology (texture), chemical composition, and crystalline structure and orientation of the materials making up the sample. In most applications, data are collected over a selected area of the surface of the sample, and a 2D image is generated that displays spatial variations in these properties. Areas ranging from approximately 1 cm to 5 μm in width can be imaged in a scanning mode using conventional SEM techniques (magnification ranging from 20× to approximately 30,000×; spatial resolution of 50–100 nm). SEM is also capable of performing analyses of selected point locations on the sample; this approach is especially useful for qualitatively or semiquantitatively determining chemical compositions (using an energy-dispersive spectrometer), crystalline structure, and crystal orientations using electron backscatter diffraction. The design and function of SEM is very similar to electron probe microanalysis; considerable overlap in capabilities exists between the two instruments (39).

4.1.6.3 Transmission electron microscopy

Transmission electron microscopy (TEM) has an advantage over SEM: Cellular structures of the specimen can be viewed at very high magnifications. However, TEM sample preparation for mollicutes is longer and more difficult than for SEM; it includes additional steps such as postfixation, embedding mollicutes in resin, sectioning samples, and staining semithin and ultrathin sections. Bozzola and Russell pointed out that perhaps the least forgiving of all the steps in TEM is the sample processing that occurs prior to sectioning. In other words, a poorly prepared specimen is useless to the investigator, whereas problems during the sectioning can be relatively easily fixed by simply cutting and staining more sections (40).

Specimen preparation of mollicutes for TEM includes eight major steps: cleaning, primary fixation, rinsing, secondary fixation, dehydration, infiltration with a transitional solvent, infiltration with resin and embedding, and sectioning with staining. The first two steps are essentially the same as those described for SEM specimen preparation (41).

4.1.6.4 Extended x-ray absorption fine structure

In this method, a monochromatic x-ray beam is directed at the sample. The photon energy of the x-rays is gradually increased such that it traverses one of the absorption edges of the elements contained within the sample. Below the absorption edge, the photons cannot excite the electrons of the relevant atomic level; thus, absorption is low. However, when the photon energy is just sufficient to excite the electrons, then a large increase in absorption occurs, known as the absorption edge. The resulting photoelectrons have a low kinetic energy and can be backscattered by the atoms surrounding the emitting atom. The probability of backscattering is dependent on the energy of the photoelectrons. The backscattering of the photoelectron affects whether the x-ray photon is absorbed in the first place. Hence, the probability of x-ray absorption will depend on the photon energy (as the photoelectron energy will depend on the photon energy). The net result is a series of oscillations on the high-photon-energy side of the absorption edge. These oscillations can be used to determine the atomic number, distance, and coordination number of the atoms surrounding the element whose absorption edge is being examined.

The necessity to sweep the photon energy implies the use of synchrotron radiation in extended x-ray absorption fine structure (EXAFS) experiments. By reflecting the x-rays from a surface at grazing incidence and detecting the resultant x-ray fluorescence with an Si(Li) detector, a more surface-sensitive signal can be obtained. This technique is known as REFLEXAFS. EXAFS spectra can be acquired in just a few seconds using the Quick EXAFS method. SEXAFS provides even greater surface sensitivity than REFLEXAFS. A related technique is NEXAFS (42).

4.2 Standardization of metallonutraceuticals

The standardization of metallonutraceuticals can be broadly categorized as follows:

1. Standardization of the raw materials
2. Standardization of the manufacturing process
3. Standardization of the final product

Because metallonutraceuticals are in the form of nanoparticles, it is necessary to determine their structure. Therefore, we have established a few guidelines as follows:

1. Establish the presence of nanoparticles in the test sample.
2. Ascertain whether the chemical compound is homogeneous.
3. Determine whether nanoparticles are crystalline or amorphous.
4. Identify the nature of defects in the sample.

Following our suggested production methodology of metallonutraceuticals as presented in **Chapter 3**, we can correlate these standardization parameters and devise the standardization strategies presented in **Table 4.1**.

Table 4.1 Standardization Strategies

Analytical Techniques	Purpose	Raw Material Standardization			In-Process Standardization					Final standardization
		Metals	Herbs	Other substances	Metal thinning into sheets, heating, and immersing into various liquids	Treatment of metal thin sheets with respective herbs, juices, decoction, or other substances	Ball milling of metal sheets for mechanical attrition	Mixing of ball milled metal with various herbs and substances	Thermal decomposition in furnace	
Electron probe microanalyzer	To identify distribution of individual elements				X	X		X	X	X
Energy-dispersive x-ray spectroscopy	To define chemical nature, size, and morphology of particles						X	X	X	X
Scanning electron microscopy	To define chemical nature, size, and morphology of particles						X			

Technique	Purpose					
Transmission electron microscopy	To define particle size and size distribution		X	X	X	X
Atomic force microscopy	To define particle size and size distribution			X	X	X
Electron probe microanalyzer	To identify distribution of individual elements					
X-ray diffraction	To analyze phase analysis					
Single-crystal x-ray diffraction	To confirm the exact molecular structure of crystalline intermediate products		X	X	X	
X-ray fluorescence	To define bulk chemical analysis		X	X	X	X
Particle induced x-ray emission analysis	To detect metal as an element	X		X	X	X

(continued)

Table 4.1 Continued

Analytical Techniques	Purpose	Raw Material Standardization			In-Process Standardization					
		Metals	Herbs	Other substances	Metal thinning into sheets, heating, and immersing into various liquids	Treatment of metal thin sheets with respective herbs, juices, decoction, or other substances	Ball milling of metal sheets for mechanical attrition	Mixing of ball milled metal with various herbs and substances	Thermal decomposition in furnace	Final standardization
High-performance liquid chromatography	To characterize organic matter		X	X				X	X	X
Nuclear magnetic resonance	To characterize organic matter							X	X	X
Infrared spectrophotometry	To characterize organic matter							X	X	X
Matrix-associated laser desorption-ionization	To characterize organic matter									
Electron spray ionization—mass spectrometry	To characterize organic matter									
Wet inorganic analysis	To characterize organic matter	X								
Atomic absorption spectrophotometry	To characterize organic matter									

AQ12 *Source:* Data from Rasheed, A., A. Marri, and M. Madhu Naik. 2011. Standardization of bhasma: Importance and prospects. *Journal of Pharmacy Research* 4:1931–1933.

4.3 The role of genomics, proteomics, metabolomics, and system biology in developing metallonutraceuticals

To study the molecular basis of health effects of specific components of the diet, nutritionists increasingly make use of the state-of-the-art "omics" technologies (43). The term *genomics* was coined by Tom Roderick in 1986 to describe the discipline of "mapping/sequencing (including analysis of the information)" of the entire genome (44). The term came to include the analysis of the genomic effects of specific biological activities of defined cells, tissues, organs, or even an organism. Instead of sporadically pinpointing differential gene expressions with limited, known, or obvious target genes, an integrated, global view of changes in gene expression (at both RNA transcription and protein translation levels) is necessary.

DNA microarray technology was developed in response to the need for a high-throughput, comprehensive, and efficient means of analyzing gene expression profiles and measuring the expression of all genes (or of a defined subset thereof) in a genome (45, 46). Based on the target DNA sequences embedded onto the arrays, hundreds, thousands, or tens of thousands of genes can be targeted and significant changes in their mRNA, microRNA, or other transcripts can be determined simultaneously. Biostatistical and bioinformatic analyses and computation of the detection signal data and functional characterization of gene patterns, clusters, and groupings, with significant changes in expression, can allow the prediction of major physiological changes in response to metallonutraceuticals (47).

Proteomics involves the large-scale study of the entire cellular array of proteins, particularly their biochemical identity, structure, and functions. The term *proteome* was proposed by Wilkins et al. (48). The main technology used in proteomics is 2D sodium dodecyl sulfate gel electrophoresis to separate different proteins according to variations in isoelectric points and relative mass (49). In addition to the 2D gel protein profiling system, other methods of systematic and/or comparative analysis of protein expression patterns include liquid chromatography (LC) followed by mass spectrometry (MS) or sequential MS analyses. Single fluorescence and fluorescence resonance energy transfer-based sensors have been successful in reporting small molecule and ion concentration, protein–ligand binding, and protein–protein interactions. Particularly exciting are protein biosensors that act as surrogate substrates for enzymes, reporting localized activities of kinases, protease, and small GTPases (50).

Because most proteins function in collaboration with other proteins in the cell, one goal of proteomics is to identify the interactions between various large key protein complexes, such as organelles or suborganelles (e.g., proteosome, nucleosome, endosome). This protein complex study is especially useful in determining potential regulatory partners in cell signaling cascades. Several methods are available to probe protein–protein interactions. One common method is yeast two-hybrid analysis. Other newer methods, including protein microarrays, immunoaffinity chromatography followed by MS, and combinations of diverse experimental methods such as phage display and computational methods, have also been developed for proteomics studies (47).

Metabolomics is an emerging field that originates from metabolite profiling, either as a targeted subset of related compounds or as a mapping of all extractable metabolites. The approach is expected to have diverse applications in toxicology, disease diagnosis, drug discovery and development, and phytomedicine research (51–54). The use of the word *metabolome* was first reported in 1998 as a way to quantitatively and qualitatively measure the defined phenotypes to assess gene functions in yeast (55) and to discuss the interplay between the global metabolite pool and specific environmental conditions in *Escherichia coli* (56). Generally, the metabolome refers to the complete set of small-molecule metabolites (e.g., metabolic hormones and other signaling molecules and secondary metabolites) to be found within a biological sample, such as a specific tissue, organ, or an organism (57).

The two major approaches in metabolomics are the targeted and the global (or unbiased) metabolite analyses. Targeted metabolite analysis or metabolite profiling, as the name implies, targets a subset (e.g., with similar solvent affinity or chemical grouping) of metabolites in a sample, instead of a complete metabolome analysis. This is usually carried out by using a particular set of analytical techniques such as gas chromatography-MS or LC-MS, together with an estimate of quantity. Various other techniques including thin layer chromatography, Fourier transform infrared spectroscopy, Raman spectroscopy, and NMR are also established as good tools of the metabolite analysis arsenal (58) (**Figure 4.1**).

Systems biology aims at simultaneous measurement of genomic, transcriptomic, proteomic, and metabolomic parameters in a given system under defined conditions. The vast amount of data generated with such omics technologies requires the application of advanced bioinformatics tools to obtain a holistic view of the effects of nutrients or nonnutrient components of foods and to identify a system of biomarkers that can predict beneficial or adverse effects of dietary nutrients or components for promoting health and preventing disease (59). Technical innovations in experimental devices, such as single-molecule measurements, femto-lasers that allow visualization of molecular interactions, confocal and three-dimensional imaging of crosstalk in cells, and nanotechnology, are also providing powerful new tools for studies of systems biology (60).

With the rapid progress of various genome sequencing and molecular biology projects that are generating in-depth knowledge of the molecular and developmental biology of many systems, we are now ready to seriously consider the possibilities offered by systems biology. A complete system-level analysis of a specific biological activity and its regulation is, however, still beyond the scope of existing experimental practices; new technologies and methodologies must be developed. Some of the newer high-throughput analytical platforms have indeed substantially enhanced the detection of a dynamic range

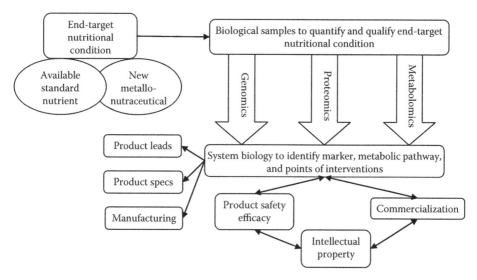

Figure 4.1 *Metallonutraceutical development pathway through systems biology.*

of metabolites, proteins, and their encoding genes (61). At the cellular scale level, it will require the development of sophisticated mathematical and computational modeling to analyze the many networks and signaling pathways, and their crosstalk and interplay that comprise the complex dynamic interactions between the chromosomes, organelles, genes, transcripts, proteins, and various metabolites (62).

4.4 Conclusion

Because metallonutraceuticals are (in one form or another) nanoparticles, one can employ various analytical techniques, alone or in combination, to ascertain the starting, intermediate, and final products as per preset standards. The optical, electrical, magnetic, and physicochemical properties of metallonutraceuticals are of utmost importance to identify structure–function relationships. To prepare reference sample processing, the methodology described in **Chapter 3** may be helpful. The final product can also be standardized on biological assays. The nutritional or therapeutic properties of metallonutraceuticals can also be anticipated in genomics, proteomics, and metabolic studies, and new products can be developed and commercialized.

References

1. Rastogi, S. 2010. Building bridges between ayurveda and modern science. *Int J Ayurveda Res* 1:41–46.
2. Burda, C., X. Chen, R. Narayanan, and M.A. El-Sayed. 2005. Chemistry and properties of nanocrystals of different shapes. *Chemical Rev* 105:1025–1102.

3. http://www.trynano.org/Nanoparticles.html
4. Roduner, E. 2006. *Nanoscopic Materials: Size-Dependent Phenomena.* Cambridge, UK: The Royal Society of Chemistry.
5. Le Bouar, Y. 2011. *An Introduction to the Stability of Nanoparticles.* Châtillon, France: LEM, UMR 104, CNRS/ONERA.
6. Edwards, P.P., R.L. Johnston, and C.N.R. Rao. 1999. In *Metal Clusters in Chemistry*, ed. P. Braunstein, G. Oro, and P.R. Raithby. Weinheim: Wiley-VCH.
7. Raether, H. 1980. *Excitations of Plasmons and Interband Transitions by Electrons.* Berlin: Springer.
8. Heip, H.M., H.M. Hiepa, T. Endob, and K. Kermana. 2007. A localized surface plasmon resonance based immunosensor for the detection of casein in milk. *Science and Technology of Advanced Materials* 8:331.
9. Sun, C.Q. *Size and Confinement Effect on Nanostructures.* Available from http://arxiv.org/ftp/cond-mat/papers/0506/0506113.pdf.
10. IHS GlobalSpec. *Nanomaterials Information.* Available from http://www.globalspec.com/learnmore/materials_chemicals_adhesives/electrical_optical_specialty_materials/nanomaterials_nanotechnology_products.
11. Huang, W.C., and J.T. Lue. 1997. Spin-glass properties of metallic nanoparticles conducted by quantum size effects. *J Chem. Solids* 58:1529.
12. Charle, K. P., W. Schultz, and B. Winter. 1989. The size dependent shift of the surface-plasmon absorption-band of small spherical metal particles. *Z Physics D* 12:471.
13. Maki, J.J., M.S. Malcuit, J.E. Sipe, and R.W. Boyd. 1991. Linear and nonlinear optical measurements of the Lorentz local field. *Phys Rev Lett* 67:972.
14. Lue, J.T. Physical properties of nanomaterials. 2007. In *Encyclopedia of Nanoscience and Nanotechnology*, ed. H.S. Nalwa, 1–46. Valencia, CA: American Scientific Publishers.
15. Rustagi, K.C., C.S. Warke, and S.S. Jha. 1967. Nonlinear electromagnetic response of Bloch electrons in a magnetic field in two-band model. *Phys Rev* 153:751.
16. Chen, L.J., Tyan, J.H., and J.T. Lue. 1994. Model analysis of temperature dependence of abnormal resistivity of a multiwalled carbon nanotube interconnection. *Phys Chem Solids* 55:871.
17. Reiss, G., J. Vancea, and H. Hoffmann. 1986. *Phys Rev Lett* 56:2100.
18. Burda, C., X. Chen, R. Narayanan, M.A. El-Sayed et al. 2005. Chemistry and properties of nanocrystals of different shapes. *Chem Rev* 105:1059.
19. Brunauer, S., P.H. Emmett, and E. Teller. 1938. Adsorption of gases in multimolecular layers. *J Am Chem Soc* 60:309.
20. Rudenberg, H.G., and P.G. Rudenberg. 2010. Origin and background of the invention of the electron microscope: Commentary and expanded notes on the memoir of Reinhold Rudenberg. *Advances in Imaging and Electron Physics* 160:207.
21. Neogy, S., R.T. Savalia, R. Tewari, and D. Shrivastva. 2006. Transmission electron microscopy of nanomaterials. *Indian Journal of Applied & Pure Physics* 44:119–124.
22. Klug, H.P., and L.E. Alexander. 1974. *X-ray Diffraction Procedures for Polycrystalline and Amorphous Materials*, 2nd ed. New York: Wiley.
23. Derry, P., and A.R. Barron. *IR Sample Preparation: A Practical Guide.* Available from http://cnx.org/content/m43564/latest/?collection=col10699/latest.
24. Hsu, C.-P.S. Infrared spectroscopy. Available from http://www.prenhall.com/settle/chapters/ch15.pdf.
25. Oliva-Chatelain, B.L., and A.R. Barron. *Basics of UV-Visible Spectroscopy.* Available from http://cnx.org/content/m34525/latest/?collection=col10699/latest.
26. Hanna, T., and A.R. Barron. *Introduction to Nuclear Magnetic Resonance Spectroscopy.* Available from http://cnx.org/content/m38356/latest/?collection=col10699/latest.
27. Bovet, C., and A.R. Barron. *EPR Spectroscopy: An Overview.* Available from http://cnx.org/content/m22370/latest/?collection=col10699/latest.
28. Payne, C., and A.R. Barron. *Raman and Surface-Enhanced Raman Spectroscopy.* Available from http://cnx.org/content/m34528/latest/?collection=col10699/latest.

29. *Auger Electron Spectroscopy*. Available from http://www.chem.qmul.ac.uk/surfaces/scc/scat5_2.htm.

30. Fisher, E., and A.R. Barron. *Introduction to Mossbauer Spectroscopy*. Available from http://cnx.org/content/m22328/latest/?collection=col10699/latest.

31. Mueller, P., and J. Vervoort. *Secondary Ion Mass Spectrometer (SIMS)*. http://serc.carleton.edu/research_education/geochemsheets/techniques/SIMS.html.

32. Benninghoven, A. 1973. Surface investigation of solids by the statical method of secondary ion mass spectroscopy (SIMS). *Surface Science* 35:427–457.

33. *Photoelectron Spectroscopy*. Available from http://www.chem.qmul.ac.uk/surfaces/scc/scat5_3.htm.

34. Murphy, E.L., and R.H. Good. 1956. Theoretical analysis of ion-enhanced thermionic emission for low-temperature, non-equilibrium gas discharges. *Phys. Rev.* 102:1464.

35. Hölzl, J., F.K. Schulte, and H. Wagner. 1979. Work function of metals. In *Solid Surface Physics*, vol. 85. Berlin: Springer-Verlag.

36. Malicki, M. 2010. *Work Function of Metals*. Available from http://photonicswiki.org/index.php?title=Work_Function_of_Metals#Work_Function_of_Metallic_Surfaces_.E2.80.93_Methods_of_Measurement.

37. Oura, K., V.G. Lifshifts, A.A. Saranin, et al. 2003. *Surface Science: An Introduction*. Berlin: Springer-Verlag.

38. Gerken, C.A., and G.A. Somorjai. 2002. Low-energy electron diffraction. In *Characterization of Materials*. New York: Wiley.

39. Swapp, S. *Scanning Electron Microscopy (SEM)*. Available from http://serc.carleton.edu/research_education/geochemsheets/techniques/SEM.html.

40. Bozzola, J.J., and L.D. Russell. 1992. *Electron Microscopy*. Boston: Jones and Bartlett.

41. Stadtländer, C.T.K.-H. 2007. Scanning electron microscopy and transmission electron microscopy of mollicutes: Challenges and opportunities. In *Modern Research and Educational Topics in Microscopy*, ed. A. Méndez-Vilas and J. Díaz. Madrid: Formatex.

42. Koningsberger, D.C., and R. Prins. 1988. *Theory of EXAFS. X-Ray Absorption: Principles, Applications, Techniques of EXAFS, SEXAFS and XANES*. Hoboken, NJ: John Wiley & Sons, Inc.

43. Zhang, X.W., Y. Yap, D. Wei, et al. 2008. Novel omics technologies in nutritional research. *Biotechnol Adv* 26:169–176.

44. McKusick, V.A., and Ruddle, F. 1997. Genomics as it enters its second decade. *Genomics* 45:243.

45. Schena, M., D. Shalon, R.W. Davis, and P.O. Brown. 1995. Quantitative monitoring of gene expression patterns with a complementary DNA microarray. *Science* 270:467–470.

46. Schena, M., D. Shalon, R. Heller, et al. 1996. Parallel human genome analysis: Microarray-based expression monitoring of 1000 genes. *Proc Natl Acad Sci USA* 93:10614–10619.

47. Aravindaram, K.H. Wilson, and N.-S. Yang. Omics for the development of novel phytomedicines. In *Genomics, Proteomics, and Metabolomics in Nutraceuticals and Functional Foods*, ed. D. Bagchi, F. Lau, and M. Bagchi. Hoboken, NJ: John Wiley & Sons, Inc.

48. Wilkins, M.R., J.C. Sanchez, A.A. Gooley, et al. 1995. Progress with proteome projects: Why all proteins expressed by a genome should be identified and how to do it. *Biotechnol Genetic Eng Rev* 13:19–50.

49. Williams, E.A., J.M. Coxhead, and J.C. Mathers. 2003. Anticancer effects of butyrate: Use of micro-array technology to investigate mechanisms. *Proc Nutri Soc* 62:107–115.

50. VanEngelenburg, S.B., and A.E. Palmer. 2008. Fluorescent biosensors of protein function. *Curr Opin Chem Biol* 12:60–65.

51. Ellis, D.I., W.B. Dunn, J.L. Griffin, J.W. Allwood, and R. Goodacre. 2007. Metabolic fingerprinting as a diagnostic tool. *Pharmacogenomics* 8:1243–1266.

52. Lindon, J.C., E. Holmes, and J.K. Nicholson. 2007. Metabonomics in pharmaceutical R&D. *FEBS J* 274:1140–1151.

53. Chen, C., F.J. Gonzalez, and J.R. Idle. 2007. LC–MS-based metabolomics in drug metabolism. *Drug Metabol Rev* 39:581–597.

54. Claudino, W.M., A. Quattrone, L. Biganzoli, et al. 2007. Metabolomics: Available results, current research projects in breast cancer, and future applications. *J Clin Oncol* 25:2840–2846.
55. Oliver, S.G., M.K. Winson, D.B. Kell, and F. Baganz. 1998. Systematic functional analysis of the yeast genome. *Trends Biotechnol* 16:373–378.
56. Tweeddale, H., L. Notley-McRobb, and T. Ferenci. 1998. Effect of slow growth on metabolism of Escherichia coli, as revealed by global metabolite pool ("metabolome") analysis. *J Bacteriol* 180:5109–5116.
57. Oliver, S.G., M.K. Winson, D.B. Kell, and F. Baganz. 1998. Systematic functional analysis of the yeast genome. *Trends Biotechnol* 16:373–378.
58. Shyur, L.F., and N.S. Yang. 2008. Metabolomics for phytomedicine research and drug development. *Curr Opin Chem Biol* 12:66–71.
59. Zhang, X., et al. Novel omics technologies in nutraceutical and functional food research. In *Genomics, Proteomics, and Metabolomics in Nutraceuticals and Functional Foods*, ed. D. Bagchi, F. Lau, and M. Bagchi. New York: Wiley Blackwell.
60. Kitano, H. 2002. Systems biology: A brief overview. *Science* 295:1662–1664.
61. Dunn, W.B., and D.I. Ellis. 2005. Metabolomics: Current analytical platforms and methodologies. *Trend Anal Chem* 24:285–294.

5

Characterization, Bioavailability, and Drug Interactions of Metallonutraceuticals

Charles Preuss

Contents

5.1 Introduction

Metallonutraceuticals, minerals, or micronutrients (e.g., iron, iodine, zinc) are essential for human health. However, individuals in the developing world often do not receive enough minerals in their diet because of poor nutrition and disease (Ramakrishnan 2002). Iron, for example, is important for hemoglobin synthesis and the healthy development of the unborn child. Iron deficiency affects approximately 50% of pregnant women in developing countries and is also an important problem in industrialized countries. Iron deficiency during pregnancy can cause anemia, intrauterine growth retardation, and small size for neonatal gestational age (Hovdenak and Haram 2012). A randomized controlled trial of pregnant women examined the effects of iron supplementation on the birth weights of their children. The trial observed higher birth weights in women who received iron supplementation versus placebo-treated women (Cogswell et al. 2003).

Several clinical studies have demonstrated that patients with type 2 diabetes mellitus have low magnesium plasma concentrations (Paolisso and Barbagallo 1997). Therefore, magnesium-rich diets may prevent the development of type 2 diabetes mellitus (Davi et al. 2010; McCarty 2005).

Bariatric weight-loss surgery is an option for obese patients. This type of surgery (e.g., Roux-en-Y gastric bypass) can cause the malabsorption of vitamins and minerals. Metallonutraceutical supplementation with calcium, zinc, iron, and magnesium can prevent mineral deficiencies from occurring in these patients (Malone 2008).

Several vitamins and minerals are important for the development and maintenance of proper bone health (Bonjour et al. 2009). For example, osteoporosis can be prevented by the adequate intake of calcium and vitamin D (Prentice 2006).

As demonstrated by these examples, evidence in the scientific literature demonstrates that the appropriate intake of metallonutraceuticals may help in the prevention of disease.

5.2 Characterization

The pharmacologic characterization of a metallonutraceutical determines its efficacy and safety. Currently, many metallonutraceuticals do not require efficacy and safety testing before marketing because they are regulated by the Dietary Supplement Health and Education Act of 1994, which considers many metallonutraceuticals as food items.

Manufacturers of metallonutraceuticals determine the contents and dose of their products. U.S. Food and Drug Administration (FDA) regulations require that the amount of metallonutraceutical listed on the label is equal to or more than the contents because of batch variation (Yetley 2007). However, there have been reports of variation in the strength of metallonutraceutical-containing products. Several dietary supplements used for prostate disease, which contained selenium, vitamin E, vitamin D, lycopene, and saw palmetto, were analyzed for accuracy and reliability (Feifer, Fleshner, and Klotz 2002). Five samples of selenium were tested and determined to be in the range of −19% to +23% of the labeled dose. Another study, which analyzed the selenium content of 15 different nutraceutical products, found that selenium content varied from −13.4% to +39.2% from the labeled amount (Veatch et al. 2005).

5.3 Bioavailability

Bioavailability is the measurement of the rate at and extent to which a drug reaches the systemic circulation. A related concept is bioequivalence, indicating a drug that has similar bioavailability to a reference drug, such as a new generic drug in comparison with an established brand name drug (Shargel and Yu 1993). Bioavailability and bioequivalence of drugs are very important parameters for drug development and regulation (Hoag and Hussain 2001). **Figure 5.1** shows the process of bioavailability for a metallonutraceutical.

Two important factors that can affect the bioavailability and bioequivalence of drugs are dissolution and excipients. Dissolution is the release of a drug from the dosage form, which is an important step in drug absorption for a solid oral dosage form (e.g., tablet, capsule). Excipients are commonly inactive ingredients that are used for the manufacturing of drugs (e.g., silicon dioxide, which is a glidant that enhances flow through a tablet machine). Excipients can affect the bioavailability by decreasing drug dissolution. The bioavailability of metallonutraceuticals lacks standard scientific and regulatory definitions (Yetley 2007). **Table 5.1** summarizes the important potential differences between prescription drugs and metallonutraceuticals (Srinivasan 2001).

Often, metallonutraceuticals are manufactured in oral solid dosage forms, such as tablets and capsules. These same manufacturing procedures are used to make prescription and nonprescription drugs. Therefore, in vitro dissolution tests can be used for metallonutraceutical oral dosage forms as a

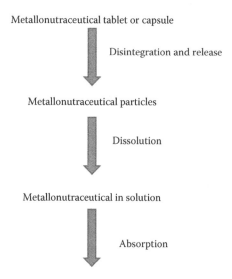

Metallonutraceutical tablet or capsule

Disintegration and release

Metallonutraceutical particles

Dissolution

Metallonutraceutical in solution

Absorption

Metallonutraceutical in systemic circulation

Figure 5.1 *Metallonutraceutical bioavailability process. (Adapted from Shargel, L., and A. Yu. 1993.* Applied Biopharmaceutics and Pharmacokinetics, *3rd ed. Norwalk, CT: Appleton and Lange.)*

Table 5.1 Differences between Prescription Drugs and Metallonutraceuticals

A metallonutraceutical's benefits may be variable and difficult to quantitate.
A metallonutraceutical's absorption lacks the scientific rigor found with prescription or OTC drug bioavailability.
Metallonutraceuticals are consumed for preventative medicine and well-being.
Metallonutraceuticals might not exhibit classic prescription or OTC drug–dose response curves.
Metallonutraceuticals might not be dosed as consistently as prescription or OTC drugs.

OTC, over the counter.

surrogate measure of absorption or bioavailability (Srinivasan 2001). When multiple metallonutraceuticals are contained in the oral solid dosage form, an index metallonutraceutical can be used as a marker for dissolution.

Table 5.2 shows the hierarchal ranking of metallonutraceuticals. Iron is ranked the highest because of its importance in maintaining public health (i.e., prevention of anemia). Calcium is ranked second for the prevention of osteoporosis (Srinivasan 2001).

The dissolution conditions for an index metallonutraceutical are as follows:

- Medium: 0.1 N HCl in 900 mL
- Apparatus: 100 rpm for capsules or 75 rpm for tablets
- Time: 1 hr

The dissolution conditions attempt to mimic the absorption of the metallonutraceutical in a tablet or capsule inside the gastrointestinal tract (Srinivasan 2001).

5.3.1 Iron

Table 5.3 lists the forms of iron in metallonutraceuticals and **Table 5.4** lists the methods used to assess the bioavailability of iron (Fairweather-Tait 2001).

Iron has two radioactive isotopes (^{59}Fe and ^{55}Fe) and four stable isotopes (^{54}Fe, ^{56}Fe, ^{57}Fe, and ^{58}Fe) (Fairweather-Tait 2001; Turnland 2006). The method of choice for determining the bioavailability of iron-containing metallonutraceuticals is hemoglobin incorporation, preferably with radioactive isotopes ^{59}Fe and ^{55}Fe because of their low sensitivities (37 kBq and 111 kBq, respectively). The radiolabeled iron metallonutraceutical is orally taken by a fasting subject

Table 5.2 Hierarchal Ranking of Metallonutraceuticals

1. Iron
2. Calcium
3. Zinc
4. Magnesium

Table 5.3 Forms of Iron in Metallonutraceuticals

Ferritin
Ferrous carbonate
Ferrous fumarate
Ferrous gluconate
Ferrous glycine sulfate
Ferrous succinate
Ferrous sulfate
Iron glycine
Polysaccharide iron complexes

Table 5.4 Methods Used to Assess the Bioavailability of Iron

Hemoglobin incorporation with isotopes
Fecal monitoring with isotopes
Whole-body counting with ^{59}Fe
Plasma concentration
Caco-2 cells

once or several times, and the percent incorporation of the radiolabeled iron into red blood cells is measured 14 days after the last dose. Stable iron isotopes can be used for hemoglobin incorporation experiments when there is a concern about exposure to radiation, but they lack the sensitivity of iron radioisotopes and thus require the use of larger doses (Fairweather-Tait 2001).

Fecal monitoring also can be used to determine the bioavailability of iron-containing metallonutraceuticals. Normally, almost an undetectable amount of absorbed iron is excreted in the feces and very little iron is excreted in the urine. Stable iron isotopes are commonly used for this procedure. A drawback of this procedure is variability of fecal collection because of differences in gastrointestinal transition times (Fairweather-Tait 2001). Whole-body counting uses ^{59}Fe, which is given to the test subject and counted 1–5 hours and 10–14 days later. The difference in readings is the absorbed ^{59}Fe. Corrections are made based on background radiation and isotope decay (Fairweather-Tait 2001).

Plasma concentrations of iron can be used to determine the bioavailability of different iron-containing metallonutraceuticals if the doses contain 25–100 mg of iron. The plasma iron concentrations increase 4–6 hours after ingestion (Fairweather-Tait 2001). Caco-2 cells can be used to rapidly predict the bioavailability of iron from metallonutraceuticals because these cell lines express Nramp2, which is a divalent transporter selective for iron. A drawback of this in vitro test is that the iron needs to be isotopically labeled (Fairweather-Tait 2001).

5.3.2 Calcium

Calcium has six stable isotopes (^{40}Ca, ^{42}Ca, ^{43}Ca, ^{44}Ca, ^{46}Ca, and ^{48}Ca) and two radioactive isotopes that are used clinically and in research (^{45}Ca and ^{47}Ca) (Boulyga 2010; Heaney and Whedon 1958; Stürup et al. 2003). **Table 5.5** lists the calcium salts found in nonprescription and prescription drugs and thus might be found in metallonutraceuticals (Cada 2004). **Table 5.6** lists the methods used to assess the bioavailability of calcium, which overlaps with methods previously discussed for iron (Heaney 2001).

The balance method refers to the difference between the amount of metallonutraceutical ingested and the amount excreted in the feces. Although it seems to be a simple method to assess the bioavailability of calcium-containing metallonutraceuticals, it has several drawbacks: It is cumbersome, imprecise, time consuming, and expensive (Heaney 2001).

Plasma concentrations of calcium can be used to determine the bioavailability from calcium-containing metallonutraceuticals. A drawback of this method is the low signal-to-noise ratio or limited sensitivity because calcium plasma concentrations are often tightly regulated and only relative bioavailability can be determined, not absolute bioavailability (Heaney 2001). For example, relative bioavailability is the oral absorption of a new generic drug compared with a well-recognized brand-name drug standard, whereas absolute bioavailability compares respective areas under the curve (AUC) after oral and intravenous administration (Shargel et al. 1993).

Table 5.5 Calcium Salts Found in Nonprescription and Prescription Drugs

Calcium glubionate
Calcium gluconate
Calcium lactate
Calcium citrate
Calcium acetate
Tricalcium phosphate
Calcium carbonate

Table 5.6 Methods Used to Assess the Bioavailability of Calcium

Balance method
Plasma concentrations
Tracer methods
Urinary excretion
Target system effects
In vitro methods

The tracer method can be very sensitive, reproducible, inexpensive, and quick. Radioactive or stable calcium isotopes may be used. This method measures the absorption from the gastrointestinal tract into the plasma after oral ingestion of a calcium-containing metallonutraceutical. This method can be very sensitive because of the high noise-to-signal ratio, because the amount of tracer in the background is in low abundance, especially with radioactive isotopes. A drawback of this method is that the calcium isotopes must be evenly distributed in the metallonutraceutical oral dosage form during the process of absorption (Heaney 2001).

Urinary excretion can be used to determine the bioavailability of a calcium-containing metallonutraceutical. As the calcium concentration in the plasma rises, it will be excreted by the kidneys into the urine. With timed urine collections, the AUC can be determined, which is similar to the plasma AUC of a drug. A drawback of this method is that there is a low signal-to-noise ratio and variability of renal clearance; in addition, only relative bioavailability can be determined (Heaney 2001).

The target system effects method refers to the desired effect of taking the calcium containing metallonutraceutical (e.g., prevention of osteoporosis), which infers bioavailability. A 5% rise in plasma calcium can cause a 40–50% decrease in plasma parathyroid hormone. Drawbacks of this method are patient variability with regard to standardizing clinical endpoints and nutritional status and parathyroid hormone assay variability, which requires large sample sizes for statistical power. Thus, the target system effects method is usually used with other methods to determine bioavailability, so that it can provide complementary information (Heaney 2001).

In vitro methods can be used to determine the bioavailability of calcium-containing metallonutraceuticals. These are quick and inexpensive screening methods that can find bioavailability problems before testing in expensive clinical trials. A drawback of these methods is that they can give inaccurate information on the bioavailability of calcium-containing metallonutraceuticals. These methods have poor correlation between solubility and absorbability of calcium salts, acid is not necessary for absorption, test conditions can create artifacts, and test conditions poorly mimic the human gastrointestinal tract (Heaney 2001).

5.3.3 Potassium

The following briefly summarized FDA guidance will provide insight into potential future regulations of dietary supplements, such as metallonutraceuticals, because of concerns about public health (Morrow et al. 2005). A drug company inquired to the FDA about designing a bioequivalence study to compare the bioavailability of their new controlled-released potassium chloride tablet and capsule versus a reference oral dosage form of potassium chloride. The FDA required that the drug company conduct a two-treatment, two-period (crossover and head-to-head), single-dose, and fasting study with at least 24 healthy human male subjects between the ages of 20 to 40 years old who were free from alcohol and medication use (FDA 1994).

The FDA indicated that human subjects are to be housed at a comfortable temperature to minimize the loss of potassium through sweating. Subjects are to be hydrated in regular intervals during the study in order to produce enough urine output. The subjects are to receive a daily diet with a set amount of calories, potassium, sodium, and water. The subjects are to be equilibrated on the daily diet for at least 4 days; then, urine should be collected at regular time intervals to obtain baseline potassium urinary excretion. On day 6, the subjects should receive an oral dosage form (i.e., tablet or capsule) that contains potassium chloride, which is either the test or reference oral dosage form of potassium chloride. Urine is collected at regular time intervals. Next, there is a 2-day washout period, during which the subjects do not receive the potassium chloride oral dosage forms. Lastly, the subjects receive crossover treatments, which are different oral dosage forms of potassium chloride than they received earlier in the study (FDA 1994). Urine potassium concentrations are used to determine the bioavailability and bioequivalence because plasma potassium concentrations are under physiologic control and are held in a narrow range of 3.5–5 mEq/L.

5.3.4 Selenium and zinc

Examples of commercially available forms of selenium and zinc are sodium selenite, seleno-L-methionine, seleno-L-cysteine, zinc sulfate, and zinc gluconate, which are known to have poor bioavailability (Mogna et al. 2012). In order to improve the bioavailability of selenium and zinc, metallonutraceutical-enriched bacterial probiotics, such as *Lactobacillus buchneri* Lb26 (DSM 16341) and *Bifidobacterium lactis* Bb1 (DSM 17850), were used; they were compared with commercially available forms of selenium and zinc in Caco-2 cells with regard to absorption (Mogna et al. 2012). Caco-2 cells are a human epithelial colorectal adenocarcinoma cell line that provides a surrogate measure of absorption [Smetanová et al. 2011]. The selenium internalized *Lactobacillus buchneri* was 6, 9, and 65 times more absorbable than sodium selenite, seleno-L-methionine, and seleno-L-cysteine, respectively. The zinc-internalized *Bifidobacterium lactis* was about 16 and 32 times more absorbable than zinc gluconate and zinc sulfate, respectively (Mogna et al. 2012). These results suggest that bacterial probiotics enriched with metallonutraceuticals may provide a better method for the absorption of these important micronutrients.

5.4 Drug interactions

Table 5.7 summarizes potential drug interactions with metallonutraceuticals.

Table 5.7 Drug Interactions with Metallonutraceuticals

Metallonutraceutical	Drug	Mechanism
Calcium	Levothyroxine forms Bisphosphonates Tetracyclines Quinolones	Decreased drug absorption from oral dosage
	Nondepolarizing neuromuscular blockers	Antagonized effects
	Sodium polystyrene sulfonate	Decreased potassium binding activity
	Thiazide diuretics	Milk-alkali syndrome
	Corticosteroids	Decreased oral calcium absorption
Magnesium	Aminoquinolones Bisphosphonates Nitrofuranatoin Penicillamine Tetracyclines Quinolones	Decreased drug absorption from oral dosage forms
	Neuromuscular blockers Barbiturates Benzodiazepines	Increased central nervous system depressive effects
Iron	Bisphosphonates Levodopa Methyldopa Levothyroxine Penicillamine Quinolones Tetracyclines	Decreased drug absorption from oral dosage forms
	Antacids H_2–Antagonists Proton pump inhibitors	Decreased oral iron absorption
Aluminum	Bisphosphonates Quinolones Tetracyclines	Decreased drug absorption from oral dosage forms
Zinc	Bisphosphonates Quinolones Tetracyclines	Decreased drug absorption from oral dosage forms

5.4.1 Calcium

The absorption of levothyroxine, bisphosphonates (e.g., risedronate), tetracyclines (e.g., doxycycline), and quinolones (e.g., levofloxacin) can be decreased with subsequent decreased plasma concentrations if taken orally concomitantly with calcium-containing metallonutraceuticals. Calcium may antagonize the effects of nondepolarizing neuromuscular blockers (e.g., pancuronium). Metabolic alkalosis and decreased potassium binding activity of sodium polystyrene sulfonate can occur with coadministration of calcium in patients with renal impairment. The therapeutic effects of verapamil can be antagonized by calcium. Patients receiving digoxin therapy should not be given intravenous calcium because of increased risk of cardiac arrhythmias. Chronic use of calcium in patients on thiazide diuretics (e.g., hydrochlorothiazide therapy) can cause milk-alkali syndrome. Patients on systemic corticosteroids (e.g., prednisone) can experience decreased intestinal absorption of calcium (Cada 2004; Gold Standard 2010).

5.4.2 Magnesium

The absorption of aminoquinolones (e.g., chloroquine), bisphosphonates, nitrofurantoin, penicillamine, tetracyclines, and quinolones can be decreased with subsequent decreased plasma concentrations if taken orally concomitantly with magnesium-containing metallonutraceuticals. At higher plasma concentrations (e.g., >4 mEq/L), magnesium can cause central nervous system (CNS) depression and thus have additive effects with CNS depressants, including neuromuscular blockers (e.g., succinylcholine), benzodiazepines (e.g., diazepam), and barbiturates (e.g., phenobarbital) (Cada 2004; Gold Standard 2010).

5.4.3 Iron

The absorption of bisphosphonates, levodopa, methyldopa, levothyroxine, penicillamine, quinolones, and tetracyclines can decrease with subsequent decreased plasma concentrations when taken orally concomitantly with iron-containing metallonutraceuticals. Antacids (e.g., aluminum hydroxide), H_2 antagonists (e.g., cimetidine), and proton-pump inhibitors (e.g., omeprazole) can decrease the absorption of iron-containing metallonutraceuticals by decreasing the acidity of the stomach (Cada 2004; Gold Standard 2010).

5.4.4 Aluminum and zinc

The absorption of bisphosphonates, quinolones, and tetracyclines can decrease with subsequent decreased plasma concentrations when taken orally concomitantly with aluminum- or zinc-containing metallonutraceuticals (Cada 2004; Gold Standard 2010).

5.5 Conclusion

Metallonutraceuticals can help with maintaining one's health and the prevention of disease, such as anemia and osteoporosis. Some metallonutraceuticals do not have the same degree of regulatory oversight as seen with prescription and nonprescription drugs. In addition, there is the concern that the commonly used oral drug delivery systems—tablets and capsules—might not effectively enter the systemic circulation. Therefore, bioavailability could potentially alter the metallonutraceutical's effectiveness. The use of dissolution testing and Caco-2 cells could help determine a metallonutraceutical's bioavailability before entering the market. A concern with metallonutraceuticals is the potential for drug interactions, of which patients might not be aware. However, metallonutraceuticals have a wonderful opportunity to improve the health and well-being of people, especially in the developing world.

References

Boulyga, S.F. 2010. Calcium isotope analysis by mass spectrometry. *Mass Spectrometry Reviews* 29:685–716.

Bonjour, J., L. Gueguen, C. Palacios, M. Shearer, and C. Weaver. 2009. Minerals and vitamins in bone health: The potential value of dietary enhancement. *British Journal of Nutrition* 101:1581–1596.

Cada, D., ed. 2004. *Drug Facts and Comparisons*, 58th ed. Philadelphia, PA: Wolters Kluwer Health.

Chemistry Explained. 2011. *Foundations and Applications: Calcium.* Available from http://www.chemistryexplained.com/elements/A-C/Calcium.html.

Cogswell, M.E., I. Parvanta, L. Ickes, R. Yip, and G.M. Brittenham. 2003. Iron supplementation during pregnancy, anemia, and birth weight: A randomized controlled trial. *American Journal of Clinical Nutrition* 78:773–781.

Davi, G., F. Santilli, C. Patrono. 2010. Nutraceuticals in diabetes and metabolic syndrome. *Cardiovascular Therapeutics* 28:216–226.

Dietary Supplement Health and Education Act. 1994. Public Law 103-417.

Fairweather-Tait, S. 2001. Iron. *Journal of Nutrition* 131:1383S–1386S.

Feifer, A., N. Fleshner, and L. Klotz. 2002. Analytical accuracy and reliability of commonly used nutritional supplements in prostate disease. *Journal of Urology* 168:150–154.

Gold Standard, Inc. 2010. *MD Consult* (online). Philadelphia, PA: Elsevier.

Heaney, R. 2001. Factors influencing the measurement of bioavailability, taking calcium as a model. *Journal of Nutrition* 131:1344S–1348S.

Heaney, R., and G.D. Whedon. 1958. Radiocalcium studies of bone formation rate in human metabolic bone disease. *Journal of Clinical Endocrinology and Metabolism* 18:1246–1267.

Hoag, S., and A. Hussain. 2001. The impact of formulation on bioavailability: Summary of workshop discussion. *Journal of Nutrition* 131:1389S–1391S.

Hovdenak, N., and K. Haram. 2012. Influence of mineral and vitamin supplements on pregnancy outcome. *European Journal of Obstetrics and Gynecology and Reproductive Biology* 164:127–132.

Malone, M. 2008. Recommended nutritional supplements for bariatric surgery patients. *The Annals of Pharmacotherapy* 42:1851–1858.

McCarty, M.F. 2005. Nutraceutical resources for diabetes prevention—an update. *Medical Hypotheses* 64:151–158.

Mogna, L., S. Nicola, M. Pane, P. Lorenzini, G. Strozzi, and G. Mogna. 2012. Selenium and zinc internalized by *Lactobacillus buchneri* Lb26 (DSM 16341) and *Bifidobacterium lactis* Bb1 (DSM 17850): Improved bioavailability using a new biological approach. *Journal of Clinical Gastroenterology* 46:S41–S45.

Morrow, J., T. Edeki, M. El Mouelhi, R. Galinsky, R. Kovelesky, and C. Preuss. 2005. American Society for Clinical Pharmacology and Therapeutics position statement on dietary supplement safety and regulation. *Clinical Pharmacology and Therapeutics* 77:113–122.

Paolisso, G., and M. Barbagallo. 1997. Hypertension, diabetes mellitus, and insulin resistance: The role of intracellular magnesium. *American Journal of Hypertension* 10:346–355.

Prentice, A., I. Schoenmakers, M. Laskey, S. de Bono, F. Ginty, and G. Goldberg. 2006. Nutrition and bone growth and development. *The Proceedings of the Nutrition Society* 65:348–360.

Ramakrishnan, U. 2002. Prevalence of micronutrient malnutrition worldwide. *Nutrition Reviews* 60:S46–S52.

Shargel, L., and A. Yu. 1993. *Applied Biopharmaceutics and Pharmacokinetics*, 3rd ed. Norwalk, CT: Appleton and Lange.

Smetanová, L., V. Štětinová, Z. Svoboda, and J. Květina. 2011. CACO-2 cells, biopharmaceutics classification system (BCS) and biowaiver. *Acta Medica (Hradec Kralove)* 54:3–8.

Srinivasan, S. 2001. Bioavailability of nutrients: A practical approach to in vitro demonstration of the availability of nutrients in multivitamin-mineral combination products. *Journal of Nutrition* 131:1349S–1350S.

Stürup, S. et al. 2003. A novel dual radio- and stable-isotope method for measuring calcium absorption in humans: Comparison with the whole-body radioisotope retention method. *American Journal of Clinical Nutrition* 77:399–405.

Turnland, J. 2006. Mineral bioavailability and metabolism determined by using stable isotope tracers. *Journal of Animal Science* 84:E73–E78.

U.S. Food and Drug Administration (FDA). 1994. *Guidance for Industry: Potassium Chloride (Slow-Release Tablets and Capsules) In Vivo Bioequivalence and In Vitro Dissolution Testing.* Available from http://www.fda.gov/ohrms/dockets/98fr/02d-0307-gdl0002.pdf.

Veatch, A., J. Brockman, V. Spate, J. Robertson, and J. Morris. 2005. Selenium and nutrition: The accuracy and variability of the selenium content in commercial supplements. *Journal of Radioanalytical and Nuclear Chemistry* 264:33–38.

Yetley, E. 2007. Multivitamin and multimineral dietary supplements: Definitions, characterization, bioavailability and drug interactions. *American Journal of Clinical Nutrition* 85:269S–276S.

6

Therapeutic Applications of Nanometals

Amy Broadwater, Bhaskar Mazumder, and Yashwant Pathak

Contents

6.1   Introduction

Nanotechnology is the synthesis and application of materials at the nanoscale. Biological nanoparticles are made up of different organic materials, such as proteins, polysaccharides, and synthetic polymers (1). Many types of synthetic nanoparticles, which have various methods of synthesis, are available on the market. Some of the most common methods include nanocrystal formation through solvent evaporation, nanosuspensions, supercritical fluid technology, polymerization of monomers, preparation of copolymerized peptide nanoparticles, and ionic gelation of hydrophilic polymers (2).

6.2   Methods of nanoparticle synthesis

Despite how different the methods of nanoparticle synthesis seem, the synthesis of nanoparticles follows a common pattern—either the "top-down" or the "bottom-up" method (3). The top-down method of synthesis is one in which a bulk of material is used as the starting material. The material is gradually cut away to reveal the final product, similar to a block of marble being sculpted into a statue. The bottom-up method of synthesis involves gradually building up to the product using small blocks of molecules or atoms—a similar process to using many bricks to build a house (3).

The synthesis of nanoparticles can also be categorized into different phases: liquid-phase, gas-phase, and/or vapor-phase synthesis. These synthesis mechanisms are quite similar to one another. Gas-phase synthesis involves supersaturation of the material by vaporizing the material into gas, then cooling the gas and condensing the nanoparticle product. Liquid-phase synthesis involves precipitating the solid nanoparticles out of an aqueous solution. Vapor-phase synthesis involves creating a supersaturated vapor under high heat conditions and condensing the nanoparticles onto a surface (3).

Nanoparticles have recently become integrated into the world of modern medicine. This area of treatment using nanoparticles, called nanomedicine, is revolutionizing the medical world with its use of biological molecules at the nanoscale to address medical problems and offer treatment solutions (2).

Many applications of nanoparticles in medicine involve pharmaceuticals, specifically in the area of drug delivery. Nanoparticles can be synthesized into specific shapes, such as nanocapsules and nanospheres, which are structurally conducive to encapsulation of drugs (1). The drug particles can be suspended within the capsule and protected by a layer of polymers or integrated throughout the sphere uniformly (1).

6.3   Applications of nanoparticles

Nanoparticles are ideal for drug therapy because of their ability to target and deliver. The nanodelivery system allows the drug particles to be released continuously, which maintains a correct level of medicine in the bloodstream (2). The nanoparticle size and surface properties allow the release of pharmacological agents at the targeted site while maintaining the correct dosage administered at the desired rate. Nanoparticles may also be coated with a hydrophilic agent that supports long-term circulation, which is a useful property when using nanotechnology for organ targeted treatment or gene therapy (1). Various coatings on the surface of nanoparticles make the particle more suitable for a particular location or for performing a particular function in the body (2).

The small size and intermolecular properties of nanoparticles penetrate tissues that are known to be relatively impermeable, such as tumors. The nanoparticles readily interact with the cell membrane and proteins, as well as with biomolecules within the cell. The interactions are possible because nanoparticles are small enough to penetrate the cell membrane and the membranes of some organelles, such as mitochondria or the nucleus. This allows high selectivity and precise targeting of the nanoparticles, and subsequently of the pharmacological agent. Because of the high specificity of the nanoparticle, the amount of drug administered can be lowered while maintaining an effective treatment (2).

The excellent targeting and delivery system using nanoparticles has given way to new methods of treating and preventing diseases at the nanoscale. Medicinal research of nanoparticles, both in vivo and in vitro, has given way to breakthroughs in diagnosis of diseases, new contrast agents, tools for analysis, and drug delivery vehicles (4). Nanometals in particular have been put into use therapeutically in various medicinal scenarios that biological nanoparticles would not be able to treat.

6.4   Trace metals in biological systems

Nanometals are nanoparticles that are made up of metallic elements, such as iron or manganese. Integrating metals and the concept of nanoparticles opens up the treatment of diseases to many possibilities that take advantage of the metallic elements' properties. Because many metals are trace elements necessary for the processes in the human body, the toxicity of nanometals is generally low, and they can be readily processed and removed from the body as waste.

Trace elements are an integral part of many of the mechanisms of life, including the most essential mechanisms, such as cellular respiration. Some of the most common metallic trace elements include iron, zinc, copper, chromium, manganese, molybdenum, cobalt, nickel, vanadium, and tin (5).

6.4.1   Iron

About two-thirds of iron in the body is found in the blood (6). The most important use of iron is in hemoglobin, the molecule used to deliver oxygen to cells throughout the body. Hemoglobin is composed of a porphyrin ring with an iron molecule contained within (6). The oxygen or carbon dioxide molecule can attach to the iron. It delivers the oxygen from the lungs to the cell, and eliminates the waste product of respiration by carrying the carbon dioxide back to the lungs to be expelled from the body. Without it, cellular respiration could not take place; thus, iron is essential for life. Iron is also present in myoglobin, cytochrome enzymes, and many oxidases—all of which are involved in various redox reactions within the cell (7).

6.4.2   Manganese

Although manganese is found in trace quantities in the human body, it is essential to life. Manganese is involved in a number of cellular processes, such as blood sugar absorption and nerve function. Manganese is also essential to the production of biological molecules, including connective tissues, blood clotting molecules, bone formation, and sex hormones. It is directly involved in the formation of bone tissue and in the processing of lipids, amino acids, and carbohydrates. A number of very important enzymes are activated by manganese, including manganese superoxide dismutase (8). A critical role of manganese is that it is involved in the prevention of damage to the cell caused by free radicals. In this way, manganese slows the rate of cell death (8).

6.4.3   Copper

Copper is found in trace quantities in the body, although large amounts of copper are toxic. Approximately 85–95% of the copper present in human blood is bound to ceruloplasmin, a ferroxidase enzyme. Copper is essential to some of the processes involved in life. It interacts with enzymes used by the cells to produce energy. Additionally, copper is involved in pigmentation of the skin and repair of connective tissues (9). One of the most important uses of copper, however, is its ability to be oxidized. Copper can collect free radical oxygen from the cell that was created by respiration. The free radical oxygen can cause damage to the cell's DNA and may result in cancer. Copper can attach to the free radical oxygen, reducing it and preventing further damage (9).

6.4.4 Zinc

Zinc is the second most common trace mineral in the human body (the adult body contains 2–3 g of zinc), with iron being the first. It is mostly confined in muscles and bones. Zinc is needed for health maintenance, but it is mainly needed for the structural integrity of proteins and regulation of gene expression (10). Zinc plays an important role in cellular division as well. It is a component of DNA transcription proteins, without which DNA could not be copied (10). It is also involved in the breakdown of carbohydrates, immune response, and fertility of males. Its deficiency causes age-related macular degeneration, night blindness, and cataracts (11). It is also involved in the senses of taste and smell in humans (10).

6.4.5 Noble metals

Two of the most versatile nanometals discussed are gold and silver nanometals. Silver, in large amounts, can be toxic to the human body. Chronic silver exposure has been shown to cause conditions such as argyria or argyrosis, which is bluish-gray discoloration of the skin or eyes, respectively (12). Worse side effects of overexposure to silver include damage to the liver, kidneys, and lungs. Silver, although toxic in large doses, is not harmful to humans in small doses (up to 0.01 mg/m^3). In fact, the various applications of silver in medical treatments have proven that silver can be very beneficial to humans. When used responsibly, the benefits of silver treatment outweigh the drawbacks (12).

Gold nanometals have many beneficial qualities as well. In small doses, gold can be processed by the kidneys and excreted from the body. Similar to silver, gold can only be taken in small doses in humans because of toxicity in high doses. However, gold nanoparticles are not only useful; they are ideal as treatments for some medical conditions.

By combining the merits of nanoparticles and the qualities of various metals, nanometals are versatile in the human body. Nanometals have therapeutic applications for treatment of cancer via hyperthermia, drug delivery, and by altering DNA within the cancerous cell. They are also powerful imaging agents for magnetic resonance imaging (MRI). They have antibacterial and antiviral properties and can be used as a treatment option for patients with diabetes. They can be used in the fields of dermatology, dentistry, and ophthalmology. Nanometals are not only versatile, but they are very effective in their various treatment applications. Their abilities to target, interact directly with membranes and macromolecules of microorganisms, release payloads, absorb and release waves, break through biological barriers, and catalyze cellular processes have made them an exciting developing tool in the world of medicine. This chapter discusses the therapeutic applications of nanometals in the body, highlighting both current techniques and developing techniques.

6.5   Therapeutic applications of nanometals

6.5.1   Nanometals as MRI imaging agents

The basic principle of MRI is based on nuclear magnetic resonance together with the relaxation of proton spins in a magnetic field (13). One of the most popular applications of metallic nanoparticles is their use as a contrast agent in MRI. MRI has the ability to read the rates of absorption of the nanometals into tissues, which a computer then can translate into an image of the inside of the human body. Advancements in nanotechnology have resulted in the development of an advanced generation of probes based on nanoparticles smaller than 100 nm, which dramatically increase relative surface area and the dominant quantum confinement effects (14).

6.5.1.1   Iron nanoparticles

Iron nanoparticles are commonly used as effective MRI contrast agents. The concept behind contrasting materials in MRI is that different tissues have different rates of uptake of the contrast materials. The difference in rates allows the contrast to be read by the MRI. Specifically, iron nanoparticles are taken into the body. Once in the bloodstream, they are absorbed at various rates by a system of vessels involved in uptake of foreign materials from the blood. The MRI reads this variation in the rate of uptake of the iron nanoparticles and translates it into an image.

Superparamagnetic iron oxide nanoparticles (SPIONS) are currently used for MRI contrast enhancement in molecular imaging. SPIONs can be produced by different methods depending on the particular application. For MRI, SPIONS should be of uniform particle size with a uniform and high superparamagnetic moment, high colloidal stability, low toxicity, and high biocompatibility. They also must possess the ability to bind a range of drugs, proteins, enzymes, antibodies, and other molecular targets (15).

The rate of uptake can be varied by creating nanoparticles of different sizes. Larger nanomolecules are removed from the bloodstream much faster than smaller ones, because the smaller nanoparticles go undetected longer than the large molecules (16). Smaller nanoparticles <30 nm in size have a longer half-life; therefore, they are used for MRI imaging (16).

The difference in uptake rates allows the MRI to differentiate between different types of tissues, specifically between normal tissues and cancerous tissues. Tumors have their own vascular system, which consequently creates a difference in the rate of uptake of the nanoparticles in comparison to the rate of uptake in normal tissue. In this way, an MRI can image which tissues are cancerous and which are not, depending on which tissues absorb the contrast agent easily and which take longer to contrast.

The small size of nanoparticles allows them to cross what is known as the blood–brain barrier, which is a method of defense by the body against foreign materials entering the brain. Because iron nanoparticles are so small, they are ideal for imaging tumors located deep within the brain. Examples of cancerous tissues that iron nanoparticles can locate and contrast in an MRI include brain tumors, liver tumors, spleen tumors, and malignant lymph nodes. Iron nanoparticles can also be used to visualize the central nervous system and the vascular system (16).

6.5.1.2 Manganese nanoparticles

Manganese nanoparticles are also used as a contrast agent in MRI. Manganese is an ideal contrast agent because of its five unpaired electrons. These electrons lend themselves to creating a long electronic relaxation time, a quality necessary in magnetic resonance imaging (17). The varying relaxation times of the different tissues can be read by the MRI to create the final image.

Manganese oxide (MnO) itself is not an ideal contrast agent. Therefore, the manganese must be altered into a nanoparticle system to create a nontoxic effective contrast agent. This is done by highly dispersing manganese nanoparticles into mesopores with silicon. By synthesizing the manganese in this state, the water molecules can easily access the highly dispersed manganese atoms in the mesopore system because of the high surface area, and use the magnetic properties of manganese to create an excellent contrasting agent for MRI (17). Manganese nanoparticles can also be formed as MnO and porous MnO, to name a few.

MnO nanoparticles treated with trioctylphosphine oxide at high temperatures (300°C) form MnO nanohollows, which have an increased nanoparticle surface percentage without decreasing the size. MnO nanoparticles coated with silica (a biocompatible material) allow easy functionality loading, either onto the particle surface or into the silica framework (18).

MnO and other colloidal nanoparticles are excellent contrast agents because of their ability to carry large loads of magnetic particles, easy penetration through biological barriers, and long circulation times. Their large load capacity is due to the large, hollow interior space of the nanoparticle. The nanosize of the nanoparticles is what allows them to be undetected by the body, keeping them in circulation for large periods of time, as well as allowing them to pass through membranes easily (19).

Manganese nanoparticles have been found to be particularly effective for anatomical brain imaging (19). Manganese nanoparticles are also particularly excellent contrast agents for characterization of early-stage tumors because of their precision (20). Manganese nanoparticles have additionally been used to contrast prostate cancer cells in vitro, as well as in vivo in animal subjects (20).

6.5.1.3 Other metallic nanoparticles as contrast agents

A study found that mesoporous silica nanoparticles infused with zinc and iron are another useful contrast agent (**Figure 6.1**). The zinc nanoparticles can be synthesized in a very simple and effective reaction. The mesoporous zinc can be used as an excellent contrast agent for MRI (21).

Gadolinium (Gd) dendrimers are useful imaging agents as well. Dendrimers are nanoparticles with repeating branches protruding from a central core (**Figure 6.2**). Metals such as Gd can be used as the central core and used as an effective magnetic contrast agent in MRI. Gd in the form of a dendrimer allows it to be excreted from the body more easily than in its molecular form. Gd dendrimers are highly effective contrasting agents for tumor tissues and lymph nodes, and they produce high-resolution MRI images (22).

Additionally, gold nanoparticles can be used as contrast agents using simpler imaging techniques. Machines as simple as an optical microscope equipped with a dark field condenser are used to image the gold nanoparticles. The gold nanoparticles interact with light waves and the conduction electrons oscillate; these oscillations create another wave that can be read as a contrast

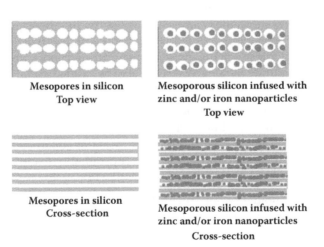

Mesopores in silicon
Top view

Mesoporous silicon infused with zinc and/or iron nanoparticles
Top view

Mesopores in silicon
Cross-section

Mesoporous silicon infused with zinc and/or iron nanoparticles
Cross-section

Figure 6.1 *A representation of mesopores infused with zinc and/or iron nanoparticles.*

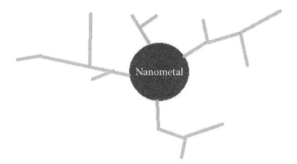

Figure 6.2 *A representation of the structure of a dendrimer.*

with a microscope. This process is known as scattering. Many other metals require MRI, lasers, or complex imaging processes to scatter the light (22). The benefit of gold nanoparticles is their simplicity in imaging techniques.

6.5.2   Cancer treatment through nanometal drug delivery

Various methods of synthesis, the shape, and the chemical makeup of nanoparticles allow some nanometals to perform drug delivery. Drug delivery is the targeting of the nanoparticle to a certain area of the body to release the pharmacological agent stored within the nanoparticle. Drug delivery via nanoparticles allows for the continuous, long-term release of a pharmacological agent at the intended site of delivery. The ability of nanoparticles to partake in drug delivery has made them not only ideal agents in delivering chemotherapy but also effective in targeted delivery of different anticancer drugs directly to the tumor.

Nanoparticles can actively or passively target a tumor. Passive targeting is when the nanoparticles in the bloodstream find the tumor based on the chemical properties of the nanoparticle as well as the pharmaceutical it is carrying. Tumors have a different vascular system than normal tissues do; this difference is often used to target the malignant tissues. Additionally, the environment of the tumor can be used to target the nanoparticle to the area. The pH and redox potential of the malignant tissue differs from normal tissues (22). Therefore, the drug carrier can flow through the blood stream until it reaches the tumor, where the difference in pH and/or the redox potential will activate the pharmaceutical.

Nanoparticles can actively target the tumors as well. The nanoparticles are equipped with surface proteins that can directly interact with the surface proteins on the malignant cells. These proteins are different from those on a normal, healthy cell. By directly interacting only with a malignant cell, the nanoparticles can be targeted to the exact type of tissue that they will interact with to deliver the carried pharmaceutical; they will not interact with normal tissues. Finally, the nanoparticles can be applied directly to the targeted area without first entering the bloodstream. This is the most accurate method of targeting because the nanoparticles are delivered to the specific area only, avoiding circulation throughout the body (22).

6.5.2.1 Manganese nanoparticles

Mesoporous manganese nanoparticles are one application of nanometals that are ideal for drug delivery. The mesopores can absorb the pharmacological agent very easily because of their large surface area and their large volume capacity of the pores.

To deliver a pharmacological agent via a mesopore, the drug is first encapsulated in the core. When the mesopore is put into the body, it targets the malignant tissues and supplies a sustained release of the pharmacological agent to the affected area. The most common pharmacological agent used in the manganese nanoparticle-infused mesopores is doxorubicin, a drug for chemotherapy. The chemical is released at the site of the cancer, and the manganese allows for a sustained release of the drug. Targeted and sustained release means less of the pharmacological agent is required to have the same effect on the cancerous tissue. Mesoporous zinc/iron complexes can also be used for drug delivery, with the same drug carriage characteristics as mesoporous manganese (21).

6.5.2.2 Gold nanoparticles

Gold nanoparticles are a promising tool for drug delivery. Gold nanoparticles have an inert core, which prevents the drug from interacting with the nanoparticle. The particles can deliver a variety of molecules, including DNA, RNA, proteins, and/or pharmaceuticals.

Gold nanometals have a variety of characteristics that allow them to effectively target and deliver the pharmaceuticals to the appropriate locations in the body. Gold nanoparticles are easily synthesized and come in a variety of sizes, from 1 nm to 150 nm. The difference in size affects where the nanoparticle will end up in the body, thereby targeting a specific location. Gold nanoparticles are also nontoxic in small quantities and have a monolayer that can be adjusted to meet certain needs. Solubility and charge can be fine-tuned to target a specific set of conditions in the body in which the pharmaceutical should unload. Gold nanoparticles can interact with thiols as well, which assists in controlled intercellular release of the contents of the nanoparticle (23).

6.5.2.3 Iron nanoparticles

Related to drug delivery techniques, iron molecules can be specialized to target certain receptors, allowing a differentiation to be made between a mutant and a normal cell. Because iron nanoparticles can target a specific type of cell, these nanoparticles can be used to deliver pharmaceuticals to malignancies (16). The ability of iron nanoparticles to target receptors on the surfaces of cells could be applied in the future to create specialized gene-targeting antibiotics that only affect the bacteria with a specific receptor (16).

Additionally, the merits of iron nanoparticles as an MRI contrast agent can be combined with their drug carriage abilities to produce a powerful tool against cancer. The iron nanoparticles carry and deliver treatments directly to brain tumors, while an MRI can be used to observe the process of the drug unloading to the tumor. The nanoparticles can be targeted to the specific area of the tumor by taking advantage of the metallic properties of iron and using magnetic targeting. Combining these factors causes an accumulation of iron nanoparticles in the brain tumor, leading to an effective treatment (24).

6.5.3   Cancer treatment by hyperthermia with nanometals

Hyperthermia is a method of killing cancerous cells by using the metallic properties of nanometals to apply heat to the cells. It is a very effective method of treating cancer, as it only affects the cells of the tumor and not the healthy surrounding tissue.

6.5.3.1   Iron nanoparticles

Iron nanoparticles are not only used in MRI visualization of tumors and drug delivery in the body; they also are involved in the treatment of tumors and cancerous cells by hyperthermia. In this treatment by hyperthermia, superparamagnetic iron oxide nanoparticles are suspended in aqueous solution and injected into cancerous tissue (25). Next, the cancerous cells are killed by applying a magnetic field or irradiating the location where the nanoparticles and tumor are located (25). This magnetic field is lethal to cancer cells because of the magnetic qualities of iron. The applied magnetic field heats up the iron molecules, and this heat kills the tumor cells. This method of treatment is formally known as magnetic fluid hyperthermia (25).

The heat is created by subjecting the nanoparticles to an alternating current field. This causes the metals in the nanoparticles to produce eddy currents, hysteresis, and Brownian rotational losses. As a result, a nanoscale source of heat is created that initiates cell death of the surrounding tissues (26).

Magnetic fluid hyperthermia can kill cancerous cells in the brain, even deep-seated tumors, as well as malignancies in different locations throughout the body (25). Alternating the magnitude and frequency of the magnetic field and/or the type of nanoparticle used for treatment can apply more or less heat to the cells. The treatment is also very selective. The heat is produced only in the iron nanoparticle-injected area, which is ideally the tumor itself. This prevents damage to the tissues surrounding the tumors.

6.5.3.2 Copper/gold nanoparticles

Copper/gold nanoparticles have also been used for phototherapy. The nanoparticle uses a combination of metals to elicit better photothermal energy conversion, as well as overcome some limitations of other nanoparticle systems (27). The copper/gold nanoparticles are heated up with a near-infrared laser and the heat produced leads to the destruction of the cancerous cells. The uses of copper/gold nanoparticles have been evaluated on human cervical cancer cells and were shown to be an effective treatment (28).

6.5.3.3 Gold nanoparticles

Gold nanoparticles are also known as effective metals to use for photothermal therapy. Although the previously mentioned metals are mainly used for tumors deep within the body, gold nanoparticles are excellent for phototherapy on tumors near or on the surface of the body. Gold nanoparticles are heated using light or lasers on cancerous cells near the surface of the skin (**Figure 6.3**). This metal has optimal characteristics for heat release, with the ability to heat to temperatures above its melting point for short periods of time before melting. The high temperature that gold can reach effectively kills cancerous cells (22). Gold nanoparticles are especially useful because of their ability to target certain tissues when coated with certain biological materials, as discussed in the section on nanometal drug delivery (29). The many forms that gold nanoparticles can take allow them to target the malignant tissues in the body.

There are many subtypes of gold nanoparticles based on size, shape, and physical properties, such as gold nanospheres, nanorods, nanoshells, and nanocages. Another type of gold-based nanoparticles has excellent surface-enhanced Raman scattering properties; thus, these nanoparticles are termed surface-enhanced Raman spectroscopy nanoparticles (**Figure 6.4**).

A study has shown that gold-coated silica nanoparticles are excellent at transferring heat to cancerous cells via hyperthermia. Again, this method of treatment is currently limited to superficial tumors, as the laser used to heat the nanoparticles does not have deep penetration into the body. Nevertheless, the qualities of gold to provide minimally invasive treatment of superficial cancerous cells have made it a strong candidate for future clinical uses (30).

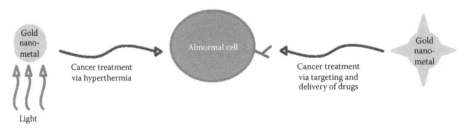

Figure 6.3 *Applications of gold nanoparticles in the treatment of cancer.*

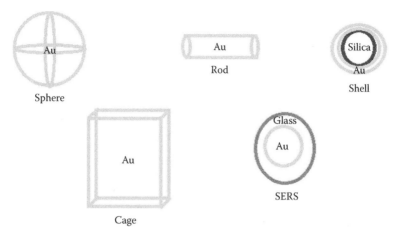

Figure 6.4 *Different forms of gold nanoparticles.*

Gold nanoparticles have the capacity to provide enormous sensitivity, output, and flexibility and thus the potential to profoundly affect cancer diagnosis and patient management. Gold nanoparticles are especially useful for cancer treatment for a number of reasons. They are easily prepared, and other biological molecules can be attached to their surfaces to enable more specific targeting of the nanoparticle. Additionally, they themselves have antiangiogenic properties (30).

6.5.4   Cancer treatment through DNA alterations with nanometals

Another method of cancer treatment is by altering DNA in cancerous cells, rendering the cancer cells inert and consequently causing cell death. Nanoparticles can be especially useful in this field because of their small size. The smaller the nanoparticle, the better it is able to penetrate through both the cellular membrane and the nuclear membrane and come into direct contact with DNA.

6.5.4.1   Copper nanoparticles

Copper nanoparticles have the ability to interact with DNA. Nanometals that are 4–5 nm in size are small enough to enter a cell. The copper nanoparticles accumulate in the nucleus. Once there, the copper binds with the DNA and breaks the double helix, rendering it inert. This substantially inhibits the growth of cancerous cells by slowing or stopping the cellular processes and leads to cell death (22). The effects of this method of DNA degradation have been studied on human histiocytic lymphoma and human cervical cancer. In both cases, the copper nanoparticle treatment was cytotoxic to the cancerous cells, causing apoptosis and cell death (31). The copper nanoparticles can be made to target cancerous cells and to leave normal cells unharmed by encapsulating the copper atoms in a liposome. Liposomes are a method of targeting the nanoparticle to a certain type of cell, in this case a cancerous cell (22). The study supports the effects of copper nanoparticles in killing cancerous cells. This information may be applied to future cancer treatments (31).

Nanoparticles can also be used to combat cancer cell growth through RNA delivery. This method delivers RNA directly to the tumor. The RNA is taken into the cells; once in the cytoplasm, it interacts with a multiprotein complex that initiates the cleavage of mRNA. Without mRNA, the cell cannot create new protein and cellular growth is halted, eventually leading to cell death (22).

6.5.4.2   Gold nanoparticles

Gold nanoparticles have been shown to be toxic to cancerous cells as well. When the gold nanoparticles are synthesized in a small enough size, they can penetrate through the cellular membrane and the nuclear membrane. Once the nanoparticles are in the nucleus, they bind to the DNA and render it inert. This causes cytokinesis arrest in the cancer cells; the cancer cells go through mitosis but are unable to complete telophase. As a result, apoptosis and cell death occur (32).

6.5.5   Nanometal antibacterial applications

Antibiotic-resistant bacteria are becoming a growing problem in today's world. Many strains of bacteria are becoming immune to the heaviest antibiotics available, and the number of these strains continues to grow daily. Silver nanoparticles in small concentrations have been proven to kill many strains of bacteria known to be antibiotic resistant, while leaving human cells unharmed (33).

6.5.5.1   Silver nanoparticles

Silver nanoparticles' toxicity to a wide range of microorganisms is dependent on a variety of qualities. First, the nanoparticles must be small enough to pass through a cellular membrane. The small size of the silver nanoparticles allows them to interact with the membranes of bacteria and viruses. They have the ability to cross the membranes of both gram-positive and gram-negative bacteria—a differentiation made by the composition of the outer membranes of the bacteria (34). The microscopic size of the nanoparticles allows them to cross the bacterial membrane and affect the processes within.

The silver nanoparticles must also be stable in the medium that the bacteria are growing in. The toxicity of the silver is dependent on how long the silver has to interact with the bacteria before it loses stability. If the silver nanoparticles are in a harsh environment, it is unlikely that they will be stable long enough to have a serious toxic effect on the bacteria. Producing stable nanoparticles that are nontoxic to humans but that significantly hinder the growth of the microorganism is a challenge that is difficult to overcome (34). Nevertheless, producing a stable silver nanoparticle that is effective as an antibiotic is a worthwhile project.

Additionally, it has been shown that interaction of the silver nanoparticles with the bacterium is dependent on the shape of the nanoparticle. A study has shown that triangular nanoparticles show significantly more interaction with the membrane of the bacteria than rod-shaped or spherical-shaped nanoparticles (35).

Although the mechanism of the interaction of silver nanoparticles is not fully understood, a general consensus is that the nanoparticles are able to bind to the surface of the bacteria, likely onto a protein. By binding to the membrane, the silver is able to slow or stop some cellular processes, resulting in cell death. Others believe that silver nanoparticles act similar to silver ions when attacking the bacteria. Silver ions have the ability to inhibit DNA replication, as well as affect the ribosomal subunits involved in the cellular respiration process, rendering them inactive and bringing the production of adenosine triphosphate to a halt (35).

Colloidal silver nanoparticles are also effective antibiotics. These nanoparticles work by hindering an enzyme in bacteria involved in cellular respiration. Without the enzyme, cellular respiration cannot occur, and the bacteria eventually shut down and die. Colloidal silver nanoparticles are lethal to bacteria, yet leave the processes of human cells unaffected (34).The antibacterial properties of silver and silver nanoparticles have opened up a variety of uses in the biomedical field, including using silver as a potential candidate for creating medicinal biomaterials that are infection resistant (36).

6.5.5.2   Zinc nanoparticles

Zinc nanoparticles have also been shown to have antibiotic effects. The zinc nanometals inhibit growth and kill both gram-positive and gram-negative bacteria, and specifically *Staphylococcus aureus* and *Escherichia coli*. *S. aureus* has been a growing problem in the field of medicine, as new antibiotic-resistant superstrains of the bacteria have emerged. Zinc nanoparticles demonstrated toxicity to resistant bacteria, such as *S. aureus*, while leaving human cells unharmed (37).

The concentration of zinc nanoparticles in a suspension has been shown to be proportional to their antibiotic properties. Additionally, the smaller the nanoparticles, the more potent they are. However, out of these two properties, it has been shown that the concentration of zinc nanoparticles is a more important factor than the shape of the nanoparticles for antibacterial effects (38).

The zinc nanoparticles kill the bacteria by directly interacting with the bacterial membrane. It has been shown that high concentrations of zinc nanoparticles can damage the membrane wall beyond repair, effectively killing the bacteria (38).

6.5.6   Treatment of surface wounds with nanometals

In order for a dermal wound to heal, the area must have sufficient blood flow along with being kept clean and infection free. Some antibiotic-resistant bacteria are hard to protect against when the skin is broken, such as the superbacteria *S. aureus*. Because of antibiotic resistance and other factors, new methods of promoting dermal wound healing must be found. Nanoparticle technology provides a method of healing wounds in a speedy and safe manner, with little to no scarring.

6.5.7   Silver nanoparticles

Silver nanoparticles are known to have a unique ability to heal wounds. The nanoparticles can be used on both burns and open wounds to disinfect the area and speed up the healing process. Gauze dressings treated with silver nanoparticles were proven to heal second-degree burns (burns that penetrate through the epidermis to the dermis of the skin) in a quick and effective manner. The nanoparticles prevented microbial growth. Without infection, the burn healed quickly with no complications. Silver nanoparticles were also found to heal the wound significantly faster than wounds treated with silver sulfadiazine cream or Vaseline (39).

Silver nanoparticles have also been shown to be effective at healing wounds because of their anti-inflammatory effects. The presence of inflammation is one of the key factors that prevent wounds from healing. Inflammation is caused by an excess of a certain kind of protein. Silver nanoparticles return proteins in the area of the wound to a normal level within a few days, which consequently lowers the inflammatory response, allowing the wound to heal. Additionally, silver nanoparticles are involved in regulation of cytokines. Cytokines are intercellular signaling molecules involved in cellular division and growth. When these are regulated, the cells divide and heal the wound as quickly as possible (40).

Diabetes is known to slow the healing process of injuries as well as make the affected area more prone to infection. An important application of silver nanoparticles is their ability to heal wounds for a patient with diabetes. Silver nanoparticles have qualities that heal diabetic wounds quickly by preventing additional infection, and they leave little to no scarring (34). These nanoparticles increase blood flow and nutrients around the area of the wound, which, combined with the antibiotic properties of silver nanoparticles, creates an ideal environment for the wound to heal.

6.5.8   Dermal protectant applications of nanometals

Many sunscreens in today's market contain nanometals, specifically zinc oxide and titanium dioxide nanoparticles (41). These nanometals are becoming increasingly common because of their clear application on the skin. Nanoparticles avoid the white, chalky consistency of older sunscreens, which can be seen as unfavorable to many. They also will not leave a white shadow on dark skin after application (42).

6.5.8.1 Zinc nanoparticles

Besides the pleasant appearance of zinc oxide nanoparticles in sunscreen applications, zinc nanoparticles are also excellent at blocking harmful ultraviolet (UV)-A, UV-B, and UV-C rays of the sun. UV rays are harmful to the skin because they penetrate through cells into the nucleus, where they can damage the DNA. If the cell with damaged DNA reproduces, it may cause skin cancer. Zinc nanoparticles can prevent this damage by reflecting the UV rays off of the skin, preventing them from penetrating the cell (43). Zinc nanoparticles protect against UV-A rays (the harmful, cancer-causing rays), as well as UV-B rays (the sunburn-causing rays).

There are a few downsides to using zinc nanoparticles as sun protectant, such as possible penetration through the skin and/or pooling under the skin. Zinc nanoparticles under the skin can cause cellular problems as well as permanent white, blotchy spots. Zinc nanoparticles, when washed off of the skin, can also seep into the water supply and cause damage to the environment.

Despite the downsides, zinc is an essential element in sun protection, and studies show that it is possible to make zinc nanoparticles safer for human use as well as for the environment. One way to reduce the toxicity of zinc nanoparticles is to combine nanoparticles with liposomes. This alteration to the nanoparticle structure helps them to better stick to the surface of the skin, as well as making them insoluble in water. Liposomes prevent the migration of zinc nanoparticles into the skin and help keep the water environment from becoming contaminated (42).

6.5.8.2 Titanium nanoparticles

Titanium dioxide nanoparticles have also been found to be effective sun-protectant agents. Titanium nanoparticles have the ability to block harmful sun rays from the skin by scattering the light. These nanoparticles are very similar to zinc oxide nanoparticles; however, the titanium nanoparticles have been shown to be a safer alternative for sun protection. Although many safety concerns have been raised involving the seeping of zinc nanoparticles into the skin, the U.S. Food and Drug Administration (FDA) has approved the use of titanium nanoparticles in sunblock. A recent study has proven that titanium dioxide nanoparticles, even when applied regularly for an extended period of time, do not significantly penetrate through the epidermis (41). Therefore, titanium dioxide nanoparticles provide a safe alternative for use in sun protection as compared to zinc nanoparticles.

6.5.8.3 Silver nanoparticles

An additional dermal application of nanometals is the antidandruff effects of silver nanoparticles. Dandruff can be caused by fungal pathogens living on the surface of the scalp, where they thrive because of the dark and damp conditions. These fungi, such as *Pityrosporum ovale*, cause the skin on the scalp to itch and flake, creating dandruff. Silver nanoparticles are well-known antifungal agents that increase the effectiveness of treatment when added to anti-dandruff shampoo (44). The silver nanoparticles interact with the fungus, hindering the growth and/or killing the microorganism.

6.5.9 Applications of nanometals in dentistry

Nanometals have great potential in the field of dentistry. One of the main causes of tooth decay is associated with bacterial growth in the mouth. The most common problem-causing bacteria include *Streptococcus mutans* and *Streptococcus lactobacilli*. Bacteria feed off of fermentable carbohydrates left in the mouth from eating and create an acidic byproduct. The acid is the cause of breakdown of hard tissues present in teeth (45).

Once damaged teeth have been restored by dental work, the probability of continuing damage and eventual tooth loss is still high. If bacteria are left in the area once the damage has been fixed, they will continue to thrive and eventually undo the tooth restoration. Typically, restorations are followed by a secondary infection, and measures have to be taken to restore the tooth a third time. Nanoparticles offer a solution to the problem of bacteria-catalyzed tooth decay.

6.5.9.1 Silver and titanium nanoparticles

Both silver and titanium nanoparticles have been introduced to the dental field as a means of preventing infection and tooth decay. Silver can be used as a general antibiotic, killing the harmful bacteria in the mouth. Possible uses of silver nanoparticles in dentistry include the topical application of the particles to combat and prevent oral infections. More effective still is the integration of silver and titanium nanoparticles into dental materials that have a high likelihood of becoming infected. These nanoparticles could be used in cements, sealants, filling materials, tooth-coating materials, restoration materials, and adhesives. Silver and titanium could also be introduced into dentures and crowns to prevent bacterial damage to the soft tissues surrounding the teeth (45). By integrating nanometals into the materials used for tooth restoration, the probability of a secondary infection or the need of a secondary restoration is significantly lessened.

6.5.10 Applications of nanometals in the treatment of diabetes

Diabetes is caused by some degree of impairment of the secretion of insulin, which is involved in the maintenance of sugar levels in the bloodstream. There are two major types of diabetes: type 1 and type 2. Type 1 diabetes typically has a juvenile onset and involves destruction of the pancreatic cells involved in insulin production. This eventually leads to a patient's complete inability to produce insulin. Type 2 diabetes develops in adults and involves resistance to the effects of insulin. It is the more common type of diabetes and is associated with obesity (46). Over time, diabetes can cause medical complications, including poor eyesight, nerve damage, lessened ability to heal wounds, and development of sores on the hands and feet (47). Nanometals can be used to slow down the process of complications as a result of diabetes.

Oxidative stress is a known cause of complications in patients with diabetes. The hyperglycemic conditions of cells affected by diabetes are known to cause an excess of free radicals from the cellular respiration process. The conditions may also cause a limitation on antioxidants, the chemicals that counteract the oxidative free radicals. The free oxygen radicals, when left uninhibited, cause damage to the interior of the cell, including the DNA. Over time, this adds to the complications of diabetes (48).

6.5.10.1 Cerium nanoparticles

Cerium oxide nanoparticles were demonstrated to be effective at counteracting the effects of the oxidative stress in diabetic rats. The cerium nanoparticles were shown to significantly protect from free radicals in the brain, heart, lungs, and kidneys created by cellular respiration. The cerium nanoparticles could lead to a breakthrough in the maintenance of health in patients with diabetes and prevent complications of diabetes from occurring (48).

Cerium is such an effective antioxidant because of its ability to participate in redox reactions. In patients with diabetes, oxidative stress in the liver is one of the early symptoms because of the liver's role in purification of the blood. Cerium nanoparticles within the liver act as a catalyst for superoxide dismutase, an enzyme found naturally in cells that reduces the free radical oxygen created by cellular respiration. Stimulating the enzyme allows it to be a much more potent antioxidant within the cell and can reduce the oxidative stress that the liver experiences (49).

Cerium nanoparticles themselves can also be effective antioxidants. Cerium nanoparticles can remain active in tissues for long periods of time, and they can spontaneously move between the oxidized and reduced state. These qualities give them the ability to collect free radical oxygen from the cell and reduce oxidative stress (49).

6.5.10.2 Gold nanoparticles

Gold nanoparticles are also effective in the treatment of diabetic complications. Gold nanoparticles can be used to deliver insulin into the bloodstream. Insulin is used by the body to regulate blood glucose levels. Typically, patients with diabetes must inject insulin into their skin in order to control their blood sugar level. Gold nanoparticles create a new method of delivering insulin into the bloodstream without having to inject it into the skin.

Gold nanoparticles bind readily to insulin molecules. These gold-bound insulin molecules take on characteristics that unbound insulin does not have. The gold nanoparticles laden with insulin can be administered via intranasal spray, where they then will absorb into the blood-rich mucous membrane. The absorption of the insulin, which normally does not easily pass through a mucous membrane, is catalyzed by the gold nanoparticles. This method of treatment shows a quick decrease in blood glucose levels because of the fast rate of absorption of the nanoparticles (50).

This method of insulin delivery is significant because insulin can be delivered into the body while avoiding undesired effects. Insulin on its own has low bioavailability because it cannot easily pass through membranes. Taking insulin orally leads to an insignificant quantity of insulin uptake because of the poor membrane absorption and the acidic conditions of the stomach. When insulin is administered through a dermal injection (the most common method used today), the insulin can stay under the skin for an extended period of time and adversely affect the diabetic patient by inducing hypoglycemia (50). The method of insulin delivery via gold nanoparticles through mucous membranes is a promising new treatment method that is safer and more effective than other methods of delivery. Gold nanoparticles provide immediate transfer of insulin into the bloodstream and will not linger in the nasal cavity for an extended period of time.

6.5.11   Nanometals in the treatment of ocular diseases

The retina is the light-sensitive tissue lining the interior of the eye. The retina is involved in the sense of sight; the cells and neurons interact with light to create vision. Because of the constant bombardment of light on the retina, these cells have the fastest rate of metabolism of oxygen out of any of the other cells in the body. This high metabolism rate results in a buildup of reactive free oxygen radicals, which can cause great damage to the neurons and the photoreceptor cells on the interior of the eye. A chronic loss of vision can occur if these free radicals are not kept in check (51).

6.5.11.1   Cerium nanoparticles

Cerium nanoparticles have been proven to lower the amount of oxidative stress found in photoreceptive cells in the eye; they can pick up the reactive oxygen that is so damaging to the cells. Studies have found that when cerium nanoparticles are injected into the eye prior to the exposure to extremely bright light, the quantity of cell death was significantly less than the amount of cell death without the cerium nanoparticles. The nanoparticles have even been found to be effective when injected into the eye after chronic bright light exposure (51). The qualities of cerium nanoparticles to lower oxidative stress in photoreceptive cells have made them a potential candidate to slow the rate of degenerative diseases not only in the eye but in other areas of the body as well.

Many of the degenerative diseases of the eye, including macular degeneration, are difficult to treat. The progression of macular degeneration can be worsened by the presence of some growth factors, which increase vascular permeability. In order to slow the progression of degeneration, these growth factors must be inhibited.

6.5.11.2 Silver nanoparticles

Silver nanoparticles have been proven to slow the progression of macular degeneration and other optical diseases. The silver nanoparticles inhibit the pathway of the growth factor that increases the permeability of endothelial cells. By inhibiting this permeability, silver nanoparticles consequently slow the progression of optical degenerative diseases and can even increase visual acuity in some patients (52).

6.5.12   Antiviral applications of nanometals

Nanoparticles can also be used as an effective antiviral agent against many known disease-causing viruses. Examples of the virus-caused diseases that nanoparticles have an effect on include human immunodeficiency virus (HIV)-1, hepatitis B virus, respiratory syncytial virus, and herpes simplex virus type 1 (53).

A virus is a DNA or RNA-carrying particle with an outer membrane composed of a lipid membrane with glycoproteins and other proteins dispersed throughout (54). The virus attaches to the surface of a normal working cell and injects its DNA, using the cell's processes to reproduce and multiply the viruses.

6.5.12.1 Silver nanoparticles

Silver nanoparticles specifically have been proven to have effective antiviral qualities. At nontoxic concentrations, silver nanoparticles express antiviral characteristics. However, it is not known exactly how these nanoparticles are able to inhibit such strong viruses as HIV. One hypothesis of the mechanism of inhibition is that the silver nanoparticles bind to the surface of the viral envelope on the glycoproteins, inhibiting it from attaching to human cells (54).

The effectiveness of the silver nanoparticles is correlated with the sizes of the nanoparticles. The nanoparticles that were within the range of 1–10 nm were shown to attach to the surface of the virus, thereby inhibiting it. Any nanoparticles larger than 10 nm were not nearly as effective in inhibition and were shown not to interact with the virus (54). Interaction with the HIV virus is a size-dependent interaction.

Silver nanoparticles show inhibition of the hepatitis virus and herpes virus as well by attaching to surface glycoproteins and preventing the virus from attaching to a cell (53). In in vitro assessments, the silver nanoparticles were shown to be effective in inhibiting the HIV virus. Although silver in large quantities is toxic to the human body, small amounts in some therapeutic applications has been approved for use by the FDA (53). Therefore, silver nanometals are a potent antiviral material that cannot be overlooked.

6.6   Potential toxicity of nanometals

Many concerns have arisen regarding the safely of nanoparticles in the human body. Major concerns include the oxidative stress and inflammatory responses caused by nanoparticles as well as their cytotoxicity (55). Nanoparticles have the ability to penetrate through the membrane of a cancerous cell or pathogen and render the processes within inert. They may also interact with the membrane of the cell, causing indirect manipulations of the cellular processes and the DNA. Although this ability is a powerful tool when applied therapeutically, it may cause harm to normal functioning human cells in the process. Nanoparticles can function on human cells similarly to how they function on pathogenic cells. It is important that the underlying mechanisms of the nanoparticle interactions as well as the genotoxic responses for specific nanometals are uncovered quickly, so that action can be taken to prevent potential harm to humans (55).

6.7   Conclusion

In conclusion, nanometals—a hybridization of both inorganic metals and organic nanoparticles—have versatile applications in the medical field. Common metals used in nanometals are zinc, copper, iron, manganese, cerium, and titanium, which are commonly found in the body, and silver and gold nanoparticles, which are not naturally found in the body.

Their applications include MRI contrast agents, cancer treatments, drug delivery agents, antibiotics, antiviral agents, and treatment/healing of skin wounds. Nanometals can also be applied in the specialties of dermatology, ophthalmology, and dentistry and additionally are used in the regulation of complications caused by diabetes.

Nanometals can be synthesized into certain forms to make them ideal for a certain use. Dendrimers, liposomes, and mesopores are a few examples of altering the structure of nanometals to create different applications. The many forms that metallic nanoparticles can take increase their possibilities for use in various treatment methods.

Nanometals are excellent at targeting a certain area of the body and penetrating into tissues and cells. These properties allow the particles to target cancerous cells and pass through the membranes to kill the cell from within. The particles can also pass into bacteria and shut down their functional mechanisms, causing bacterial death. Because of this, nanometals are powerful antibiotics. Nanometals can also block receptors on the surface of viruses and prevent them from binding with human cells and multiplying, thus making powerful antiviral treatments as well.

The magnetic properties of metallic nanoparticles can allow them to be powerful MRI contrast agents and powerful cancer treatments. The uptake of the nanometal in the body creates a contrast that can be read by an MRI. Additionally, nanometals injected into cancer tissues can be heated up by a laser or infrared ray. The heat that the metallic particles create kills the cancerous cells while leaving the normal surrounding tissues unharmed.

Many therapeutic applications of nanometals are still being developed and tested. Although nanometals are a relatively new technology, their applications promise to revolutionize the field of medicine. Because of their properties, nanometals will continue to be integrated into various treatments in the field of medicine for years to come.

References

1. Maruthi, G., A. Smith, and R. Manavalan. 2011. Nanoparticles: A review. *Journal of Advanced Scientific Research* 2:12–19.
2. Jin, S., and K. Ye. 2007. Nanoparticle-medicated drug delivery and gene therapy. *Biotechnology Progress* 23:32–41.
3. Hu, X.L., G.S. Li, and J.C. Yu. 2010. Design, fabrication and modification of nanostructured materials for environmental and energy applications. *Langmuir* 26:3031–3039.
4. Mritunjai, S., S. Singh, S. Prasad, and I.S. Gambhir. 2008. Nanotechnology in medicine and antibacterial effect of silver nanoparticles. *Digest Journal of Nanomaterials and Biostructures* 3:115–122.
5. Mason, J.B. 2012. Vitamins, trace minerals, and other micronutrients. *Goldman's Cecil Medicine* 2:1–5.
6. Institute of Medicine. 2001. "Iron," in *Dietary Reference Intakes for Vitamin A, Vitamin K, Arsenic, Boron, Chromium, Copper, Iodine, Iron, Manganese, Molybdenum, Nickel, Silicon, Vanadium, and Zinc*. Washington, DC: The National Academies Press, 290–293.
7. Dallman, P.R., and H.C. Schwartz. 1965. Myoglobin and cytochrome response during repair of iron deficiency in the rat. *Journal of Clinical Investigation* 44:1631–1638.
8. Institute of Medicine. 2001. "Manganese," in *Dietary Reference Intakes for Vitamin A, Vitamin K, Arsenic, Boron, Chromium, Copper, Iodine, Iron, Manganese, Molybdenum, Nickel, Silicon, Vanadium, and Zinc*. Washington, DC: The National Academies Press, 394–419.
9. Institute of Medicine. 2001. "Copper," in *Dietary Reference Intakes for Vitamin A, Vitamin K, Arsenic, Boron, Chromium, Copper, Iodine, Iron, Manganese, Molybdenum, Nickel, Silicon, Vanadium, and Zinc*. Washington, DC: The National Academies Press, 224–257.
10. Institute of Medicine. 2001. "Zinc," in *Dietary Reference Intakes for Vitamin A, Vitamin K, Arsenic, Boron, Chromium, Copper, Iodine, Iron, Manganese, Molybdenum, Nickel, Silicon, Vanadium, and Zinc*. Washington, DC: The National Academies Press, 442–501.

11. International Zinc Association. 2011. *Zinc: Essential for Human Health*. Durham, NC: International Zinc Association.
12. Drake, P.L., and K.J. Hazelwood. 2005. Exposure-related health effects of silver and silver compounds: A review. *Annals of Occupational Hygiene* 49:575–585.
13. Brown, M.A., and R.C. Semelka. 2003. *MRI: Basic Principles and Applications*. New York, NY: Wiley-Liss.
14. Na, H.B., I.C. Song, and T. Hyeon. 2009. Inorganic nanoparticles for MRI contrast agents. *Advanced Materials* 21:2133–2148.
15. Lodhia, J., G. Mandarano, N.J. Ferris, P. Eu, and S.F. Cowell. 2010. Development and use of iron oxide nanoparticles (Part 1): Synthesis of iron oxide nanoparticles for MRI. *Biomedical Imaging and Intervention Journal* 6:e12.
16. Pankhurst, Q.A., J. Connolly, S.K. Jones, and J. Dobson. 2006. Applications of magnetic nanoparticles in biomedicine. *Journal of Physics D: Applied Physics* 36:167–181.
17. Chen, Y., H. Chen, S. Xhang, F. Chen, S. Sun, Q. He, et al. 2012. Structure-property relationships in manganese oxide-mesoporous silica nanoparticles used for T-weighted MRI and simultaneous anti-cancer drug delivery. *Biomaterials* 33:2388–2398.
18. Zhen, Z., and J. Xie. 2012. Development of manganese-based nanoparticles as contrast probes for magnetic resonance imaging. *Theranostics* 2:45–54.
19. Shin, J., R.M. Anisur, M.K. Ko, G.H. Im, J.H. Lee, and I.S. Lee. 2008. Hollow manganese oxide nanoparticles as multifunctional agents for magnetic resonance imaging and drug delivery. *Angewandte Chemie International Edition* 48:321–324.
20. Sterenczak, K.A., M. Meier, S. Glage, M. Meyer, S. Willenbrock, P. Wefstaedt, et al. 2012. Longitudinal MRI contrast enhanced monitoring of early tumor development with manganese chloride (MnCl2) and superparamagnetic iron oxide nanoparticles (SPIOS) in a CT1258 based in vivo model of prostate cancer. *BMC Cancer* 12:284.
21. Kim, H., K. Shin, M. Han, K. Ana, J.-K. Lee, I. Honma, et al. 2009. One-pot synthesis of multifunctional mesoporous silica nanoparticle incorporated with zinc(II) phthalocyanine and iron oxide. *Scripta Materialia* 61:1137–1140.
22. Barreto, J.A., W. O'Malley, M. Kubeil, B. Graham, H Stephan, and L. Spiccia. 2011. Nanomaterials: Applications in cancer imaging and therapy. *Advanced Materials Weinheim* 23:H18–H40.
23. Ghosh, P., G. Han, M. De, C.K. Kim, and V.M. Rotello. 2008. Gold nanoparticles in delivery applications. *Advanced Drug Delivery Reviews* 60:1307–1315.
24. Chertok, B., B.A. Moffat, A.E. David, F. Yu, C. Bergemann, B.D. Ross, et al. 2008. Iron oxide nanoparticles as a drug delivery vehicle for MRI monitored magnetic targeting of brain tumors. *Biomaterials* 29:487–496.
25. Berger, M. 2013. Metal oxide based breath nanosensors for diagnosis of diabetes. *Nanowerk*. Available at http://www.nanowerk.com/spotlight/spotid=28303.php.
26. Hadjipanayis, C.G., M.J. Bonder, S. Balakrishnan, X. Wang, H. Mao, and G.C. Hadjipanayis. 2008. Metallic iron nanoparticles for MRI contrast enhancement and local hyperthermia. *Small* 4:1925–1929.
27. Lakshmanan, S. 2011. *Gold/Copper Sulfide and Gold Nanoparticles for Application in Cancer Therapy*. Arlington, TX: University of Texas at Arlington.
28. Li, Y., W. Lu, Q. Huang, M. Huang, C. Li, and W. Chen. 2010. Copper sulfide nanoparticles for photothermal ablation of tumor cells. *Nanomedicine (London)* 5:1161–1171.
29. Pissuwan, D., S.M. Valenzuela, and M.B. Cortie. 2006. Therapeutic possibilities of plasmonically heated gold nanoparticles. *Trends in Biotechnology* 24:62–67.
30. Gannon, C.J., C.R. Patra, R. Bhattacharya, P. Mukherjee, and S.A. Curley. 2008. Intracellular gold nanoparticles enhance non-invasive radiofrequency thermal destruction of human gastrointestinal cancer cells. *Journal of Nanobiotechnology* 6:2.
31. Jose, G.P., S. Santra, S.K. Mandal, and T.K. Sengupta. 2011. Singlet oxygen mediated DNA degradation by copper nanoparticles: potential towards cytotoxic effect on cancer cells. *Journal of Nanobiotechnology* 9:9.

32. B. Kang, M.A. Mackey, and M.A. El-Sayed. 2010. Nuclear targeting of gold nanoparticles in cancer cells induces DNA damage, causing cytokinesis arrest and apoptosis. *Journal of the American Chemical Society* 132:1517–1519.

33. Ravindran, A., P. Chandran, and S.S. Khan. 2013. Biofunctionalized silver nanoparticles: Advances and prospects. *Colloids and Surfaces B: Biointerfaces* 105:342–352.

34. Singh, M., S. Singh, S. Prasad, and L.S. Gambhir. 2008. Nanotechnology in medicine and anitbacterial effects of silver nanoparticles. *Digest Journal of Nanomaterials and Biostructures* 3:115–122.

35. Pal, S., Y.K. Tak, and J.M. Song. 2007. Does the antibacterial activity of silver nanoparticles depend on the shape of the nanoparticle? A study of the gram-negative bacterium *Escherichia coli*. *Applied and Environmental Microbiology* 73:1712–1720.

36. Nair, L.S., and C.T. Laurencin. 2008. Nanofibers and nanoparticles for orthopaedic surgery applications. *Journal of Bone and Joint Surgery* 90(Suppl 1):128–131.

37. Reddy, K.M., K. Feris, J. Bell, D.G. Wingett, C. Hanley, and A. Punnoose. 2007. Selective toxicity of zinc oxide nanoparticles to prokaryotic and eukaryotic systems. *Applied Physics Letters* 90:2139021–2139023.

38. Zhang, L., Y. Jiang, Y. Ding, M. Povey, and D. York. 2007. Investigation into the antibacterial behaviour of suspensions of ZnO nanoparticles (ZnO nanofluids). *Journal of Nanoparticle Research.* 9:479–489.

39. Chen, J., C.M. Han, X.W. Lin, Z.J. Tang, and S.J. Su. 2006. Effect of silver nanoparticle dressing on second degree burn wound. *Zhonghua Wai Ke Za Zhi* 44:50–52.

40. Tian, J., K.K. Wong, C.M. Ho, C.N. Lok, W.Y. Yu, C.M. Che, et al. 2007. Topical delivery of silver nanoparticles promotes wound healing. *ChemMedChem* 2:129–136.

41. Sadrieh, N., A.M. Wokovich, N.V. Gopee, J. Zheng, D. Haines, D. Parmiter, et al. 2010. Lack of significant dermal penetration of titanium dioxide from sunscreen formulations containing nano- and submicron-size TiO_2 particles. *Toxicology Sciences* 115:156–166.

42. Gulson, B., M. McCall, M. Korsch, L. Gomez, P. Casey, Y. Oytam, et al. 2010. Small amounts of zinc from zinc oxide particles in sunscreens applied outdoors are absorbed through human skin. *Toxicology Sciences* 118:140–149.

43. Connecticut Department of Public Health. 2012. *Fact Sheet: Playing It Safe in the Sun.* Hartford, CT: Environmental Health Section, Connecticut Department of Public Health.

44. Pant, G., N. Nayak, and R. Prasuna. 2013. Enhancement of antidandruff activity of shampoo by biosynthesized silver nanoparticles from Solanum trilobatum plant leaf. *Applied Nanosciences* 3:431–439.

45. Hamouda, I.M. 2012. Current perspectives of nanoparticles in medical and dental biomaterials. *Journal of Biomedical Research* 26:143–151.

46. Cuthbertson, D., and R. Holman. 2008. Synopsis of causation: Diabetes mellitus. Ministry of Defense, London.

47. American Diabetes Association. 2013. Standards of medical care in diabetes—2013. *Diabetes Care* 36:S11–S66.

48. Navaei-Nigjeh, M., M. Rahimifard, N. Pourkhalili, A. Nili-Ahmadabadi, M. Pakzad, M. Baeeri, et al. 2012. Multiorgan protective effects of cerium oxide nanoparticle/selenium in diabetic rats: Evidence for more efficiency of nanocerium in comparison to metal form of cerium. *Asian Journal of Animal and Veterinary Advances* 7:605–612.

49. Pourkhalili, N., A. Hosseini, and M. Abdollahi. 2011. Biochemical and cellular evidence of the benefit of a combination of cerium oxide nanoparticles and selenium to diabetic rats. *World Journal of Diabetes* 2:204–210.

50. Joshi, H.M., D.R. Bhumkar, K. Joshi, V. Pokharkar, and M. Sastry. 2006. Gold nanoparticles as carriers for efficient transmucosal insulin delivery. *Langmuir* 22:300–305.

51. Chen, J., S. Patil, S. Seal, and J.F. McGinnis. 2006. Rare earth nanoparticles prevent retinal degeneration induced by intracellular peroxides. *Nature Nanotechnology* 1:142–150.

52. Sheikpranbabu, S., K. Kalishwaralal, D. Venkataraman, S.H. Eom, J. Park, and S. Gurunathan. 2009. Silver nanoparticles inhibit VEGF-and IL-1beta-induced vascular permeability via Src dependent pathway in porcine retinal endothelial cells. *Journal of Nanobiotechnology* 7:8.
53. Lara, H.H., N.V. Ayala-Nuñez, L. Ixtepan-Turrent, and C. Rodriguez-Padilla. 2010. Mode of antiviral action of silver nanoparticles against HIV-1. *Journal of Nanobiotechnology* 8:1.
54. Elechiguerra, J.L., J.L. Burt, J.R. Morones, A. Camacho-Bragado, X. Gao, H.H. Lara, et al. 2005. Interaction of silver nanoparticles with HIV-1. *Journal of Nanobiotechnology* 3:6.
55. Singh, N., B. Manshian, G.J. Jenkins, S.M. Griffiths, P.M. Williams, T.G. Maffeis, et al. 2009. Nanogenotoxicology: The DNA damaging potential of engineered nanomaterials. *Biomaterials* 30:3891–3914.

7

Metallonanotherapeutics for Neurodegenerative Diseases

Komal Vig and Shivani Soni

Contents

7.1 Neurodegenerative diseases

Alzheimer disease (AD) and Parkinson disease (PD) are the most devastating central nervous system (CNS) disorders, affecting more than 6 million Americans (Nutt 2005). Worldwide, approximately 33.9 million people have AD (Banks 2012), which is followed by PD (Kumar 2010). Both of these debilitating neurodegenerative disorders cause worldwide medical and socioeconomic burden. As these diseases advance, the increasing disability interferes with a patient's capacity to perform day-to-day activities, which increases the burden on families and caregivers. AD is characterized by dementia or memory dysfunction, loss of cognitive function, depression, loss of judgment, and violent behaviors with progression of the disease (Banks 2012). The rate of AD increases in the population with aging: 5–10% of individuals who are 60–80 years of age have AD, increasing to 40–50% for individuals older than 85 years (Sahni 2012). Thus, institutionalizing patients in their final years of life accounts for large share of direct costs from the health care system. In United States, it is estimated that the cost of AD will exceed $500 billion annually by 2020 (Smith 1998); the cost of PD is predicted to exceed $50 million by 2040 (Findley 2007).

Besides a genetic component, other conditions that have been associated with an increased risk of AD are cardiovascular factors, including diabetes and hypertension; obesity; depression and other psychosocial factors; and behavioral factors, such as a less active lifestyle and smoking (Barnes 2011). The three main types of pathologies that have been related to AD are tau pathology, amyloid pathology, and neuronal injury (Querferth 2010). AD pathology is characterized by plaques of amyloid β (Aβ) and neurofibrillary tangles of hyperphosphorylated tau, both of which are considered to be neurotoxic along with elevated oxidative stress and the amount of metal ions (Srikanth 2012). In amyloid pathology, amyloid forms plaques in the brain, thus decreasing the concentration of Aβ proteins in cerebrospinal fluid (CSF). Conversely, due to tau pathology, the amount of tau and phosphorylated tau increases in CSF, thus forming very important groups of biomarkers for identifying people who have a high risk of developing dementia over time (Holland 2012). The major risk for sporadic AD is inheritance of the apolipoprotein (Apo) Ee4 allele of the *ApoE* gene located on chromosome 19q13, which is thought to be associated with formation of amyloid plaques and neurofibrillary tangles due to failure of phosphorylated tau, and can lead to early onset and also affect the rate of progression of disease (Poierier 1993). The mutations of three genes— amyloid precursor proteins encoding genes on chromosome 21, *presenilin 1* on chromosome 14, and *presenilin 2* on chromosome 1—are considered to be responsible for genetic predisposition to familial AD (Schaffer 2011).

PD is the second most common chronic, progressive, and age-related neurodegenerative disease (ND) following AD (Nussubaum 2003). It is a disease of aging brain, with up to 2% probability of risk for individuals over the age of 50 (Park 2012) and 1.5 times higher prevalence in men than women older than 70 years (Twelves, 2003). PD is a heterogeneous disorder characterized by bradykinesia, hypokinesia, rigidity, resting tremors, postural instability, hallucinations, depression, dysphagia, and urinary incontinence (Hely 2005; Exner 2012). The hallmark of PD pathogenesis is loss of dopaminergic neurons in substantia nigra pars compacta and accumulation of α-synuclein (SNCA) proteins inside Lewy bodies and Lewy neuritis (Fahn 2004). Oxidative stress related to mitochondrial dysfunction plays a crucial role in disease progression (Pimentel 2012).

Environmental toxins, such as 1-methyl-4-phenyl-1,2,3,6-tetrahydropyridine, toluene, and carbon disulfide are thought to induce PD (Chao 2012).

Herbicides and pesticides including Paraquat (dipyridylium), organophos-phates, and rotenone exposure have also been implicated in the epidemiology of PD (Ascherio 2006). Familial PD accounts for more than 10% of cases. A plethora of genes related to PD have been identified; for example, *PARK1-10*, present in either the autosomal dominant or recessive form, was linked to PD (Fahn 2004). Other genetic factors related to disease include mutations in the *SNCA* gene, leucine–rich repeat kinase 2 (LRRK2), ubiquitin carboxyl-termi-nal esterase L1 (ubiquitin thiolesterase), and PTEN-induced putative kinase 1 (Lachenmayer 2012). Heterozygous glucocerebrosidase (GBA) gene mutation and many polymorphisms related to LRRK2 gene are known to be associated with predisposition to sporadic PD (Tsuboi 2012).

7.2 Blood–brain barrier

For several years, progress has been made in the understanding of the genetic basis of ND disorders and pathology related to it, but advances in neuro-therapeutic efficacy have been hampered due to a very selective and efficient barrier known as the blood–brain barrier (BBB). Paul Ehlrich was the first scientist to speculate about the possibility of a BBB when, after injecting a water-soluble dye intravenously, he observed staining of all tissues except the brain and spinal cord. Since then, the concept of the BBB has been refined and well accepted (Ronaldson 2011).

The BBB is the composition of anatomical features and physiological pro-cesses (**Figure 7.1**) that allows the controlled movement of substances between blood and fluids present in the CNS. The BBB maintains the homeo-stasis of the CNS, which is critical for neuronal and glial functions, by pre-venting bloodborne pathogens and substances from entering. However, this protective barrier also limits the delivery of neurotherapeutic and diagnostic neuroimaging agents to the brain (Jalalai 2011; Koffie 2011).

Structurally, the BBB is mainly constituted of polarized brain capillary endo-thelial cells with tight junctions as well as lack of fenestrea, increased num-ber of mitochondria and decreased macropinocytosis, astrocytes forming close associations with endothelial cells and providing biochemical support, closely placed neuronal endings, and pericytes with efficient phagocytic capacity (Nunes 2012). Astrocytes, pericytes, and neurons form the neu-rovascular units (NVUs), which play an important role in maintaining the immune-privileged environment of the CNS and also control cellular traf-ficking (Zozulya 2007). Furthermore, NVUs are also continuously involved in crosstalk with components of the blood side of the BBB (Neuwelt 2008). Endothelial cells and pericytes are covered by a 30- to 40-nm-thick base-ment membrane constituted of extracellular components, such as collagen type IV, heparin sulfate proteoglycan, laminin, and fibronectin, arranged in trilaminar structure composed of lamina rara externa, lamina rara interna, and a transitory layer (Farkas 2001). Lamina rara externa of the basement

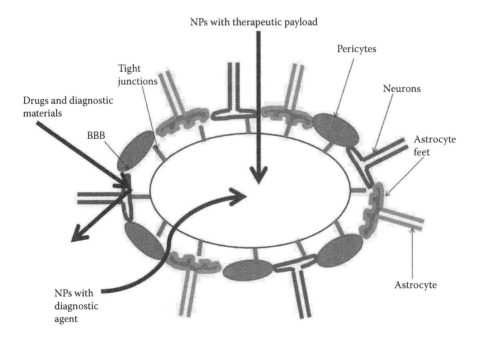

Figure 7.1 *Blood–brain barrier structure showing its main components and possible delivery of molecules with nanoparticles.*

membrane is closely associated with the plasma membrane of the astrocytic end-feet processes ensheathing cerebral capillaries (Naik 2011). Moreover, astrocytes–neuron signaling is speculated to be involved in controlling brain vasculature via calcium-dependent communication between the endothelium and astrocytes through intracellular IP3 and gap junctions or through extracellular transmission of the purinergic messenger (Ballabh 2004). The least studied pericytes are embedded in the basement membrane; they are thought to have contractile properties and may have functional role in regulating blood flow in capillaries as well in angiogenesis (Bhandopadhyay et al. 2001).

The tight junctions of BBB endothelial cells are composed of various protein complexes, such as junctional adhesion molecules, occludin, claudins (claudin 1, 3, and 5), membrane-associated guanylate kinase-like proteins, and transmembrane proteins, such as zonula occluden 1, 2, and 3 linked to cytoskeleton via accessory proteins (i.e., cingulin, AF-6, 7H6, and EMP-1). Additionally, pericyte-derived angiopoietin plays a key role in expression of occludin, a major component of BBB tight junctions (Hawkins 2005). Thus, these tight junctions restrict the paracellular flux of materials through the BBB. However, transport across the barrier does take place through certain transporters, such as ABC transporters, specifically P-glycoprotein (P-gp) and solute carrier transporters (Ronaldson 2011). P-gp is a 170-KDa membrane protein that was first discovered in colchicine-resistant Chinese hamster ovary cells. It plays a protective role against harmful xenobiotics in circulation, especially for sensitive organs, such as the brain (Kivisto et al. 2004). P-gp transports a wide variety of substrates, mainly weakly amphipathic and relatively hydrophobic; the majority of them contain aromatic rings and a positively charged tertiary nitrogen atom (Sharom 2006).

Multidrug resistance proteins (MRPs) are other members of the C branch of the ABC transporter subfamily. They are responsible for the transport of a broad spectrum of endobiotics and xenobiotics (Bandler et al. 2008). Involvement of MRPs (especially MRP2) in transporting anionic xenobiotics in brain capillaries using confocal microscopy was shown by Miller et al. (2011). Another example of an ABC transporter is the breast cancer resistance protein, which plays an important role as a part of the BBB by effluxing unwanted molecules back into the blood, thus reducing the neurotoxic effects of certain drugs (Agarwal et al. 2011). An additional major group of efflux transporters in CNS circulation that are involved in efflux transport of organic anions is the solute carrier transporters (SLCs). Out of 43 SLC transporter families, SLC21/SLCO, including OATP and SLC22 members such as organic cation transporters, are known to be expressed in the BBB, playing vital part in removing xenobiotics from the brain (Kusuhara et al. 2005). Although these transporters are a defense mechanism of BBB to prevent penetration of drugs and xenobiotics from entering CNS, they also pose limitations to the delivery of therapeutic agents to target diseases of the CNS, such as AD and PD.

One big challenge of neurotherapeutics is the constraint of the BBB and drug release kinetics that cause peripheral side effects. High molecular-weight drugs cannot cross the BBB (Pardridge 2007); therefore, many neurological disorders remain untreated. Due to these difficulties in the delivery of drugs, recent research has focused on developing new strategies for drug delivery. One such promising strategy is using nanoparticles (NPs) for drug delivery. Various NPs are being employed as therapeutics for NDs. Polymeric NPs seem to be promising candidates because they are capable of opening the tight junctions of the BBB; they can also enhance prolonged drug release and therefore prevent drug degradation by the enzyme system (Roney et al. 2005).

7.3 Nanoparticles for theranostics

NPs have been increasingly used for the treatment of various diseases, including cancer, human immunodeficiency virus, cardiovascular diseases, and NDs (Friese et al. 2000; Jain et al. 2008; Mitra et al. 2001; Wilson et al. 2008a). NPs offer high surface-to-volume ratio, which often results in a reduction in the amount of drug required for the targeted effect, as well as a decrease in the frequency of drug administration (**Figure 7.2**).

The properties of NPs, such as their size and surface charge, play a significant role in their use as drug delivery vehicles. The surface charge of NPs not only influences the NP's stability but also indicates the interaction of NPs with cells

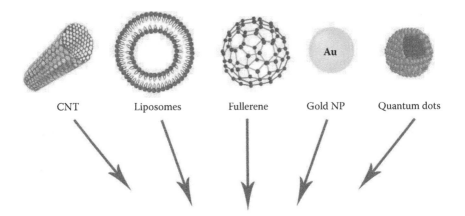

Figure 7.2 *Use of nanopaticles as theranostics for neurodegenerative diseases.*

(Wilson et al. 2008b). It has been reported that a physically stable nanosuspension that is only stabilized by electrostatic repulsion between the particles should have a minimum zeta potential of 30 mV (Muller et al. 2001). Smaller NPs are internalized more efficiently compared to larger particles. However, smaller NPs (<5 nm) have a greater chance of being eliminated by the reticuloendothelial system; hence, they may not deliver the desired drug or allow sufficient residence time for the effect of the drug. Larger particles (≥100 nm), on the other hand, are easily engulfed by phagocytic immune cells, such as Kupffer cells, thus restricting their biodistribution (Wissee and Leeuw 1984). Studies on the biodistribution of NPs in mice have shown that the distribution of NPs in organs and tissues depends on size (Balasubraminium et al. 2010; Sonovane et al. 2008). NPs of 15 nm in size were detected in various tissues including blood, liver, lung, spleen, kidney, brain, heart, and stomach. However, NPs of 200 nm in size were present in much lower quantities in the blood, brain, stomach, and liver (Sonovane et al. 2008).

7.4 Uptake of nanoparticles by brain cells

There are several different mechanisms by which nanomaterials can be internalized. Perhaps the most prominent method is diffusion across the plasma membrane (either directly across the membrane or through membrane channels) or endocytosis. The absence of toxicity at the BBB both in vitro and in situ suggests that NPs may be transported through the barrier by endocytosis or transcytosis or by passive diffusion in the absence of a barrier opening. Uptake by the endothelial cells of the brain can be manipulated by designing NPs to mimic low-density lipoproteins so as to interact with their receptors (Faraji and Wipf 2009; Wagner et al. 2012). Knowledge of the uptake mechanism of NPs enables future developments to rationally create very specific and effective carriers to overcome the BBB.

7.5 Use of nanomaterials as delivery vehicles

NPs can be loaded with drugs, bioactive molecules, antioxidant species or molecules that act like antioxidant sponges for the targeted effect (**Figure 7.3**). Various nanomaterials are gaining use both as diagnostic agents and therapeutics for detection and treatment of ND.

7.5.1 Nanoparticles conjugated with chelating ligands

Transition metal ions, such as Cu^{2+} and Zn^{2+}, are known to increase in the brain with aging. In patients with AD, the number of metal ions is often higher than in healthy individuals. Oxidative reactions catalyzed by Fe^{2+}, Cu^{2+}, Al^{3+}, and Zn^{2+} increase the local concentration of these metals, which is linked with triggering AD (Castellani et al. 2007). Treatments for AD thus focus on lowering brain metal ions and targeting the interaction between the Aβ peptide and metal ions. Chelators reduce the metal levels in neuronal tissue and therefore protect the brain from the harmful effect of the oxidative stress. They also increase uptake of therapeutic agents in the brain, along with enhancing their bioavailability and decreasing their toxic side effects. Metal chelating compounds, such as ethylenediaminetetraacetic acid salts, desferrioxamine, and

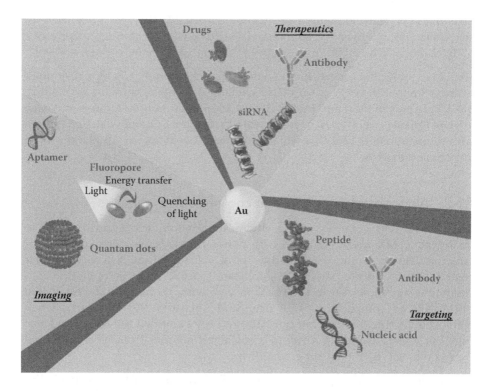

Figure 7.3 Nanoparticles designed for therapeutics, targeting, and imaging.

clioquinol are recruited to improve the clinical conditions in patients with AD (Crapper et al. 1991). However, due to their low BBB permeability and neurotoxicity, chelators are restricted in their usage (Liu et al. 2005). This drawback of chelators can be overcome by using NPs for therapies of brain diseases. D-penicillamine (cuprimine NPs, Merck), a Cu(I) chelator, is currently approved by the U.S. Food and Drug Administration (FDA) for the treatment of Wilson's disease and rheumatoid arthritis. It has been covalently conjugated to 1,2-dioleoyl-snglycero-3-phosphoethanolamine-N-[4-(p-maleimidophenyl) butyramide] (sodium salt) and 1,2-dioleoylsn-glycero-3-phosphoethanolamine-N-[3-(2-pyridyldithio)-propionate] (sodium salt) NPs (Cui et al. 2005). Uptake of these NPs by the brain in situ and in vivo without significant changes in BBB integrity or permeability was observed (Cui et al. 2005; Koziara et al. 2003; Lockman et al. 2003a, 2003b). These NPs could also solubilize Cu^{2+}-amyloid-β aggregates in the reduced environment in vitro (Cui et al. 2005). Likewise, phosphoethanolamine constructs with Cu(II) ions were internalized by cells easily, along with weaker Aβ turnovers (Treiber et al. 2009).

In the same manner, deferoxamine and 2-methyl-N-(2-aminoethyl)-3-hydroxyl-4-pyridone), when conjugated to polysorbate 80-coated NPs, have demonstrated ability to chetale ions in brain sections of patients with AD (Liu et al. 2006). In another study, Liu et al. (2009) demonstrated the ability of nanoparticle–chelator conjugate (Nano-N2PY) to deplete excess metals without affecting essential ions, thus showing a great potential for AD therapy. They used an iron chelator (2-methyl-N-(2′-aminoethyl)-3-hydroxyl-4-pyridinone), a derivative of deferiprone, which had high affinities for iron, aluminum, copper, and zinc.

Similarly, quinoline derivatives, such as clioquinol (5-chloro-7-iodo-8-hydroxyquinoline) (CQ), a Cu^{2+}/Zn^{2+} chelator, has been known to solubilize the Aβ plaques in vitro and inhibit the Aβ accumulation in AD transgenic mice in vivo. Kulkarni et al. (2010) encapsulated n-butyl-2-cyanoacrylate (BCA) NPs with radioiodinated 125I-CQ, which exhibited specificity for Aβ plaques both in vitro and in vivo.

7.5.2 Nanoparticles loaded with cholinesterase inhibitors

Tacrine, the first centrally acting cholinesterase inhibitor approved for the treatment of AD, was conjugated with poly(n-butylcyanoacrylate) (PBCA) for targeted delivery to the brain (Wilson et al. 2008a). These NPs were further coated with 1% polysorbate 80. A significant fourfold increase of tacrine concentration was observed in the brain in vivo with 1% polysorbate 80 coated PBCA NPs as compared to free drug and uncoated NPs. Similarly, when employed, tacrine-loaded chitosan NPs not only extended the drug residence time but also released the drug in a sustained manner from NPs that could freely cross the BBB (Wilson et al. 2010).

Rivastigmine, a noncompetitive and reversible inhibitor of both acetylcholinesterase (AChE) and butyrylcholinesterase that is approved by the FDA, was targeted in the brain using PBCA NPs and 1% polysorbate 80 coated PBCA NPs (Wilson et al. 2008b). A 3.8-fold increase in rivastigmine in the brain was achieved after its administration in vivo as 1% polysorbate 80 coated PBCA NPs compared to that conjugated to PBCA NPs or free drug. Likewise, rivastigmine encapuslated in poly(lactic-co-glycolic) acid (PLGA) and PBCA NPs resulted in positive therapeutic outcomes in amnesic mice (Joshi et al. 2010).

Aβ plaques were also targeted in vivo by an AChE PE154 following its release by polyglycidylmethacrylate or polystyrene/polybutylcyanoacrylate nanoparticles after intrahippocampal injections (Hartig et al. 2010). Single-walled carbon nanotubes (CNTs) were also employed to deliver acetylcholine (ACh) to the brain for AD treatment (Yang et al. 2010); however, its usage is restricted due to lack of toxicity data.

7.5.3 Nanoparticles with payloads of drugs

Thioflavin-T (ThT), a benzothiazole dye, is used for staining and identifying amyloid fibrils both in vitro and in vivo because it emits fluorescence upon binding to amyloid fibrils. However, due to its hydrophilicity, ThT on its own cannot cross the BBB and enter the brain (Klunk et al. 2001). To overcome this, ThT-loaded core-shell latex NPs were prepared using butyl-cyanoacrylate on thioflavin-filled polystyrene (Hartig et al. 2003). ThT was delivered from nanospheres to neurons and microglia, suggesting that NPs are good carriers for drug delivery against Aβ accumulation in the AD brain. Equally, when ThT and Thioflavin-S were encapsulated in a polystyrene core and PBCA shell, they exhibited stronger fluorescence as compared to free fluorophore (Siegemund et al. 2006).

The efficacy of the brain-targeted effects of trimethylated chitosan (TMC)/ PLGA NPs prepared by loading Co-Q10 was tested in vivo for brain uptake and showed that Co-Q10 could dissolve senile plaques in vivo (Wang et al. 2010). Likewise, chitosan-based nanocarriers with tripolyphosphate cores, functionalized with a polyamine-modified F(ab') portion of IgG4.1, an anti-amyloid antibody, demonstrated the ability to cross the BBB and target the brain amyloid deposits (Agyare et al. 2008).

Similarly, vasoactive intestinal neuroprotective peptide—a widely distributed peptide in the nervous system (Fahrenkrug et al. 1979)—showed an increase in brain uptake in mice with a neuroprotective effect when encapsulated in polyethylene glycol-polylactic NPs modified with wheat germ agglutinin (WGA) (Gao et al. 2007). Another NP, VP025 (Vasogen; Ontario, Canada), a phosphatidylglycerol-based phospholipid NP, is currently enrolled in phase II clinical development for PD and has shown potential to induce an anti-inflammatory response of neural tissue, regulating cytokine productions and controlling inflammation reactivity of brain tissue (Crotty et al. 2008). In addition, polymeric NPs carrying nerve growth factor (Kurakhmaeva et al. 2009) or urocortin (Hu et al. 2011) also enhanced BBB penetration and elicited behavior recovery of Parkinson models when administrated systemically. Equally, odorranalectin conjugated to PEG-PLGA nanoparticles could lead to functional and motor recovery of 6-hydroxydopamine (6-OHDA) Parkinson models (Wen et al. 2011). Similarly, human serum albumin NPs bound to ApoE were detected in brain capillary endothelial cells and neurons, whereas no uptake of NPs without Apo into the brain was observed (Zensi et al. 2009).

7.5.4 Hormone-loaded nanoparticles

Estrogen is a well-known hormone in neurotransmission that is involved in cognitive processing (Al-Ghananeem et al. 2002; Bodor et al. 2002), which makes it a good candidate molecule for neurodegenerative therapy. Significant improvement in estradiol (E2) levels transported to the brain was observed when E2-loaded chitosan NPs were administrated to male Wistar rats (Wang et al. 2008). Mittal et al. (2007) also observed improved oral bioavailability and sustained release of E2 when E2 loaded-PLGA NPs were administrated in male Sprague-Dawley rats.

Similarly, melatonin-loaded NPs using the Eudragit S100 polymer (Degussa Pharma, Darmstadt, Germany) provided an increase in the antioxidant effect of melatonin against lipid peroxidation (Schaffazick et al. 2005). Melatonin is known to inhibit oxidative damage in a number of neurological disease models and diseases such as AD due to its ability to scavenge free radicals (Reiter et al. 1997, 1998, 2002, 2003). Schaffazick et al. (2008) also observed a marked reduction on lipid peroxidation in mice tissues with melatonin-loaded polysorbate 80-coated nanocapsules compared to melatonin aqueous solution.

7.5.5 Nanoparticles loaded with polyphenolic phytochemicals

Polyphenolic compounds are known to protect cells against Aβ-induced toxicity. However, they are poorly absorbed from the intestine and get either quickly metabolized or eliminated rapidly from the body. NPs are therefore now being used to encapsulate poorly absorbed polyphenols, such as curcumin, catechins, or resveratrol, to increase their bioavailability.

Epigallotechin-3-gallate (EGCG) and catechins, promising antioxidant polyphenols that promote nonamyloidogenic processing, were encapsulated in chitosan NPs to enhance their intestinal absorption (Dube et al. 2010). Similarly, nanolipidic EGCG particles promoted enhanced levels of α-secretase activity. In vivo administration of these NPs showed enhanced bioavailability of EGCG compared to free EGCG (Smith et al. 2010).

Curcuminoids obtained from *Curcuma longa* (turmeric) have been widely known for their anti-inflammatory, antioxidant, neuroprotective, anticarcinogenic, and antiviral properties; they also reduce Aβ aggregated-related toxicity to neurons (Ringman et al. 2005; Osawa 2007; Sharma et al. 2007; Ishrat et al. 2009). However, due to its poor stability and ability for photodegradation or oxidation, it shows low brain uptake (Anand et al. 2007). Cumurin half-life and brain concentration was increased when encapsulated into polymeric PBCA NPs (Mulik et al. 2009;

Sun et al. 2010). Likewise, when PBCA NPs containing curcumin functionalized with ApoE3 (ApoE3-C-PBCA) were synthesized, it enhanced its cellular uptake and stability (Mulik et al. 2010). Ferulic acid [3-(4-hydroxy-3-methoxyphenyl)-2-propenoic acid], an antioxidant obtained from curcumin when loaded on solid-lipid NPs, demonstrated the ability to cross the BBB with enhanced inhibition of neuronal oxidative stress (Picone et al. 2009). When LAN5 human neuroblastoma cells were treated with Aβ142 oligomers, these NPs blocked the reactions leading to cellular death (Picone et al. 2009).

Nanoliposomes (NLs) with curcumin, or the curcumin derivative, could inhibit the formation of fibrillar and/or oligomeric Aβ in vitro; hence, they could be developed as a treatment for AD (Taylor et al. 2011). NLs adorned with a curcumin derivative were synthesized and showed extremely high affinity for Aβ1-42 fibrils (Mourtas et al. 2011).

Resveratrol, another phytochemical, has demonstrated antioxidant and anti-inflammatory effects. Resveratrol-loaded NPs, using poly-ε-caprolactone as a core and PEG shell, could protect PC12 cells (a model system for neuronal differentiation) against Aβ, while free resveratrol did not (Liu et al. 2009).

7.5.6 Nanoparticles conjugated with antioxidant molecules

PEG-glutathione (GSH) NPs were synthesized to increase GSH, a water-soluble endogenous antioxidant that can protect cells from oxidative stress levels in the brain (Williams et al. 2009). In another study, superoxide dismutase (SOD), another antioxidant molecule involved in defense against oxidative stress, was encapsulated into PLGA NPs to increase its circulating half-life, cell membrane permeability, and brain uptake (Reddy et al. 2008). Similarly, hydralazine-loaded chitosan NPs were employed to reduce membrane integrity damage, secondary oxidative stress, and lipid peroxidation (Cho et al. 2010). Cerium and yttrium NPs were used in vitro to reduce oxidative stress by the inhibition of reactive oxygen species production and the consequent reduction of neuronal death associated with γ-irradiation (Schubert et al. 2006). Likewise, Batrakova et al. (2007) developed PEI-PEG NPs to deliver catalase enzymes to injured brain regions in preinduced PD models.

7.5.7 Metallic nanoparticles

Gold NPs have been employed to redissolve amyloid fibrils remotely using the local heat dissipated by gold NPs after being irradiated electromagnetically (Kogan et al. 2006). Likewise, gold NPs along with weak microwave fields were employed to dissolve toxic protein deposits of Aβ1–42 involved in AD (Bastus et al. 2007). TiO$_2$ NPs were also engaged to promote Aβ fibrillation by accelerating the nucleation process after the peptide assembled to the surface of the NPs (Wu et al. 2008).

Copolymer NPs, cerium oxide particles, hydrophilic polymer-coated quantum dots (QDs), and CNTs could also accelerate the nucleation step of protein fibrils from human ß2 microglobulin (Linse et al. 2007). Similarly, Xiao et al. (2010) demonstrated that the amyloid fibrillation can be inhibited by quenching the nucleation and elongation processes with water-dispersed N-acetyl-L-cysteine capped CdTe QDs (NAC-QDs), which displayed a threshold response. Fluorescent and nonfluorescent maghemite (γ-Fe_2O_3) NPs have also been employed for the removal of amyloid plaques derived from different amyloidogenic proteins that lead to NDs (Skaat and Margel 2009).

7.5.8 Polymeric nanoparticles

Dendrimers have been employed for therapeutic applications against diseases such as PD. Poly-L-lysine dendrimers complexed with human glial cell line-derived neurotrophic factor gene (h-GDNF) have promoted the recovery of 6-OHDA in a preinduced Parkinson model (Gonzalez-Barrios et al. 2006). Correspondingly, PEGylated nanoparticles conjugated to lactoferrin and loaded with h-GDNF for treatment of different preinduced Parkinson models were developed (Huang et al. 2009, 2010). For the treatment of PD, delivery and release of dopamine (DA) in the brain have been reported using nanomaterials (Modi et al. 2010). Chitosan NPs with DA adsorbed onto the external surface were administrated to rats; they showed that loaded DA could reach the brain and increases in the striatum more than DA alone (Trapani et al. 2011).

Nanobased approaches are also used for the treatment of AD. lysine-leucine-valine-phenylalanine-phenylalanine-functionalized dendrimeric scaffolds were developed, which had an inhibitory effect on Aβ1-42 aggregation, along with the ability to disassemble preexisting amyloid aggregates (Chafekar et al. 2007). In the same way, low concentrations of third-generation poly-amidoamine (PAMAM) dendrimers could reduce Aβ1-28 peptide aggregation (Klajnert et al. 2006). Sialic acid-conjugated PAMAM dendrimers were also engaged to reduce Aβ-induced toxicity (Patel et al. 2006).

Carboxyfullerenes (C60) can also be employed as a therapy for AD (Arispe et al. 1993). When administered as an intracerebroventricular injection, they resulted in the improved performance of cognitive tasks in rats induced by amyloid-β peptide. C60 has been engaged to trap multiple radicals (Krusic et al. 1991). Additionally, polyhydroxylated fullerenes act like powerful radical scavengers, preventing mitochondrial oxidative damage in an acute cellular Parkinson model (Cai et al. 2008). Water-soluble C60 carboxylic acid derivatives containing malonic acid groups were used to reduce damage caused by Aβ toxicity (Dugan et al. 1997). Similarly, fullerenol-1 has been engaged for reduction of Aβ-related toxicity (Huang et al. 2000). Fullerenes C60 were used to inhibit Aβ fibrillation, which could salvage rats from Aβ-induced cognitive impairment (Podolski et al. 2007).

Aβ fibril formation was hindered and strongly dependent on the physicochemical characteristics of the NP surface and concentration, when poly(N-isopropylacrylamide)-co-poly(N-tert-butylacrylamide) NPs were used (Cabaleiro-Lago et al. 2008). Additionally, Aβ fibril aggregation behavior was affected when amino-functionalized polystyrene nanoparticles were employed (Cabaleiro-Lago et al. 2010). Aβ oligomerization was also suppressed after sulfonated, sulfated, and fluorinated PS were employed in cultured neurons (Rocha et al. 2008; Saraiva et al. 2010). PEGylated poly(alkyl cyanoacrylate) NPs have also been used for suppression of Aβ aggregation (Brambilla et al. 2010). Furthermore, cholesterol-bearing pullulan nanogels were engaged for controlling the aggregation and cytotoxicity of Aβ1-42 (Ikeda et al. 2006). NLs functionalized with phosphatidic acid or cardiolipin were also developed for reducing the brain level of Aβ aggregation (Gobbi et al. 2010).

7.6 Nanoparticles for imaging and diagnosis

NPs, such as superparamagnetic particles, are being employed for specific tissue targeting during MRI. Fluorescent and nonfluorescent-maghemite (γ-Fe$_2$O$_3$) NPs could be used as multimodal imaging agents, as these have a combination of the magnetic and fluorescence imaging in one nanostructured system (Skaat and Margel 2009). These can be used for Aβ40 fibrils labeling and for early detection of plaques using both magnetic resonance imaging (MRI) and fluorescence microscopy, and therefore may be applied for in vivo AD diagnostic studies. Yang et al. (2011) detected amyloid deposition by magnetic resonance microscopy in AD transgenic mice using Aβ-coupled iron oxide NP.

Surface properties of gold NPs have been used for optical applications, such as probes to detect neuronal cell activity (Zhang et al. 2009). The electron configuration on the surface of the NP was used to measure the activity of neurons. Gold NPs have also been used for the diagnosis of AD. Georganopoulou et al. (2005) developed an ultrasensitive biobarcode assay by means of a sandwich process involving oligonucleotide-modified gold NPs and magnetic microparticles, both functionalized with antibodies specific to the antigen of interest to measure the concentration of amyloid-derived diffusible ligands (ADDLs), a potential soluble pathogenic AD marker, in the CSF of individuals. Correspondingly, monoclonal anti-tau antibody-coated gold NPs were used for the detection of AD's tau protein in CSF (Neely et al. 2009). Additionally, gold NP-generated photothermal bubbles were employed, which enhanced the imaging sensitivities of cancerous cells and atherosclerotic plaques, aiding the diagnosis and therapy (Lapotko 2009). Equally, gold-SiO$_2$ nanoshells were used, which had a substantial scattering cross-section that makes them ideal for image enhancement applications (Cole et al. 2009). Furthermore, the optical properties of silver nanotriangles were used to develop localized surface plasmon resonance nanosensors to understand the interaction between ADDL and the anti-ADDL antibody (Haes et al. 2004).

Another nanomaterial that is gaining immense usage in diagnostics is QDs. TAT or transferrin peptide-conjugated QDs have been employed for brain imaging (Santra et al. 2005; Xu et al. 2008). Likewise, after encapsulating CdSe/ZnS QDs within PEG-poly(lactic acid) and functionalizing them with the lectin WGA, Gao et al. (2008) observed peak uptake of the QDs in the brain at approximately 3–4 hours following its intranasal administration. Equally, peglayted-QDs crosslinked to Aβ peptides were employed to monitor formation of fibrils in the cellular system (Tokuraku et al. 2009). In vivo Aβ peptide aggregation could be studied using this approach. Gupta et al. (2010) developed biphenyl ether conjugated QDs, which acted as nanoscale devices for the detection of amyloid deposits in the brain and can be useful in probing their formation. In another study, QDs conjugated with streptavidin can effectively recognize amyloid precursor proteins with high sensibility compared with conventional fluoroimmunoassay (Feng et al. 2010).

Relatedly, novel water-soluble and luminescent ZnO QDs capped by (3-aminopropyl)triethoxysilane were developed for sensitive and selective detection of DA in biological fluids, such as serum (Zhao et al. 2013); thus, this probe has potential for practical applications in clinical analysis. In the same way, polymeric n-butyl-2-cyanoacrylate NPs were developed as biological markers for the early diagnosis of AD by encapsulating the radioiodinated amyloid affinity drug125I-CQ (Kulkarni et al. 2010). The drug-loaded NPs exhibited specificity for Aβ plaques both in vitro and in vivo; they could be used as a promising delivery vehicle for in vivo single photon emission tomography [(123)I] or PET [(124)I] amyloid imaging agents.

Thioflavin-T-conjugated NPs have been used in the detection of plaques in AD. Thioflavin-T is a hydrophilic and fluorescent marker. When it was conjugated to a PBCA NP, it enabled visualization of various neuronal areas (Siegemund et al. 2006). Likewise, NL decorated with an anti-Aβ monoclonal antibody showed binding to both Aβ monomers and fibrils with high affinity (Canovi et al. 2011). Further studies showed that the same Aβ-MAb-liposomes bound amyloid deposits in postmortem AD brain samples, confirming the potential of these NPs for the diagnosis and therapy of AD.

Aβ plaques were also targeted in vivo by an AChE PE154 following its release by poly(glycidyl methacrylate) or PS/PBCANPs in vivo after intrahippocampal injections (Hartig et al. 2010). PE154 was shown to act not only as an AChE inhibitor but also as a useful fluorescent probe for Aβ in tissue from transgenic mice autopsy case with AD (Elsinghorst et al. 2009). Likewise, Chen et al. (2011) constructed a simple and sensitive biosensor for detection of choline and ACh based on the fluorescence quenching of the QDs by hydrogen peroxide (H_2O_2), which was produced from the enzymatic reaction of choline or ACh.

In vitro quantitative assays for neurotransmitters involved in PD pathology, which exploit plasmon absorbance and gold NPs, have been developed (Baron et al. 2005). Efficient techniques are also being developed for in vitro diagnosis of PD. Gold-doped TiO_2 nanotube arrays were used to design a high-sensitivity photoelectrochemical immunosensor for SNCA detection (An et al. 2010).

7.7 Toxicity issues of nanoparticles

Although nanomaterials are gaining immense usage for targeted delivery of drugs, proteins, and other molecules, there is growing concern about their safety in living systems. NP toxicity depends on various parameters of NPs, such as morphology, size, surface area, material solubility, dose, route, and duration of administration. Commonly, NPs cause oxidative stress (Nel et al. 2006) or generate reactive oxygen species (Hussain et al. 2006). NPs are also known to induce apoptosis in neural stem cells (Deng et al. 2009). NPs are generally eliminated from the body through the reticuloendothelial system. However, long-term exposure of NPs has resulted in a variety of toxicological changes with accumulations in kidney, liver, and brain after oral administration (Kim et al. 2008). For example, Fe_2O_3 NPs were found to adversely affect neuronal fatty acid degeneration in the hippocampus when administered intranasally (Kircher et al. 2003).

Various studies have been reviewed related to toxicology of nanomaterials (Lacerda et al. 2006; Gomez–Gualdrón et al. 2011; Tiwari et al. 2011; Kumar et al. 2012). Strategies towards reducing the toxicity should therefore be considered when developing nanomaterials for biological use. The use of biodegradable and biocompatible polymers such as chitosan, PBCA, polylactic acid, and poly(ε-caprolactone) is encouraged in the preparation of NPs to reduce the toxicity and aid the drug release (Hans and Lownan 2002). These polymers produce biodegradable products that are either absorbed or eliminated from the body (Shirahama et al. 1994). Chitosan is one such nanomaterial that is considered as a Generally Recognized as Safe material by the FDA and has been accepted as safe for dietary usage in Japan, Italy, and Finland. The FDA allows chitosan for applications in wound dressings (Dash et al. 2011). To date, clinical trials using chitosan or chitosan-based biomaterials have not demonstrated any immunological side effects in the human body (Cadete et al. 2012).

Despite adverse and long-term effects, the special surface and physical properties of NPs make them unique and offer attractive applicability in therapeutics and diagnostics. However, extremely limited data are available on possible NP neurotoxicity, and there is an urgent need for future studies to determine guidelines for the adverse effects in question.

7.8 Conclusions

Nanotechnology offers opportunities in understanding the mechanisms of NDs, along with detection and treatment. NP-based systems can help overcome the BBB barrier. NPs thus seem promising for treating ND diseases by enhancing drug delivery and offering new approaches for early diagnosis. A reduction in NP toxicity along with target specificity will be the future of NP-based systems.

Acknowledgments

The authors would like to acknowledge funding support by the National Science Foundation Centers of Research Excellence in Science and Technology (HRD1241701) and Historically Black Colleges and Universities Undergraduate Program (HRD1135863) grants to Prof. Shree R. Singh, Director, Center for Nanobiotechnology Research, Alabama State University.

References

Agarwal S., A.M. Hartz, W.F. Elmquist, and B. Bauer. 2011. Breast cancer resistance protein and P-glycoprotein in brain cancer: Two gatekeepers team up. *Curr Pharm Des* 17:2793–2802.

Agyare E.K., G.L. Curran, M. Ramakrishnan, C.C. Yu, J.F. Poduslo, and K.K. Kandimalla. 2008. Development of a smart nano-vehicle to target cerebrovascular amyloid deposits and brain parenchymal plaques observed in Alzheimer's disease and cerebral amyloid angiopathy. *Pharm Res* 25: 2674–2684.

Al-Ghananeem A.M., A.A. Traboulsi, L.W. Dittert, and A.A. Hussain. 2002. Targeted brain delivery of 17 beta-estradiol via nasally administered water soluble prodrugs. *AAPS Pharm Sci Tech* 3: E5.

Anand P., A. B. Kunnumakkara, R. A. Newman, and B.B. Aggarwal. 2007. Bioavailability of curcumin: problems and promises. *Mol Pharm* 4:807–818.

Arispe N., H.B. Pollard, and E. Rojas. 1993. Giant multilevel cation channels formed by Alzheimer disease amyloid β-protein [A β P-(1-40)] in bilayer membranes. *Proc Natl Acad Sci USA* 90:10573–10577.

Ascherio A., H. Chen, M.G. Weisskopf, E. O'Reilly, M.L. Mccullough, E.E. Calle, et al. 2006. Pesticide exposure and risk for Parkinson's disease. *Ann Neurol* 60:197–203.

Balasubramanian, S.K., J. Jittiwat, J. Manikandan, C.N. Ong, L.E. Yu, and W.Y. Ong. 2010. Biodistribution of gold nanoparticles and gene expression changes in the liver and spleen after intravenous administration in rats. *Biomaterials* 31:2034–2042.

Ballabh, P., A. Braun, and M. Nedergaard. 2004.The blood-brain barrier: An overview structure, regulation, and clinical implications. *Neurobiol Dis* 16:1–13.

Bandler, P.E., C.J. Westlake, C.E. Grant, S.P. Cole, and R.G. Deelev. 2008. Identification of regions required for apical membrane localization of human multidrug resistance protein 2. *Mol Pharm* 74:9–19.

Banks, W.A. 2012. Drug delivery to the brain in Alzheimer's disease: Consideration of the blood-brain barrier. *Adv Drug Del Review* 64:629–639.

Barnes, D.E., and K. Yaffe. 2011. The projected effect of risk factor reduction on Alzheimer's disease prevalence. *Lancet Neurol* 10:819–828.

Bastus, N., M. Kogan, R. Amigo, D. Bosch, E. Araya, A. Turiel, A., et al. 2007. Gold nanoparticles for selective and remote heating of β-amyloid protein aggregates. *Mater Sci Eng C* 27:1246–1240.

Batrakova, E.V., S. Li, A.D. Reynolds, R.L. Mosley, T.K. Bronich, A.V. Kabanov, et al. 2007. A macrophage-nanozyme delivery system for Parkinson's disease. *Bioconjug Chem* 18:1498–1506.

Bhandopadhyay, R., C. Orte, J.G. Lawrenson, A.R. Reid, S. De Silva, and G. Allt. 2001. Contractile proteins in pericytes at blood-brain and blood-retinal barriers. *J Neurocytol* 30: 35–44.

Bodor, N., and P. Buchwald. 2002. Barriers to remember: Brain-targeting chemical delivery systems and Alzheimer's disease. *Drug Discov Today* 7: 766–774.

Brambilla, D., R. Verpillot, M. Taverna, L. De Kimpe, B. Le Droumaguet, J. Nicolas, M. Canovi, et al. 2010. New method based on capillary electrophoresis with laser-induced fluorescence detection (CE-LIF) to monitor interaction between nanoparticles and the amyloid-β Peptide. *Anal Chem* 82: 10083–10089.

Cabaleiro-Lago C., F. Quinlan-Pluck, I. Lynch, K. Dawson, and S. Linse S. 2010. Dual effect of amino modified polystyrene nanoparticles on amyloid protein fibrillation. *ACS Chem Neurosci* 1: 279–287.

Cabaleiro-Lago, C., F. Quinlan-Pluck, I. Lynch, S. Lindman, A.M. Minogue, E. Thulin, et al. 2008. Inhibition of amyloid β protein fibrillation by polymeric nanoparticles. *J Am Chem Soc* 130:15437–15443.

Cadete, A., L. Figueiredo, R. Lopes, C.C. Calado, A.J. Almeids, and L.M. Goncalves. 2012. Development and characterization of a new plasmid delivery system based on chitosan-sodium deoxcholate nanoparticles. *Eur J Pharm Sci* 45: 451–458.

Cai, X., H. Jia, Z. Liu, B. Hou, C. Luo, W. Li, et al. 2008. Polyhydroxylated fullerene derivative C(60)(OH)(24) prevents mitochondrial dysfunction and oxidative damage in an MPP(+)-induced cellular model of Parkinson's disease. *J Neurosci Res* 86: 3622–3634.

Calvo, P., B. Gouritin, H. Chacun, D. Desmaele, J. D'Angelo, J.P. Noel, et al. 2001. Long-circulating PEGylated polycyanoacrylate nanoparticles as new drug carrier for brain delivery. *Pharm Res* 18:1157–1166.

Canovi, M., E. Markoutsa, A.N. Lazar, G. Pampalakis, C. Clement, F. Re, et al. 2011. The binding affinity of anti-Aβ1-42 peptides in vitro and to amyloid deposits in post-mortem tissue. *Biomaterials* 32:5489–5497.

Castellani, R.J., P.I. Moreira, G.Liu, J. Dobson, G. Perry, M.A. Smith, et al. 2007. Iron: The redox-active center of oxidative stress in Alzheimer disease. *Neurochem Res* 32:1640–1645.

Chafekar, S.M., H. Malda, M. Merkx, E.W. Meijer, D. Viertl, H.A. Lashuel, et al. 2007. Branched KLVFF tetramers strongly potentiate inhibition of β-amyloid aggregation. *Chembiochem* 8:1857–1864.

Chao, J., Y. Leung, M. Wang, and R.C.C. Chang. 2012. Neutraceuticals and their preventive or potential therapeutic value in Parkinson's disease. *Nutrition Rev* 70:373–386.

Chen, Z., X. Ren, X. Meng, D. Chen, C. Yan, J. Ren, et al. 2011. Optical detection of choline and acetylcholine based on H_2O_2-sensitive quantam dots. *Biosens Bioelectron* 28:50–55.

Cho, Y., R. Shi, and R. Ben Borgens. 2010. Chitosan nanoparticle-based neuronal membrane sealing and neuroprotection following acrolein-induced cell injury. *J Biol Eng* 4:2.

Cole, J. R., N.A. Mirin, M.W. Knight, G.P. Goodrich, and N.J. Halas. 2009. Photothermal efficiencies of nanoshells and nanorods for clinical therapeutic applications. *J Phys Chem C* 113:12090–12094.

Crapper McLachlan, D.R., A.J. Dalton, T.P. Kruck, M.Y. Bell, W.L. Smith, W. Kalow, et al. 1991. Intramuscular desferrioxamine in patients with Alzheimer's disease. *Lancet* 337:1304–1308.

Crotty, S., P. Fitzgerald, E. Tuohy, D.M. Harris, A. Fisher, A. Mandel, et al. 2008. Neuroprotective effects of novel phosphatidylglycerol based phospholipids in the 6-hydroxydopamine model of Parkinson's disease. *Eur J Neurosci* 27:294–300.

Cui, Z., P.R. Lockman, C.S. Atwood, C.H. Hsu, A. Gupte, D.D. Allen, et al. 2005. Novel D-penicillamine carrying nanoparticles for metal chelation therapy in Alzheimer's and other CNS diseases. *Eur J Pharm Biopharm* 59:263–272.

Dash, M., F. Chiellini, R.M. Ottenbrite, and E. Chiellini. 2011. Chitosan—a versatile semi-synthetic polymer in biomedical applications. *Prog Polym Sci* 36:981–1014.

Deng, X., Q. Luan, W. Chen, Y. Wang, M. Wu, H. Zhang, and Z. Jiao. 2009. Nanosized zinc oxide particles induce neural stem cell apoptosis. *Nanotechnology* 20:115101.

Dube, A., J.A. Nicolazzo, and I. Larson. 2010. Chitosan nanoparticles enhance the intestinal absorption of the green tea catechins (+)-catechin and (−)-epigallocatechin gallate. *Eur J Pharm Sci* 41:219–225.

Dugan, L.L., D.M. Turetsky, C. Du, D. Lobner, M. Wheeler, C.R. Almli, et al. 1997. Carboxyfullerenes as neuroprotective agents. *Proc Natl Acad Sci USA* 94:9434–9439.

Elsinghorst, P.W., W. Härtig, S. Goldhammer, J. Grosche, and M. Gutschow. 2009. A gorge-spanning, high-affinity cholinesterase inhibitor to explore β-amyloid plaques. *Org Biomol Chem* 7:3940–3946.

Exner, N., A.K. Lutz, C. Haass, and K.F. Winkhofer. 2012. Mitochondrial dysfunction in Parkinson's disease: Molecular mechanisms and pathophysiological consequences. *Embo J* 31:3038–3062

Fahn, S., D. Oakes, I. Shoulson, K. Kieburtz, A. Rudolph, A. Lang, et al. 2004. Levodopa and the progression of Parkinson's disease. *N Engl J Med* 351:2498–2508.

Fahn S., and D. Sulzer. 2004. Neurodegeneration and neuroprotection in Parkinson disease. *Neuro Rx* 1:139–154.

Fahrenkrug, J. 1979. Vasoactive intestinal polypeptide: Measurement, distribution and putative neurotransmitter function. *Digestion* 19: 149–169.

Faraji, A.H., and P. Wipf. 2009. Nanoparticles in cellular drug delivery. *Bioorg Med Chem* 17: 2950–2962.

Farkas, E., and P.G.M. Luiten. 2001. Cerebral microvascular pathology in aging and Alzheimer's disease. *Prog Neurobiol* 64:575–611.

Feng, L., S. Li, B. Xiao, S. Chen, R. Liu, and Y. Zhang. 2010. Fluorescence imaging of APP in Alzneimer's disease with quantam dot or cy3: A comparative study. *J Cent South Univ* 35:903–909.

Findley L.J. 2007. The economic impact of Parkinson's disease. *Parkinsonism Relat Disord* 13: S8–S12.

Friese, A., E. Seiller, G. Quack, B. Lorenz, and J. Kreuter. 2000. Increase of the duration of the anticonvulsive activity of a novel NMDA receptor antagonist using poly (butyl-cyanoacrylate) nanoparticles as a parenteral controlled release system. *Eur J Pharm Biopharm* 49:103–109.

Gao, X., J. Chen, B. Wu, H. Chen, and X. Jiang. 2008. Quantam dots bearing lectin-functionlized nanoparticles as a platform for in vivo brain imaging. *Bioconjug Chem* 19:2189–2195.

Gao, X., B. Wu, Q. Zhang, J. Chen, J. Zhu, W. Zhang, Z., et al. 2007. Brain delivery of vasoactive intestinal peptide enhanced with the nanoparticles conjugated with wheat germ agglutinin following intranasal administration. *J Con Release* 121:156–167.

Georganopoulou, D.G., L. Chang, J.M. Nam, C.S. Thaxton, E.J. Mufson, W.L. Klein, et al. 2005. Nanoparticle-based detection in cerebral spinal fluid of a soluble pathogenic biomarker for Alzheimer's disease. *Proc Natl Acad Sci USA* 102: 2273–2276.

Gomez-Gualdron, D.A., J.C. Burgos, J. Yu, and P.B. Balbuena. 2011. Carbon nanotubes: Engineering biomedical applications. *Prog Mol Biol Transl Sci* 104:175–245.

Gonzalez-Barrios, J.A., M. Lindahl, M.J. Bannon, V. Anaya-Martinez, G. Flores, I. Navarro-Quiroga, et al. 2006. Neurotensin polyplex as an efficient carrier for delivering the human GDNF gene into nigral dopamine neurons of hemiparkinsonian rats. *Mol Ther* 14:857–865.

Gupta, S., P. Babu, and A. Surolia. 2010. Biphenyl ethers conjugated CdSe/ZnS core/shell quantam dots and interpretation of the mechanism of amyloid fibril disruption. *Biomaterials* 31:6809–6822

Haes, A.J., W.P. Hall, L. Chang, W.L. Klein, and R.P. Van Duyne. 2004. A localized surface plasmon resonance biosensor: First steps toward an assay for Alzheimer's disease. *Nano Lett* 4:1029–1034.

Hans, M.L., and A.M. Lownan. 2002. Biodegradable nanoparticles for drug delivery and targeting. *Curr Opin Solid State Mat Sci* 6:319–327.

Hartig, W., J. Kacza, B.R. Paulke, J. Grosche, U. Bauer, and A. Hoffmann. 2010. In vivo labelling of hippocampal beta-amyloid in triple-transgenic mice with a fluorescent acetylcholinesterase inhibitor released from nanoparticles. *Eur J Neurosci* 31: 99–109.

Hartig, W., B.R. Paulke, C. Varga, J. Seeger, T. Harkany, and J. Kacza. 2003. Electron microscopic analysis of nanoparticles delivering thioflavin-T after intrahippocampal injection in mouse: Implications for targeting beta-amyloid in Alzheimer's disease. *Neurosci Lett* 338:174–176.

Hawkins, B.T., and T.P Davis. 2005. The blood-brain barrier/neurovascular unit in health and disease. *Pharmacol Rev* 57:173–185.

Hely, M.A., J.G Morris, W.G Reid, and R. Trafficante. 2005. Sydney multicenter study of Parkinson's disease: Non-L-dopa-responsive problems dominate at 15 years. *Mov Disord* 20:190–199.

Holland, D., L.K. McEvoy, R.D. Desikan, and A.M. Dale. 2012. Enrichment and stratification for predementia Alzheimer disease clinical trials. *PLoS One* 7:e47739.

Hu, K.L., Y.B. Shi, W.M. Jiang, J.Y. Han, S.X. Huang, and X.G. Jiang. 2011. Lactoferrin conjugated PEG–PLGA nanoparticles for brain delivery: Preparation, characterization and efficacy in Parkinson's disease. *Int J Pharm* 415:273–283.

Huang, H.M., H.C. Ou, S.J. Hsieh, L.Y. Chiang. 2000. Blockage of amyloid β peptide-induced cytosolic free calcium by fullerenol-1, carboxylate C60 in PC12 cells. *Life Sci* 66:1525–1533.

Huang, R., L. Han, J. Li, F. Ren, W. Ke, C. Jiang, et al. 2009. Neuroprotection in a 6-hydroxy-dopamine-lesioned Parkinson model using lactoferrin-modified nanoparticles. *J Gene Med* 11:754–763.

Huang, R., W. Ke, Y. Liu, D. Wu, L. Feng, C. Jiang, et al. 2010. Gene therapy using lactoferrin-modified nanoparticles in a rotenone-induced chronic Parkinson model. *J Neurol Sci* 290:123–130.

Hussain, S.M., K.L. Hess, J.M. Gearhart, K.T. Geiss, and J.J. Schlager. 2005. In vitro toxicity of nanoparticles in BRL3A rat liver cells. *Toxicol In Vitro* 19:975–983.

Ikeda, K., T. Okada, S. Sawada, K. Akiyoshi, and K. Matsuzaki. 2006. Inhibition of the formation of amyloid beta-protein fibrils using biocompatible nanogels as artificial chaperones. *FEBS Lett* 580: 6587–6595.

Ishrat, T., M.N. Hoda, M.B. Khan, S. Yousuf, M. Ahmad, M.M. Khan, et al. 2009. Amelioration of cognitive deficits and neurodegeneration by curcumin in rat model of sporadic dementia of Alzheimer's type (SDAT). *Eur Neuropsychopharmacol* 19:636–647.

Jain, S.K., Y. Gupta, A. Jain, A.R. Saxena, and P. Khare. 2008. Mannosylated gelatin nanoparticles bearing an anti-HIV drug didanosine for site-specific delivery. *Nanomedicine* 4:41–48.

Jalali, N., F. Moztarzadeh, M. Mozafari, S. Asgari, M. Motevalian, and S. N. Alhosseini. 2011. Surface modification of poly(lactide-coglycolide) nanoparticles by d-alpha-tocopheryl polyethylene glycol 1000 succinate as potential carrier for the delivery of drugs to the brain. *Physiochem Eng Aspects* 392:335–342.

Joshi, S.A., S.S. Chavhan, and K.K. Sawant. 2010. Rivastigmine-loaded PLGA and PBCA nanoparticles: Preparation, optimization, characterization, in vitro and pharmacodynamic studies. *Eur J Pharm Biopharm* 76:189–199.

Kim, I.D., C.M. Lim, J.B. Kim, H.Y. Nam, K. Nam, S.W. Kim, et al. 2010. Neu siRNA delivery in primary cortical cultures and in the postischemic brain. *J Con Release* 142:422–430.

Kircher, M.F., U. Mahmood, R.S. King, R. Weissleder, and L. Josephson. 2003. A multimodel nanoparticle for preoperative magnetic resonance imaging and intraoperative optical brain tumor delineation. *Cancer Res* 63:8122–8125.

Kivisto, K.T., M. Niemi, and M.F. Fromm. 2004. Functional interaction of intestinal CYP3A4 and P-glycoprotein. *Fund Clin Pharmacol* 18: 621–626.

Klajnert, B., M. Cortijo-Arellano, M. Bryszewska, and J. Cladera. 2006. Influence of heparin and dendrimers on the aggregation of two amyloid peptides related to Alzheimer's and prion diseases. *Biochem Biophys Res Commun* 339:577–582.

Klunk, W.E., Y. Wang, G.F. Huang, M.L. Debnath, D.P. Holt, and C.A. Mathis. 2001. Uncharged thioflavin-T derivatives bind to amyloid-beta protein with high affinity and readily enter the brain. *Life Sci* 69:1471–1484.

Koffie, R.M., C.T. Farrar, L.J. Saidi, C.M. Willain, B.T. Hyman, and T.L. Spires-Jones. 2011. Nanoparticles enhance brain delivery of blood-brain-impermeable probes for in vivo optical and magnetic resonance imaging. *Proc Natl Acad Sci USA* 108:18837–18842.

Kogan, M.J., N.G. Bastus, R. Amigo, D. Grillo-Bosch, E. Araya, A. Turiel, et al. 2006. Nanoparticle-mediated local and remote manipulation of protein aggregation. *Nano Lett* 6:110–115.

Koziara, J.M., P.R. Lockman, D.D. Allen, and R.J. Mumper. 2003. In situ blood-brain barrier transport of nanoparticles. *Pharm Res* 20: 1772–1778.

Krusic, P.J., E. Wasserman, P.N. Keizer, J.R. Morton, and K.F. Preston. 1991. Radical reactions of C60. *Science* 254:1183–1185.

Kulkarni, P.V., C.A. Roney, P.P. Antich, F.J. Bonte, A.V. Raghu, and T.M. Aminabhavi. 2010. Quinoline-n-butylcyanoacrylate-based nanoparticles for brain targeting for the diagnosis of Alzheimer's disease. *Wiley Interdiscip Rev Nanomed Nanobiotechnol* 2: 35–47.

Kumar, S., G. Ho, and Y.L. Zhang. 2010. In vivo imaging of retinal gliosis: A platform for diagnosis of PD and screening of anti-PD compounds. *Conf Proc IEEE Eng Med Biol Soc* 2010: 3049–3052.

Kumar, V., A. Kumari, P. Guleria, and S.K. Yadav. 2012. Evaluating the toxicity of selected types of nanochemicals. *Rev Environ Contam Toxicol* 215:39–121.

Kurakhmaeva, K.B., I.A. Djindjikhashvili, V.E. Petrov, V.U. Balabanyan, T.A. Voronina, S.S. Trofimov, et al. 2009. Brain targeting of nerve growth factor using poly(butyl cyanoacrylate) nanoparticles. *J Drug Target* 17:564–574.

Kusuhara H., and Y. Sugiyama. 2005. Active efflux across the blood-brain barrier: Role of the solute carrier family. *NeuroRx* 2:73–85.

Lacerda, L., A. Bianco, P. Maurizio, K. Kostarelos. 2006. Carbon nanotubes as nanomedicine. From toxicology to pharmacology. *Adv Drug Deliv Rev* 58:1460–1470.

Lachenmayer, M.L., and Z. Yue. 2012. Genetic animal models for evaluating the role of autophagy in etiopathogenesis of Parkinson disease. *Autophagy* 8:1837–1838.

Lapotko, D. 2009. Plasmonic nanoparticle-generated photothermal bubbles and their biomedical applications. *Nanomedicine (Lond)* 4:813–845.

Linse, S., C. Cabaleiro-Lago, W.F. Xue, I. Lynch, S. Lindman, E. Thulin, et al. 2007. Nucleation of protein fibrillation by nanoparticles. *Proc Natl Acad Sci USA* 104:8691–8696.

Liu, G., M.R. Garrett, P. Men, X. Zhu, G. Perry, and M.A. Smith. 2005. Nanoparticle and other metal chelation therapeutics in Alzheimer disease. *Biochim Biophys Acta* 1741:246–252.

Liu, G., P. Men, P.L.R. Harris, R.K Rolston, G. Perry, and M.A. Smith. 2006. Nanoparticle iron chelators: A new therapeutic approach in Alzheimer disease and other neurologic disorders associated with trace metal imbalance. *Neurosci Lett* 406:189–193.

Liu, G., P. Men, W. Kudo, G. Perry, and M.A. Smith. 2009. Nanoparticle-chelator conjugates as inhibitors of amyloid-[beta] aggregation and neurotoxicity: A novel therapeutic approach for Alzheimer disease. *Neurosci Lett* 455:187–190.

Lockman, P.R., J. Koziara, K.E. Roder, J. Paulson, T. J. Abbruscato, R.J. Mumper, et al. 2003. In vivo and in vitro assessment of baseline blood-brain barrier parameters in the presence of novel nanoparticles. *Pharm Res* 20:705–713.

Lockman, P.R., M.O. Oyewumi, J.M. Koziara, K.E. Roder, R.J. Mumper, and D.D. Allen. 2003. Brain uptake of thiamine-coated nanoparticles. *J Con Release* 93:271–282.

Miller, D.W., M. Hinton, and F. Chen. 2011. Evaluation of drug efflux transporter liabilities of darifenacin in cell culture models of the blood-brain and blood-ocular barriers. *Neurourol Urodyn* 30:1633–1638.

Mitra, S., U. Gaur, P.C. Ghosh, and A.N. Maitra. 2001. Tumour targeted delivery of encapsulated dextran-doxorubicin conjugate using chitosan nanoparticles as carrier. *J Con Release* 74:317–323.

Mittal, G., D.K. Sahana, V. Bhardwaj, and M.N. Ravi Kumar. 2007. Estradiol loaded PLGA nanoparticles for oral administration: Effect of polymer molecular weight and copolymer composition on release behavior in vitro and in vivo. *J Con Release* 119:77–85.

Modi, G., V. Pillay, and E. Yahya. 2010. Advances in the treatment of neurodegenerative disorders employing nanotechnology. *Ann NY Acad Sci* 1184:154–172.

Mourtas, S., M. Canovi, C. Zona, D. Aurilia, A. Niarakis, B. LaFerla, et al. 2011. Curcumin-decorated nanoliposomes with very high affinity for amyloid-β1-42. *Biomaterials* 32:1635–1645.

Mulik, R., K. Mahadik, and A. Paradkar. 2009. Development of curcuminoids loaded poly(butyl cyanoacrylate) nanoparticles: Physicochemical characterization and stability study. *Eur J Pharm Sci* 37:395–404.

Mulik, R.S., J. Monkkönen, R.O. Juvonen, K. R. Mahadik, and A.R. Paradkar. 2010. ApoE3 mediated poly(butyl) cyanoacrylate nanoparticles containing curcumin: Study of enhanced activity of curcumin against β-amyloid-induced cytotoxicity using in vitro cell culture model. *Mol Pharm* 7:815–825.

Muller, R.H., C. Jacobs, and O. Kayser. 2001. Nanosuspensions as particulate drug formulations in therapy. Rationale for development and what we can expect for the future. *Adv Drug Deliv Rev* 47:3–19.

Naik, P., and L. Cucullo. 2012. In vitro blood-brain barrier models: Current and perspective technologies. *J Pharm Sci* 101:1337–1354.

Neely, A., C. Perry, B. Varisli, A.K. Singh, T. Arbneshi, D. Senapati, et al. 2009. Ultrasensitive and highly selective detection of Alzheimer's disease biomarker using two-photon Rayleigh scattering properties of gold nanoparticle. *ACS Nano* 3:2834–2840.

Nel, A., T. Xia, and L. Madler. 2006. Toxic potential of materials at the nanolevel. *Science* 311:622–627.

Neuwelt, E., N.J. Abbott, L. Abrey, W.A. Banks, B. Blakley, T. Davis, et al. 2008. Strategies to advance translational research into brain barriers. *Lancet Neurol* 7:84–89.

Nunes, A., K.T. Al-Jamal, and K. Kostarelos. 2012. Therapeutics, imaging and toxicity of nanomaterials in the central nervous system. *J Con Release* 161:290–306.

Nussbaum, R.L., and C.E. Ellis. 2003. Alzheimer's disease and Parkinson's disease. *N Eng J Med* 348:1356–1364.

Nutt, J.G., and G.F. Wooten. 2005. Clinical practice. Diagnosis and initial management of Parkinson's disease. *N Engl J Med* 353:1021–1027.

Osawa, T. 2007. Nephroprotective and hepatoprotective effects of curcuminoids. *Adv Exp Med Biol* 595:407–423.

Pardridge, W.M. 2007. Brain drug development and brain drug targeting. *Pharm Res* 24:1729–1732.

Park, N.G. 2012. Parkinson disease. *JAAPA* 25:73.

Patel, D.A., J.E. Henry, and T.A. Good. 2006. Attenuation of β-amyloid induced toxicity by sialic acid-conjugated dendrimeric polymers. *Biochim Biophys Acta* 1760:1802–1809.

Patel, D.A., J.E. Henry, and T.A. Good. 2007. Attenuation of β-amyloid-induced toxicity by sialic acid-conjugated dendrimers: Role of sialic acid attachment. *Brain Res* 1161:95–105.

Picone, P., M.L. Bondi, G. Montana, A. Bruno, G. Pitarresi, G. Giammona, et al. 2009. Ferulic acid inhibits oxidative stress and cell death induced by Ab oligomers: Improved delivery by solid lipid nanoparticles. *Free Rad Res* 43:1133–1145.

Podolski, I.Y., Z.A. Podlubnaya, E.A. Kosenko, E.A. Mugantseva, E G. Makarova, L.G. Marsagishvili, et al. 2007. Effects of hydrated forms of C60 fullerene on amyloid 1-peptide fibrillization in vitro and performance of the cognitive task. *J Nanosci Nanotechnol* 7:1479–1485.

Poirier, J., J. Davignon, D. Bouthilier, S. Kogan, P. Bertrand, and S. Gauthier. 1993. Apolipoprotein E polymorphism and Alzheimer's disease. *Lancet* 342:697–699.

Querferth, H.W., and F.M. LaFerla. 2010. Alzheimer's disease. *N Eng J Med* 362:329–344.

Reddy, M.K., L. Wu, W. Kou, A. Ghorpade, and V. Labhasetwar. 2008. Superoxide dismutase-loaded PLGA nanoparticles protect cultured human neurons under oxidative stress. *Appl Biochem Biotechnol* 151:565–577.

Reiter, R., L. Tang, J.J. Garcia, and Munoz-Hoyos A. 1997. Pharmacological actions of melatonin in oxygen radical pathophysiology. *Life Sci* 60:2255–2271.

Reiter, R.J. 2003. Melatonin: Clinical relevance. *Best Pract Res Clin Endocrinol Metab* 17:273–285.

Reiter, R.J. 1998. Oxidative damage in the central nervous system: Protection by melatonin. *Prog Neurobiol* 56:359–384.

Reiter, R.J., D.X. Tan, and S. Burkhardt. 2002. Reactive oxygen and nitrogen species and cellular and organismal decline: Amelioration with melatonin. *Mech Ageing Dev* 123:1007–1019.

Ringman, J.M., S.A. Frautschy, G.M. Cole, D.L. Masterman, and J.L. Cummings. 2005. A potential role of the curry spice curcumin in Alzheimer's disease. *Curr Alzheimer Res* 2:131–136.

Rocha, S., A.F. Thunemann, M. Pereira, C. Pereira Mdo, M. Coelho, H. Mohwald, et al. 2008. Influence of fluorinated and hydrogenated nanoparticles on the structure and fibrillogenesis of amyloid β-peptide. *Biophys Chem* 137:35–42.

Ronaldson, P.T., and T.P. Davis. 2011. Targeting blood-brain barrier changes during inflammatory pain: An opportunity for optimizing CNS drug delivery. *Ther Deliv* 2:1015–1041.

Roney, C., P. Kulkarni, V. Arora, P. Antich, F. Bonte, A. Wu, et al. 2005. Targeted nanoparticles for drug delivery through the blood-brain barrier for Alzheimer's disease. *J Con Release* 108:193–214.

Sahni, J.K., S. Doggui, J. Ali, S. Baboota, L. Dao, and C. Ramassamy. 2011. Neurotherapeutic applications of nanoparticles in Alzheimer's disease. *J Con Release* 152:208–231.

Santra, S., H. Yang, J.T. Stanley, P.H. Holloway, B.M. Moudgil, G. Walter, et al. 2005. Rapid and effective labeling of brain tissue using TAT-conjugated CdS:Mn/ZnS quantum dots. *Chem Commun* 25:3144–3146.

Saraiva, A.M., I. Cardoso, M.J. Saraiva, K. Tauer, M.C. Pereira, M.A. Coelho, et al. 2010. Randomization of amyloid-β-peptide(1-42) conformation by sulfonated and sulfated nanoparticles reduces aggregation and cytotoxicity. *Macromol Biosci* 10:1152–1163.

Schaeffer, E.L., M. Figueiro, and W.F. Gattaz. 2011. Insights into Alzheimer disease pathogenesis from studies in transgenic animal models. *Clinics* 1:45–54.

Schaffazick, S.R., A.R. Pohlmann, C.A.S. de Cordova, T.B. Creczynski-Pasa, and S.S.Guterres. 2005. Protective properties of melatonin-loaded nanoparticles against lipid peroxidation. *Int J Pharm* 289:209–213.

Schaffazick, S.R., I.R. Siqueira, A.S. Badejo, D.S. Jornada, A.R. Pohlmann, C.A. Netto, et al. 2008. Incorporation in polymeric nanocapsules improves the antioxidant effect of melatonin against lipid peroxidation in mice brain and liver. *Eur J Pharm Biopharm* 69:64–71.

Schubert, D., R. Dargusch, J. Raitano, and Chan S.W. 2006. Cerium and yttrium oxide nanoparticles are neuroprotective. *Biochem Biophys Res Commun* 342:86–91.

Sharma, R.A., W.P. Steward, and A.J. Gescher. 2007. Pharmacokinetics and pharmacodynamics of curcumin. *Adv Exp Med Biol* 595:453–470.

Sharom, F.J. 2006. Shedding light on drug transport: Structure and function of the P-glycoprotein multidrug transporter (ABCB1). *Biochem Cell Biol* 84:979–992.

Shirahama, H., and H. Yasuda. 1994. Biodegradation of optically active new-polyesters. In *Studies in Polymer Science*, eds. Y. Doi and K. Fukuda, pp. 541–548. Amsterdam: Elsevier.

Siegemund, T., B.R. Paulke, H. Schmiedel, N. Bordag, A. Hoffmann, T. Harkany, et al. 2006. Thioflavins released from nanoparticles target fibrillar amyloid β in the hippocampus of APP/PS1 transgenic mice. *Int J Dev Neurosci* 24:195–201.

Skaat, H., and S. Margel. 2009. Synthesis of fluorescent-maghemite nanoparticles as multimodal imaging agents for amyloid-beta fibrils detection and removal by a magnetic field. *Biochem Biophys Res Commun* 386:645–649.

Smith, A., B. Giunta, P.C. Bickford, M. Fountain, J. Tan, and R.D. Shytle. 2010. Nanolipidic particles improve the bioavailability and α-secretase inducing ability of epigallocatechin-3-gallate (EGCG) for the treatment of Alzheimer's disease. *Int J Pharm* 389:207–212.

Smith, M.A. 1998. Alzheimer disease. *Int Rev Neurobiol* 42:1–54.

Sonavane, G., K. Tomoda, and K. Makino. 2008. Biodistribution of colloidal gold nanoparticles after intravenous administration: Effect of particle size. *Colloids Surf B Biointerfaces* 66:274–280.

Taylor, M., S. Moore, S. Mourtas, A. Niarakis, F. Re, C. Zona, B. La Ferla, et al. 2011. Effect of curcumin-associated and lipid ligand-functionalized nanoliposomes on aggregation of the Alzheimer's Aβ peptide. *Nanomedicine* 7:541–550.

Tiwari, P.M., K. Vig, V.A. Dennis, and S.R. Singh. 2011. Functionalized gold nanoparticles and their biomedical applications. *Nanomaterials* 1:31–63.

Tokuraku, K., M. Marquardt, and T. Ikezu. 2009. Real-time imaging and quantification of amyloid-β peptide aggregates by novel quantum-dot nanoprobes. *PLoS One* 4:e8492.

Trapani, A., E. De Giglio, D. Cafagna, N. Denora, G. Agrimi, T. Cassano, et al. 2011. Characterization and evaluation of chitosan nanoparticles for dopamine brain delivery. *Int J Pharm* 419:296–307.

Treiber, C., M.A. Quadir, P. Voigt, M. Radowski, S. Xu, L. M. Munter, et al. 2009. Cellular copper import by nanocarrier systems, intracellular availability, and effects on amyloid β peptide secretion. *Biochemistry* 48:4273–4284.

Tsuboi, Y. 2012. Environmental-genetic interactions in the pathogenesis of Parkinson's disease. *Exp Neurobiol* 21:123–128.

Twelves, D., K.S. Perkins, and C. Counsell. 2003. Systematic review of incidence studies of Parkinson's disease. *Mov Disord* 18:19–31.

Wagner, S., A. Zensi, S.L. Wien, S.E. Tschickardt, W. Maier, T. Vogel, et al. 2012. Uptake mechanism of ApoE-modified nanoparticles on brain capillary endothelial cells as a blood-brain barrier model. *PLoS One* 7:e32568.

Wang, X., N. Chi, and X. Tang. 2008. Preparation of estradiol chitosan nanoparticles for improving nasal absorption and brain targeting. *Eur J Pharm Biopharm* 70:735–740.

Wang, Z.H., Z.Y. Wang, C.S. Sun, C.Y. Wang, T.Y. Jiang, and S.L. Wang. 2010. Trimethylated chitosan-conjugated PLGA nanoparticles for the delivery of drugs to the brain. *Biomaterials* 31:908–915.

Wen, Z.Y., Z.Q. Yan, K.L. Hu, Z.Q. Pang, X.F. Cheng, L.R. Guo, et al. 2011. Odorranalectin-conjugated nanoparticles: Preparation, brain delivery and pharmacodynamic study on Parkinson's disease following intranasal administration. *J Con Release* 151:131–138.

Williams, S., B. Lepene, C. Thatcher, and T. Long. 2009. Synthesis and characterization of poly(ethylene glycol)-glutathione conjugate selfassembled nanoparticles for antioxidant delivery. *Biomacromolecules* 10:155–161.

Wilson, B., M.K. Samanta, K. Santhi, K.P. Kumar, N. Paramakrishnan, and B. Suresh. 2008a. Targeted delivery of tacrine into the brain with polysorbate 80-coated poly(n-butylcyanoacrylate) nanoparticles. *Eur J Pharm Biopharm* 70:75–84.

Wilson, B., M.K. Samanta, K. Santhi, K.P. Kumar, N. Paramakrishnan, and B. Suresh. 2008b. Poly(n-butylcyanoacrylate) nanoparticles coated with polysorbate 80 for the targeted delivery of rivastigmine into the brain to treat Alzheimer's disease. *Brain Res* 1200:159–168.

Wilson, B., M.K. Samanta, K. Santhi, K.P.S. Kumar, M. Ramasamy, and B. Suresh. 2010. Chitosan nanoparticles as a new delivery system for the anti-Alzheimer drug tacrine. *Nanomed Nanotechnol Biol Med* 6:144–152.

Wisse, E., and A.M. De Leeuw. 1984. Structural elements determining transport and exchange process in the liver. In *Microspheres and Drug Therapy: Pharmaceutical, Immunological and Medical Aspects*, eds. S.S. Davis, L. Illum, J.G. McVie, and E. Tomlinson, pp. 1–23. Elsevier: Amsterdam.

Xiao, S.J., P.P. Hu, X.D. Wu, Y.L. Zou, L.Q. Chen, L. Peng, et al. 2010. Sensitive discrimination and detection of prion disease-associated isoform with a dual-aptamer strategy by developing a sandwich structure of magnetic microparticles and quantum dots. *Anal Chem* 82:9736–9742.

Xu, G., K.T. Yong, I. Roy, S.D. Mahajan, H. Ding, S.A. Schwartz, et al. 2008. Bioconjugated quantum rods as targeted probes for efficient transmigration across an in vitro blood-brain barrier. *Bioconjug Chem* 19:1179–1185.

Yang, J., Y.Z.W Adghiri, D.M. Hoang, W. Tsui, Y. Sun, E. Chung, et al. 2011. Detection of amyloid plaques targeted by USPIO-Aβ1-42 in Alzheimer's disease transgenic mice using magnetic resonance microimaging. *Neuroimage* 55:1600–1609.

Yang, Z., Y. Zhang, Y. Yang, L. Sun, D. Han, H. Li, et al. 2010. Pharmacological and toxicological target organelles and safe use of single-walled carbon nanotubes as drug carriers in treating Alzheimer disease. *Nanomed Nanotechnol Biol Med* 6:427–441.

Zensi, A., D. Begley, C. Pontikis, C. Legros, L. Mihoreanu, S. Wagner, et al. 2009. Albumin nanoparticles targeted with Apo E enter the CNS by transcytosis and are delivered to neurons. *J Con Release* 137:78–86.

Zhang, J., T. Atay, and A.V. Nurmikko. 2009. Optical detection of brain cell activity using plasmonic gold nanoparticles. *Nanoletters* 9:519–524.

Zhao, D., H. Song, H. Liying, X. Liu, L. Zhang, and Y. Lv. 2013. Luminescent ZnO quantum dots for sensitive and selective detection of dopamine. *Talanta* 107:133–139.

Zozulya, A.L., E. Reinke, D.C. Baju, J. Karman, M. Sandor, and Z. Fabry. 2007. Dendritic cell transmigration through brain microvessel endothelium is regulated by MIP-1 alpha chemokine and matrix metalloproteinases. *J Immunol* 178:520–529.

Nanometals and Complexes in Cancer Diagnosis and Therapy

Sajid Bashir, Srinath Palakurthi, Hong-Cai Zhou, and Jingbo Liu

Contents

8.1 A new frontier of cancer theranostics using nanomaterials

Cancer nanotheranostics have advanced with the advent of nanotechnology due to their unique physicochemical properties, such as quantum confinement in quantum dots, superparamagnetism in certain oxide nanoparticles, and surface-enhanced Raman scattering in metallic nanoparticles (Wang 2012), resulting in the emergence of sensitive and cost-effective imaging agents (Lee 2009). Similarly, their properties, such as large surface area-to-volume ratio, capability to control size, hydrophobicity, and a surface charge according to the intended application, make them valuable carriers for therapeutic drugs and genes (Liu 2007). Many nanocarriers for anticancer drugs (e.g., immuno-lipsomes of doxorubicin and polymeric micelles with paclitaxel) are being investigated clinically for their targeting capabilities (Matsumura 2004, 2008). Thus, nanoparticles have the required attributes to house therapeutic payloads along with diagnostic imaging agents for real-time monitoring of treatment response (Peer 2007).

In this chapter, three classifications of nanomaterials in cancer nanotheranostics—nanometals (NMs), nanometal oxides (MOs), and nanometal-organic frameworks (MOFs)—are introduced, followed by an overview of synthetic methods and characterization, and a discussion of the efficacy of a nanomaterial when loaded with an anticancer drug. The related results, including the advantages and disadvantages of each type of nanomaterial, are discussed, with a comparison to the current literature of similar works. Lastly, a new approach using MOFs is described, followed by a conclusion and future trends in the area of cancer nanotheranostics science.

8.2 Introduction to cancer nanotheranostics

Since its introduction in 1997, pharmacogenomics has underscored the central dogma of the inextricably complicated field of personalized medicine (Marshall 1997). However, the pharmacogenomics approach poses a risk of compartmentalization, particularly for the reliance on biomarkers, because the strategy relies excessively on singular biotechnology (Naranjo 1997). This led to the emergence of a new concept called theragnostics, which synthesizes information based on various biotechnologies involving genomics, transcriptomics, proteomics, and metabolomics (Ozdemir 2006). The term *theragnostics* (or *theranostics*) was first coined by John Funkhouser, Chief Executive Officer of PharmaNetics, while he was describing his company's business model, which involved developing diagnostic tests directly linked to the application of specific therapies (Wall Street Transcript 2004). Theragnostics encompasses a wide array of topics, including predictive medicine, personalized medicine, integrated medicine, and pharmacodiagnostics. The principal applications of theragnostics for personalized medicine are as follows:

1. *Efficacy:* To profile subgroups of patients based on the likelihood of occurrence of a positive outcome to a given treatment so as to provide them with targeted therapies (Ginsburg 2009)

2. *Safety:* To identify subgroups of patients who are at risk of side effects during a treatment using pharmacogenomics (Huang 2009)
3. *Efficacy and safety:* Monitoring therapeutic response after treatment (Pene 2009)

The process of fabrication of a nanotheranostic system has multiple components, with each component having unique properties and advantages. For example, a signal emitter emits a signal upon excitation by external source. The therapeutic moiety can be a drug (Park, 2007), gene, or nucleic acid, such as small interfering RNA (Hamilton 1999). The nanocarrier is a polymeric carrier capable of carrying a high drug payload (Rhee 2011). The targeting ligand is the entity that can bind to a disease marker with high specificity, so as to deliver the entire system to the target cell (Misra 2010). The response of a signal emitter can be captured in real time by non-invasive imaging techniques, such as magnetic resonance imaging (MRI; Bentzen 2005), computed tomography (CT; Janib 2010), positron emission tomography (Grégoire 2008), single-photon emission computed tomography (Sevick-Muraca 2012), and ultrasound (Koizumi 2008). The therapeutic moiety and the signal emitter can be encapsulated or covalently attached to the polymeric carrier.

Many synthetic and natural polymers were proven to be effective carriers. Polymers that are approved for clinical applications or are currently in clinical trials are polyethylene glycol (PEG; $[-CH_2CH_2O-]_n$), dextran ($[C_6H_{10}O_5]_n$), carboxydextran ($[C_7H_{10}O_7]_n$), β-cyclodextrin ($C_{42}H_{70}O_{35}$), poly(lactic-co-glycolic) acid (PLGA; $[C_3H_4O_2]_x[C_2H_2O_2]_y$), and poly(L-lysine) (PLL; $[D-Lys-(D-Lys)_n-D-Lys]$). The most important component of a theragnostic is its targeting ligand (Wang 2010). The targeting ligand includes small-molecule ligands, such as short peptides and aptamers, and large-molecule ligands, such as antibodies (Xie 2010).

8.2.1 Nanometals

Metal-based nanoparticles (NPs) have been widely used in various applications, including biological diagnostics, cell labeling, targeted drug delivery, cancer therapy, and biological sensors and as antiviral, antibacterial, and antifungal agents (Medina 2007). An understanding of the potential toxicity induced by these NPs to human health and the environment is of prime importance in the clinical development of these NPs for biomedical applications (Wildgoose 2005).

Metal NPs can enter the body via routes such as the gastrointestinal tract, lungs, intravenous injection, and exposure to skin. When NPs come to contact with biological membranes, they pose a threat by affecting physiology of the body (Kohen 2002). For example, silver (Ercal 2001), copper (Gupte 2009), and aluminum (Shulaev 2006) NPs may induce oxidative stress and generate free radicals that could disrupt the endothelial cell membrane. For example, in utero exposure to NPs that are present in diesel exhaust was reported to affect the testicular function of the male fetus by inhibiting testosterone production (Fowler 2002; Li 2009). In the following sections, the toxicity of each metal NP—gold, silicon, copper, titania, and ceria—will be discussed.

Aluminum (Al) NPs are widely used in military applications, such as fuels, propellants, and coatings (Carlson 2008). Thus, exposure of soldiers and other defense personnel to aluminum is on the rise. Wagner et al. (2007) showed that Al NPs exhibit higher toxicity in rat alveolar macrophages and their phagocytic ability is diminished after 24 h of exposure. Al NPs have produced a significant increase in lactate dehydrogenase leakage and were shown to induce apoptosis after exposure to mammalian germline stem cells (Kawata 2009). In vivo toxicity experiments of aluminum oxide NPs in imprinting control region strained mice indicated that nanoalumina impaired neurobehavioral functions. Furthermore, these defects in neurobehavioral functions were shown to be mediated by mitochondrial impairment, oxidative damage, and neural cell loss (Andersen 2004).

Gold (Au) NPs (**Figure 8.1a**) are now widely used in cellular imaging and photodynamic therapy (Huang 2007). Au NPs exhibit size-dependent toxicity, with smaller-sized particles showing more toxicity than larger-sized particles in various cell lines in vitro; a similar pattern of size-dependent toxicity was observed in vivo (Johnston 2010). The effect of shape of the Au NPs on toxicity was also assessed. It was reported that Au nanorods were more toxic than spherical Au NPs (Qiu 2010). The effect of surface chemistry of Au NPs was investigated in monkey kidney cells (CV-1), cells carrying SV40 genetic material (Cos-1 cells), and *Escherichia coli* bacteria (Medina-Ramirez 2012). The results indicated that cationic Au NPs were more toxic compared to their anionic counterparts (Shukla 2005).

Biofunctionalization of Au NPs has been explored in an attempt to reduce their toxicity. Lysine-capped Au NPs were not toxic to macrophages at concentrations up to 100 µM after 72 hours of exposure. Moreover, they did not elicit the secretion of the proinflammatory cytokines tumor necrosis factor (TNF)-α or interleukin (IL)-1β (Johnston 2010). In our research, to facilitate "smart" release, we engineered a biodegradable polymer to decompose upon near-infrared radiation (NIR). The system was composed of doxorubicin (DOX)-loaded PEG-PLGA-gold half-shell nanoparticles by depositing Au films on DOX-loaded PEG-PLGA NPs. As the PEG-PLGA NPs biodegraded, DOX was released, and heat was locally generated upon NIR irradiation due to the NIR resonance of DOX-loaded PLGA half-shell NPs in a mouse model (**Figure 8.1b–d**).

This study used tumor-bearing mice and implanted adherent growth mode human epidermoid carcinoma cell line 431 (A431; 7.5×10^5 cells per animal) into the dorsal subcutis of the Bagg Albino (inbred research mouse strain) generation 100 (Balb/c) nude mouse. A431 tumor cells allow the skin cancer to be exposed by NIR light. When the tumor size became approximately

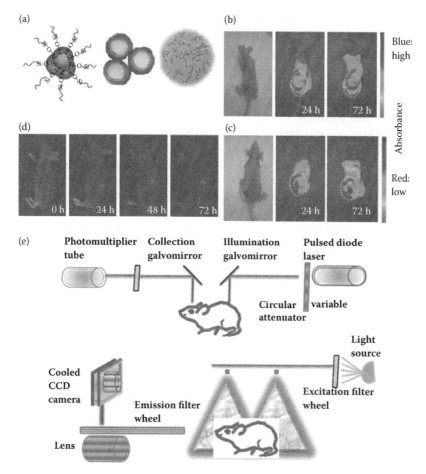

Figure 8.1 *Various nanometals used in the cancer theranostics. (a) Schematic fabrication process of doxorubicin (DOX)-loaded polyethylene glycol (PEG)–poly(lactic-co-glycolic) acid (PLGA)–gold half-shell nanoparticles (NPs). (b) Time-lapse in vivo near-infrared (NIR) images of the mouse after intraveneous injection of 200 µg DOX-loaded PEG-PLGA-gold half-shell NPs. (c) Time-lapse in vivo NIR images of the mouse after intratumoral injection of 200 µg DOX-loaded PEG-PLGA-gold half-shell NPs. (d) Time-lapse in vivo fluorescence images of the mouse treated with intratumorally injected NPs before and after NIR irradiation. (e) Illustration of eXplore Optix TD (GE Healthcare; Port Washington, NY) and planar imaging systems, respectively. (**Figure 8.1b–d** from Lee, S.-M., H. Park, and K.-H. Yoo. 2010. Synergistic cancer therapeutic effects of locally delivered drug and heat using multifunctional nanoparticles. Advanced Materials 22:4049. With permission.)*

100 mm³, 200 µL of DOX-loaded PEG-PLGA-Au half-shell NPs (1 mg/mL stock solution) was administered into the mice via intravenous or intratumoral injection. The injected NPs were monitored by measuring time-lapse in vivo NIR images using the eXPLORE Optics system (GE Healthcare, London, Ontario, Canada) (Lee 2011).

Metals can be grouped into three categories: metals that are essential for metabolic processes in trace amounts, metals that are toxic in higher amounts, and metals that are toxic and not involved in biochemical processes (Shukla 2005). Copper (Cu) has historically been used by bakers as a sterilization agent in chimney fumes to inactivate any yeast; when accumulated, it can be toxic (Heinlaan 2008). Silver (Ag) was used in ancient times either as metal silver or in the form of 1% silver sulfadiazine solution to prevent wound infection or surface sterilization. In recent times, silver has been reevaluated in the form of nanoparticles (Johnston 2010). Metal ions diffuse into the membrane and disrupt cell membrane proteins, promoting metal-assisted catalysis of reactive oxygen species, such as hydroxyl radicals, in addition to nucleic acid binding, which is general to metal ions. In addition, silver is known to irreversibly bind to cysteine, causing protein denaturation, changes in permeability of the cell, and inhibition of respiration through disruption of cytochromes, the proton motive force of the membrane, and oxidative phosphorylation. In the form of silver sulfadiazine, additional wall damage is observed in the form of microvesicle formation and nucleic acid denaturation (Foldbjerg 2009).

8.2.2 Nanometal oxides

Titania is a transition metal oxide that has been extensively used in water purification and disinfection, particularly with gram-negative microbes through photocatalytic oxidation and generation of reactive oxygen species, usually coupled with ultraviolet radiation. This section discusses different strategies to inactivate microbes through mechanical damage of the cell wall, plasma membrane, denaturation of membrane proteins, or cytoplasmic proteins and nucleic acids.

Iron oxide (Fe_3O_4) has low toxicity, is biocompatible and biodegradable, and has a controllable surface area; therefore, it is a good candidate for the above-mentioned applications and has been extensively studied (Lu 2007). These magnetic nanoparticles (MNPs) can be redispersed when the external magnetic field withdraws (Jordan 1999). Therefore, it is critical to fabricate the Fe_3O_4 materials with polymer grafting to functionalize the magnetic properties (Han 2007).

Iron oxide NPs, in particular, have received significant attention because of their proven biocompatibility and biodegradability. Iron from degraded NPs is used in the body's natural iron stores, such as hemoglobin in red blood cells (Sun 2010). These NPs develop superparamagnetism at the nanoscale, which can be used in MRI (Weissleder 1989). The superparamagnetic NPs generate local inhomogeneities in the magnetic field, and their signals will be decreased when an external magnetic field is applied (Arbab 2003). Therefore, regions in the body that have iron oxide NPs appear to be darker on MRI as a result of the negative contrast.

The relaxivity of Fe_3O_4 NPs and their ability to provide contrast on MRI can be improved by tuning the size, shape, and defect of the NP cores (Ferrucci 1990). However, detection of these negative-contrast NPs is difficult in low-signal-intensity tissues, such as lungs and blood clots.

Positive contrast can be achieved with magnetic NPs, which can improve detection in low-signal body regions (Kooi 2003). For example, manganese oxide (Na 2009) and gadolinium oxide (Bridot 2007) can be used. Furthermore, iron oxide NPs with core sizes less than 10 nm can provide positive contrast in MRI (Longmire 2008).

Hyperthermia can be achieved with iron oxide NPs using a rapidly changing magnetic field (Veiseh 2010). High-frequency alternating magnetic fields cause the magnetic moment of the superparamagnetic NPs to quickly shift through Néel fluctuations, which creates very high local temperatures. This mechanism can be used for tumor cell destruction after iron oxide NPs are internalized by the target cells (Goya 2008).

The applications for magnetic metal oxides in cancer theranostics are summarized in **Figure 8.2**. Synthesis of various oxides with different formulations may be achieved using a feasible wet-chemistry method; modification of transition metal oxides (TMOs), which can be conducted to adjust the hydrophobicity by introducing functional groups; polymer grafting or anticancer drug loading, which can be implemented to target the tumor cells of interest; and binding of the cancer cells to the functionalized TMOs through specific binding.

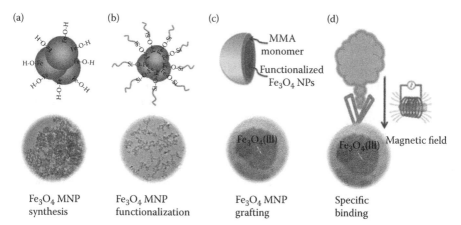

Figure 8.2 *Nanostructured metal oxides used in cancer theranositics (with Fe_3O_4 as an example), (a) Nanothesis via wet-chemistry approach. (b) Functionalization of transition metal oxide (TMO) nanoparticles to adjust the hydrophobicity. (c) Polymer grafting to improve biocompatibility of the TMOs or loading the anticancer drug to target the cell lines. (d) Binding to tumor cells via specific binding under magnetic fields.*

8.2.3 Nanometal-organic frameworks

Metal-organic molecular architectures, such as capsules, boxes, polyhedral cages, and polygonal grids/rings, have attracted significant research interest in recent decades (Ye 2005; Li 2010). This family of materials has drawn significant attention due to their unique structures, from which different applications can be derived, including mechanistic study of self-assembly in nature, materials science, and host-guest related chemistries (Steel 2005). However, there are several challenges in the design and construction of metal-containing supramolecular assemblies with desired structures and functions, which hinder the motif development (summarized in **Figure 8.3**).

After a decade of exploration, chemists and engineers gained more control over synthetic approaches and the underlying assembly rules involving both inorganic and organic components (Yaghi 2003; Li 2009). In some cases, not only the shape and size but also the function of the target assembly can be predesigned, thereby fabricating a particular molecular host for a special application, such as acting as a molecular reactor and a molecular catalytic box (Barth 2007; Kuppler 2009).

Among the structurally varied metal-organic molecular architectures, polygonal rings and polyhedral cages are desirable because of their fixed geometric shape and highly cooperative stability (Ni 2005). A variety of highly symmetrical metal-organic polygons and polyhedra have been synthesized by using highly directional organic bridging ligands to bind geometrically prefixed metal-containing nodes (Chen 2007; Sun 2006). Most of these reported molecular assemblies were constructed with single metal ions as connecting nodes and neutral organic ligands as linkers (Macgillivray 2008; Zhao 2010). The assembly of molecular architectures with multimetal entities or metal clusters acting as nodes has not been widely explored, particularly for molecular polygons and polyhedra (Perry Iv 2009).

In contrast, multimetal node-based MOFs are pervasive, and most of the reported MOFs contain molecular cages or rings as higher-level building units (James 2003; Sun 2006). A concentrated study of molecular metal-organic assemblies should result in a greater control in the construction of MOFs, and certain molecular assemblies have already been utilized directly as precursors for the preparation of MOFs. (Janiak 2003; Zhao 2009). Mining of the inorganic small-molecule library reveals an enormous range of multimetal entities (e.g., multinuclear metal carboxylates) that are potentially adoptable as nodes for the construction of diverse molecular architectures (Qiu 2009). Another additional advantage of using multimetal nodes comes from the inclusion of their inherent properties, such as stability and magnetism, into the supramolecular systems, resulting in additional interesting properties and functions (Kurmoo 2009). These characteristics can widen the spectrum of metal-containing supramolecular chemistry (Eddaoudi 2001).

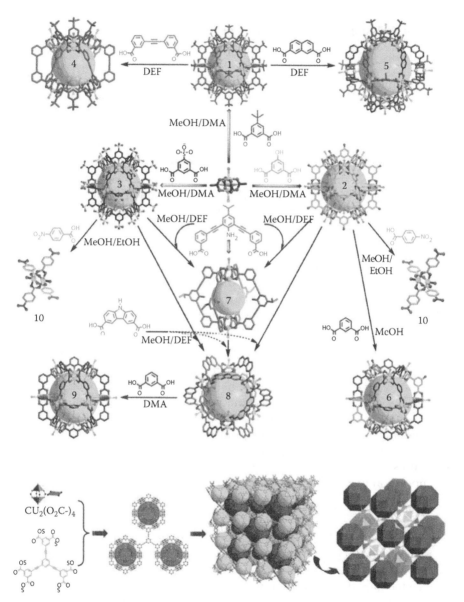

Figure 8.3 (See color insert.) *Nanostructured metal-organic frameworks: bridging ligand substitution reactions. (Reprinted from Li, J.R., and H.C. Zhou. 2010. Bridging-ligand-substitution strategy for the preparation of metal-organic polyhedra.* Nature Chemistry *2:893; Zhao, D., D. Yuan, D. Sun, and H.C. Zhou. 2009. Stabilization of metal-organic frameworks with high surface areas by the incorporation of meso-cavities with microwindows.* Journal of the American Chemical Society *131:9186; Yuan, D., D. Zhao, D. Sun, and H.C. Zhou. 2010. An isoreticular series of metal-organic frameworks with dendritic hexacarboxylate ligands and exceptionally high gas-uptake capacity.* Angewandte Chemie International Edition *49:5357. With permission.)*

8.3 Methodology: wet-chemistry synthesis

8.3.1 Preparation of nanometals

The development of cost-effective environmentally friendly methods for large-scale synthesis of benign, highly efficient nanomaterials represents a critical challenge to their practical applications in biomedical research (Gao 2009). Some nanoporous materials with regular shapes, such as porous nanowires, nanotubes, spheres, and nanoparticles, have been successfully prepared by chemical or physical methods; these methods can be carried out in a variety of ways, such as in the gas phase, in solution, supported on a substrate, or in a matrix (Xia 2008). Although a comprehensive comparison of these approaches does not exist, significant differences in the physicochemical properties (and therefore, the performance) of the resulting materials do exist, allowing for some quantitative assessment (Borm 2006).

In general, physical methods (also known as "top-down" techniques) have high energy requirements; in addition, it is difficult to control the size and composition of the fabricated materials (Zhang 2003). In the top-down method, the bulk is broken down to the nanometer-length scale by lithographic or laser ablation-condensation techniques (Dretler 2005).

Chemical methods (also known as "bottom-up" techniques) are the most popular methods of manufacturing nanomaterials (Mijatovic 2005). These techniques are characterized by narrow nanoparticle size distribution, relative simplicity of control over synthesis, and reliable stabilization of nanoparticles in the systems. In addition, kinetically controlled mixing of elements using low-temperature approaches might yield nanocrystalline phases that are not otherwise accessible (Yan 2003). The methods are based on various reduction procedures involving surfactants or templating molecules, as well as thermal decomposition of metal or metal-organic precursors (Vayssieres 2003).

The sol-gel process (shown in **Figure 8.4**) has proven to be very effective in the preparation of diverse metal oxide nanomaterials, such as films, particles, and monoliths (Su 2004). The sol-gel process consists of the hydrolysis of metal alkoxides and subsequent polycondensation to form the metal oxide gel (Niederberger 2007). One means of achieving shape control is by using a static template to enhance the growth rate of one crystallographic phase over another (Vayssieres 2004). The organic surfactants may be undesired for many applications, and a relatively high temperature is needed to decompose the material (Liong 2008).

Nanometals, such as silver and gold nanoparticles, have been studied and prepared using the bottom-up colloidal method (Medina-Ramirez 2009). In this study, silver nitrate ($AgNO_3$, 0.0071 mol) was dissolved in 30.00 mL of Milli-Q water (EMD Millipore Corporation, Billerica, MA). Gum arabic gum (3%) solution was added to prevent the agglomeration of the particles. Various reducing agents, such as ascorbic acid ($C_6H_8O_6$), sodium citrate ($C_6H_5ONa_3$), sodium borohydride ($NaBH_4$), and dimethylamine borane ($C_2H_{10}NB$), were used for the

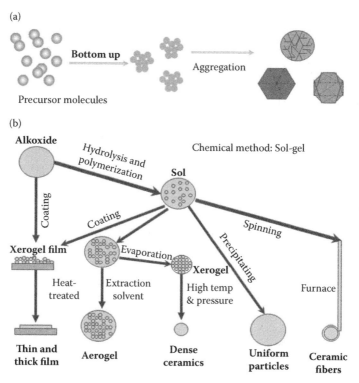

Figure 8.4 *Bottom-up synthesis of nanomaterials. (a) The simplified concept of the bottom-up method in general. (b) The selected colloid synthesis to product series of nanomaterials.*

complete chemical reduction of silver cation (Ag^+). Titanium butoxide was used as the starting material and formed a colloid suspension, which was modified using an Ag–NP suspension. Then, the mixture was stirred for 30 minutes at 60°C to derive the stable colloidal suspension, which was used as disinfectant. In this method, the particle agglomeration was successfully prevented using a green chemical, with gum arabic as the surfactant. The synthetic process consists of sol synthesis starting from the appropriate metal alkoxides, followed by the reduction of the metal ion, which is extracted using wet chemistry. Finally, the nanocolloid is characterized using electron microscopy and x-ray spectroscopy.

8.3.2   Preparation of nanostructured metal oxides

A feasible colloidal wet-chemistry method (a bottom-up technique; see **Figure 8.5**) is used to produce metal oxide nanoparticles. From our research, two oxides (TiO_2 and Fe_3O_4) with different geometries were found to actively interact with cell surfaces of National Institutes of Health fibroblast cell line in a 3-day transfer of 3×10^5 cells per 20-cm² dish (NIH/3T3) in a controllable uptake rate. This study demonstrated a reverse microemulsion method, which was employed to prepare TiO_2-base NP and Fe_3O_4

MNPs, which were selected because of their cost-effectiveness, flexibility, and homogeneity control at a molecular level (Zeng 2004). This reverse microemulsion also provided the advantages of a composition-control, cost-effective, and green-synthesis process compared with other reported approaches (Capek 2004).

This preparation mainly focuses on two major steps: producing ultrafine MNPs with tunable size and architecture and grafting the MNPs with monomers to form a core-shelled structure. Via this approach, the monodispersive MNPs with uniform shapes were synthesized under ambient conditions. The MNPs were then functionalized using silane coupling agent 3-glycidoxypropyltrimethoxysilane (KH-570; Ogden et al. 2008). After being modified, the hydrophilic MNPs were then converted to be hydrophobic in nature, which allows for grafting of methyl methacrylate monomers due to the introduction of chemically active sites (Babu 2008). Importantly, the initial molar ratio of the metal

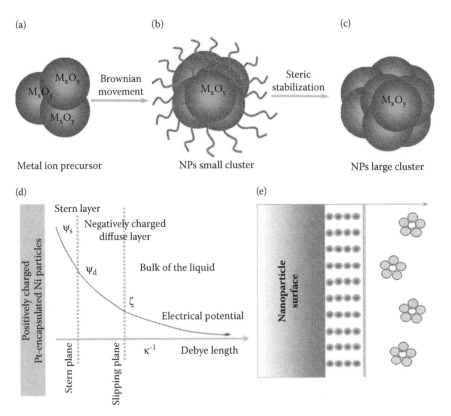

Figure 8.5 *Bottom-up synthesis of nanometal oxides. (a) TiO_2 colloid synthesis using $Ti(O^nBu)_4$ metal alkoxide as starting materials. (b) Ag colloid synthesis through redox reaction. (c) Ag-TiO_2 nanoparticle synthesis via extraction of solvents. (d) Nanostructural characterization of the nanoparticles, such as crystalline structure, particle size, and size distribution study.*

precursors such as Fe^{3+} can be maintained in the final product, which ensures the controllable composition of the core-shelled MNPs (Bourgeat-Lami 2002).

In summary, hydrophilic Fe_3O_4 MNPs were prepared by reverse microemulsion to achieve a highly dispersive and uniform structure. To adjust the hydrophobicity of the MNP surface, a coupling agent was employed, leading to functionalization of the NP surfaces. Lastly, the so-functionalized Fe_3O_4 was grafted via polymerization to increase the biocompatibility of NPs and biological cell lines.

8.3.3 Preparation of metal-organic complexes

Synthetic method development has been playing a major role in the advancement of metal-organic frameworks. These diversified synthetic methods include microwave, electrochemical, mechanochemical, ultrasonic, and high-throughput syntheses (Zhou 2012). In one study, the metal-organic complexes were generated through a hydrosolvothermal approach to produce well-defined and highly crystallized organometallic compounds (Zhuang 2012). This allows for great flexibility in terms of physical, optical, or electrical properties, as well as various parameters in the design, synthesis, and fabrication of the metal-organic frameworks.

The transition metals, such as iron, cobalt, nickel, and silver, serve central element and ligand as organic linker. The nucleation of MOFs crystal units occurred at elevated temperatures. In this approach, the fabrication variables were optimized and simplified. The size and stability of MOFs crystals were further regulated (**Figure 8.6**). Fabrication is performed using a hydrosolvothermal approach (due to the short reaction time and lower reaction temperature) via a nucleation reaction between the active site of the ligand and metal through a coordinative covalent bond. In this way, crystal geometry, porosity, and atomic size can be controlled (Liu 2012).

8.4 Instrumentation and magnetism characterization

Briefly, nanomaterials and treated cancer cells have been examined using fluorescent microscopy (Horisberger 1981) and high-resolution transmission electron microscopy (TEM; Smith 1989), equipped with electron diffraction (Spence 1989) to determine wall/membrane integrity. To obtain morphological images and elemental distribution, electron energy loss spectroscopy (Spence 1989) was also employed to determine the distribution of the elemental composition, as described in the next section. Crystalline phase structure was evaluated using APEX-II diffractometer (Bruker AXS, Madison, WI) and Ultima III x-ray powder diffraction (XRD; Rigaku, The Woodlands, TX) with a copper diffractometer. Additionally, the stability of the nanocolloid suspension and particle size distribution were estimated using ZetaPALS (Brookhaven Instruments, Holtsville, NY). Finally, the magnetism of nanoparticles was

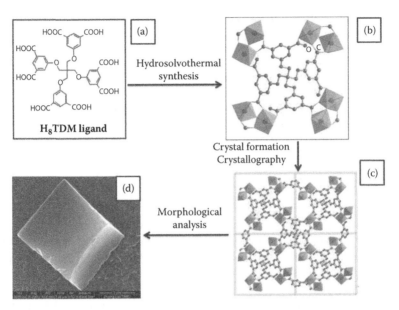

Figure 8.6 *Schematic of metal organic-framework (MOF) synthesis using a flexible chemistry approach. (a) An octa-topic ligand. (b) The connection mode of MOFs. (c) The dimeric carboxylate cluster. (d) Scanning electron microscopic image of MOFs. (Reprinted from Zhuang, W., D. Yuan, J.R. Li, et al. 2012. Highly potent bactericidal activity of porous metal-organic frameworks. Advanced Healthcare Materials 1:225. With permission.)*

measured using the Lakeshore 7304 VSM instrument (Westerville, OH) with an applied magnetic field up to 12 kOe at ambient temperature.

8.4.1 Morphology and elemental composition analyses

A transmission electron microscope (FEI Tecnai G2-F20, Houston, TX) equipped with Oxford electron diffraction (FEI Company 2002) was used as previously described (Chamakura 2011) to obtain nanostructure information and crystalline phase of the materials, including metal nanoparticles, metal oxides, and metal-organic frameworks (**Figure 8.7**). The instrument was operated in the scanning TEM modes to obtain high-resolution chemical information. Due to their higher magnification, larger depth of focus, greater resolution, and ease of sample observation, both modes (STEM and TEM) have been employed widely to determine the crystalline phase, defects, and texture of materials (Traversa 2001).

High-resolution field-emission TEM was employed to achieve high spatial resolution, high contrast, and unsurpassed versatility. Particularly, the Tecnai F20 G2 TEM used in this study, which includes a Schottky field emission source, provides ultrahigh brightness, low energy spread, and very small probe sizes. The Tecnai G2 F20 used in this study was designed and preconfigured

Figure 8.7 *Transmission electron microscopy (TEM) morphological analyses. (a) TEM image and ring pattern of Ag nanoparticles. (b) TEM image and ring pattern of Fe₃O₄ nanoparticles. (c) TEM image and ring pattern of Fe metal-organic frameworks.*

specifically to meet the strict requirements of nanomaterials. This TEM is also equipped with a robust high-brightness field emission gun, allowing for a wide range of applications, from morphological analysis to defect characterization (Fujishima 1972; Traversa 2001). The morphological images indicated that the semispherical particles with tunable size and shape can be obtained from bottom-up chemical synthesis. The ring pattern from the polycrystals suggested that the employed method also allowed for high crystallinity. High crystallinity is important in allowing generation of uniform layering of fixed dispersity.

8.4.2 Crystallography analyses

Data were collected on a Bruker SMART APEX-II diffractometer with a low-temperature device and a fine-focus sealed-tube x-ray source (Mo-K_α radiation, $\lambda = 0.71073$ Å). Suitable single crystals were directly picked up from the mother liquor, attached to a glass loop, and quickly transferred to a cold stream of liquid nitrogen (110 K) for data collection. Raw data collection and reduction were manipulated using APEX2 software from Bruker. The semi-empirical method SADABS (version 2.05; SADABS 2000) was applied for absorption corrections. The structures were solved by direct methods and refined by the full-matrix least-squares technique against F^2 with anisotropic temperature parameters for all nonhydrogen atoms using the SHELXTL software package (Sheldrick 2002).

All hydrogen atoms were geometrically placed and refined in riding model approximation. There are large pore volumes in the crystals of some compounds, which are occupied by heavily disordered solvent molecules. In some cases, the positions of solvent molecules can be found and a satisfactory disorder model could be achieved. However, it cannot be done for all solvent molecules; therefore, the SQUEEZE program implemented in PLATON was used to model this electron density (Farrugia 1999). Ultima III XRD was used to identify the crystalline phase of the engineering nanomaterial due to its capability to provide rapid, nondestructive analysis of multicomponent mixtures without the need for extensive sample preparation (Wang 2008). The electromagnetic radiation of wavelength provided by XRD is about 1 Å (10^{-10} m), which is essentially similar to the size of an atom (Rothen-Rutishauser 2009). A direct combination of nanoparticle fabrication and exposure to lung cell cultures in a closed setup was used as a method to simulate accidental nanoparticle exposure of humans.

The x-ray interacted with the nanoparticles, allowing quick and accurate analysis of phase, crystallinity, lattice parameters, expansion tensors, bulk modules, and periodical arranged clusters. Based on the peak broadening, the crystallite size can also be calculated using the Scherrer equation (Zhang 2010). XRD characterization is widely used in various fields, such as metallurgy, mineralogy, forensic science, archaeology, condensed matter physics, and the biological and pharmaceutical sciences. The single-crystal structure (**Figure 8.8**) showed three-dimensional framework, the dumbbell-shaped open channels of 10.2×5.6 Å in the c direction, and iron centers in its polyhedral geometry. The XRD patterns of selected samples (**Figure 8.9**) indicated that the face-centered cubic Ag nanoparticles (**Figure 8.9a**; powder diffraction file: 01-089-3722, space group: Fm3m, lattice constants: $a = 4.0855$ Å, a = 90°) with crystallinity (>1) were derived from the colloidal method. The inverse spinel Fe_3O_4 was obtained and its crystalline phase was well indexed with the standard samples (**Figure 8.9b**, powder diffraction file: 65-3107, space group: Fd3m [Gratias 1979], lattice constants: $a = 8.3905$ Å, a = 90°). The XRD of single-crystal Cu-benzene-1,3,5-tricarboxylic acid indicated that cubic phase was formed (**Figure 8.9c**; space group: Fm3m, lattice constant: $a = 26.3336$ Å, a = 90°).

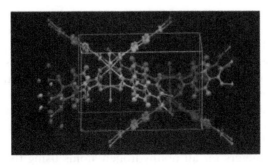

Figure 8.8 *Single crystal structure of Fe metal-organic frameworks.*

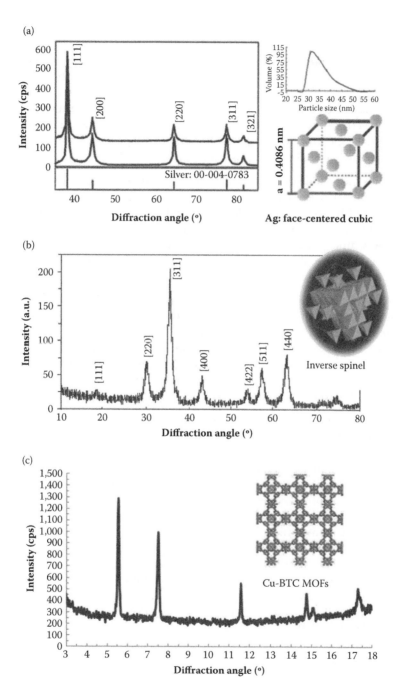

Figure 8.9 *X-ray diffraction (XRD) of selected samples. (a) Silver. (b) Fe₃O₄. (c) metal-organic framework.*

8.4.3 Electrokinetics analyses

A NanoBrook ZetaPALS analyzer (Brookhaven Instruments Corporation, Holtsville, NY) was employed to measure the particle size distribution and the zeta-potential to evaluate the stability and electrokinetic value of polyatomic nanoparticles (PANs) ZetaPALS uses phase-analysis light scattering to measure the electrophoretic mobility of charged colloid. ZetaPALS (31311) then converts these electrophoretic mobility data to zetapotential values based on the Smoluchowski mathematical model (McConville 2010).

To evaluate the stability and electrokinetic behavior of the PAN colloid, the zeta-potential (also known as the effective surface potential, denoted as ζ) was measured by imposing an electrical field (20 mV) across the colloid particles (**Figure 8.10**). The measured ζ allowed an estimation of the degree of aggregation: fine particles have high surface tension during the fabrication procedure and large absolute zeta-potential, from which the degree of aggregation can be estimated.

The particle size distribution and zetapotential analysis of an Au colloidal suspension are summarized in **Figure 8.10**. The particle size ranges from approximately 10 to 14.5 nm (**Figure 8.10a**). This observation is different from XRD and scanning/transmission electron microscopy studies, which indicated that the agglomeration occurs after dehydrolysis of gum arabic (the dispersing agent) at 85°C. After removal of the gum arabic, the Au cluster agglomerates due to its high surface tension.

The zetapotential essentially ranged from −40 to −45 mV (**Figure 8.10b**), depending on the particle size. The significance of the measured zetapotential indicated the high stability of the colloid. The particles with large absolute zetapotential (arising from negatively charged particles) were able to minimize their agglomeration due to electrostatic repulsion. The particle interaction and double-layer formation are shown in **Figure 8.10c**. From the schematic, it can be seen that the zetapotential corresponds to the electric potential at the conceptual diffusion layer, which is negatively charged. If this electric potential exhibits high absolute (magnitude) value, the repulsion between particles exceeds the van der Waals long-range attractive forces and fine particles will be formed. When the zetapotential approaches zero, particles tend to aggregate when gum arabic is absent due to dehydrolysis. Under these conditions, the van der Waals attraction exceeds the electrostatic repulsion (**Figure 8.10d**).

8.4.4 Magnetism analyses

Magnetism is an electronically driven physical phenomenon, although it is weak compared with electrostatic interactions. The concept originated from the Pauli exclusion principle and the existence of electron spin (Krishnan 2006). This magnetism depends highly on the structural or morphological arrangements of phases, grains, or individual atoms.

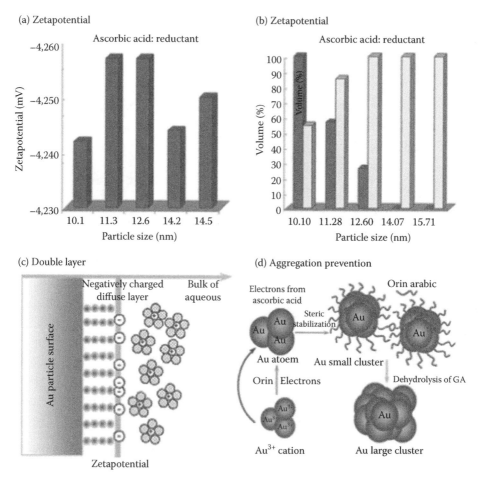

Figure 8.10 *Electrokinetic analysis of selected samples (Au nanoparticles). (a) Zetapotential analysis of Au nanoparticles (NPs), indicating that NPs are highly electrokinetically stable. (b) Particle size averaged 10.3 nm (this size can be varied according to the dispersing agents and preparation conditions). (c) The schematic demonstration of the double layer of negatively charged Au NPs. (d) The electrostatic prevention of nanoparticles agglomeration. (Reprinted from Medina-Ramírez, I., M. Gonzalez-Garcia, and J. L. Liu. 2009. Nanostructure characterization of polymer-stabilized gold nanoparticles and nanofilms derived from green synthesis.* Journal of Materials Science *44:6325. With permission.)*

The magnetic behavior of materials is complex and poorly understood, and many fundamental questions remain unanswered. To the best of our knowledge, the size of domain wall-widths (nanostructures), lateral confinement (shape and size), and interparticle interactions are critical factors for determining the magnetism of prepared materials. Magnetization curves (**Figure 8.11**) dominated by rotation and magnetocrystalline anisotropy were evaluated systematically by Li et al. (2012).

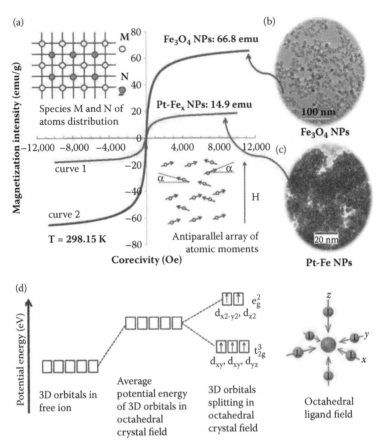

Figure 8.11. *Magnetism analyses. (a) The magnetization intensity of Fe-Pt nanoalloys (curve 1) under magnetic field and Fe$_3$O$_4$ nano-oxides (curve 2). Note that in the top insert, which shows the random distribution of different atoms from different species, the surroundings of the atoms may be vastly different and the approximation of a single molecular field representing the action of the surroundings for all the sites must be very poor. The bottom insert indicates the motion of an antiparallel array of atomic moments acted upon by a magnetic field H. (b) Transmission electron microscopy (TEM) image of nanospherical particles of Fe$_3$O$_4$. (c) The TEM image of nanospherical particles of Fe-Pt nanoalloys. (d) Demonstration of d electrons and d-orbital splitting under octahedral fields. (**Figure 8.11d** adapted from Bashir, S., J. Liu, H. Zhang, X. Sun, and J. Guo. 2013. Band gap evaluations of metal-inserted titania nanomaterials, Journal of Nanoparticle Research 15:1572. With permission.)*

Based on Weiss's hypothesis, the domain wall motion is a more important reversal mechanism in real engineering materials because defects, such as grain boundaries and impurities, serve as nucleation sites for reversed-magnetization domains. The role of domain walls in determining coercivity is complex because defects may pin domain walls in addition to nucleating them. The dynamics of domain walls in ferromagnetism is similar to that of grain boundaries and plasticity in metallurgy because both domain walls and grain boundaries are planar defects (Néel, 1971).

Intrinsic coercivity is highly dependent on the chemically ordered crystalline structure of nanomaterials. The hysteresis loops of iron platinum (**Figure 8.11c**) allow the formation of Fe-Pt$_x$ nanoalloys (**Figure 8.11a**, curve 1), indicating that FePt$_x$ NPs exhibit soft ferromagnetic behavior due to near zero coercivity; however, the coercivity of annealed Fe-Pt$_x$ nanoalloys was 275 Oe (Li 2012). It is logical to identify the mechanism of the coercivity of the FePt$_x$ NPs. The four unpaired electrons in the Fe 3d ([Ar] 4s23d6) orbital lay the foundation for its magnetism. Upon formation of the NPs, the magnetization will take place at the surfaces of hard magnetic grains or at the grain boundaries and intergranular phases, which are also regions of reduced magneto-crystalline anisotropy (**Figure 8.11d**). In thermally demagnetized materials, the large initial susceptibility is due to a domain wall motion in grains that are largely free of defects. The coercivity can be related to nucleation of reversed domains in regions where the anisotropy constant has been reduced from the bulk value of the SmCo$_5$ phase.

The magnetic property study indicated that the intrinsic coercivity of the Fe$_3$O$_4$ MNPs (**Figure 8.11c**) was subject to no major changes upon modification. The coercivity was determined by measurement of the hysteresis loop (magnetization curve). These hysteresis loops of the Fe$_3$O$_4$ MNPs (**Figure 8.11a**, curve 2) indicated that both Fe$_3$O$_4$ and functionalized MNPs exhibited soft ferromagnetic behavior due to its near-zero coercivity, H_c. This magnetic hysteresis is mainly attributed to the inverse spinel Fe$_3$O$_4$ phase. From the magnetization-coercivity loop, it can also be concluded that Fe$_3$O$_4$ MNPs displayed high magnetization intensity (66.8 A·m^2·kg^{-1}).

8.5 Potency evaluation of nanomaterials

Cancer is a complex and multifactorial disease that accounts for more than 10 million new cases every year. With advancements in cellular, molecular, and genetic techniques, novel biomarkers and new therapeutic targets for cancer diagnosis, therapy, and therapy management have been developed. However, effective application of the bioactive molecules needs suitable technology for delivery in vivo.

In the early 21st century, rapid advances in nanotechnology led to the development of new medical devices and delivery techniques to bypass the biological barriers for targeted delivery of bioactive molecules for monitoring the status of biomarkers in the prevention and prognosis of cancer. Nanotechnology platforms with high target specificity and minimum collateral damage and immune reactions are currently under development. Nanomaterials—especially metals (Au, Ag), metal oxide (Fe$_3$O$_4$) metal-organic frameworks, and metal (single or bimetallic) supramolecular polymers—are versatile vectors for theranostics applications, including simultaneous diagnostic imaging and controlled drug/gene delivery, hyperthermia, radiotherapy enhancement, and resistance reversal.

We hypothesize that the targeted delivery of 7-ethyl-10-hydroxy-camptothecin [SN-38] as an inclusion complex will generate a significantly high cytotoxicity to breast cancer cells. In addition, incorporating an imaging module should facilitate the monitoring of the fate of the cyclodextrin complexes in the pharmacodynamics activity of the chemotherapeutic. To fully understand the therapeutic efficacy and mechanism of breast cancer damage, we examined nanomaterial synthesis, fine structural characterization of the nanomaterial, measurement of the nanomaterial's properties (including electrokinetic and magnetic properties), and the efficiency of the theranostics analysis. In this section, our discussion centers on the efficient study of various nanomaterials prepared by our group and collaborators' groups.

8.5.1 Efficiency evaluation

8.5.1.1 Nanoparticle–cell interactions and cellular uptake

Despite a rapid surge in the use of nanoparticles in diagnostic and therapeutic applications, interactions of NPs with the cell membrane are not clear yet. A small change in the physicochemical properties can result in drastic changes in the mechanism and magnitude of NP uptake and their intracellular disposition. Several endocytic routes exist; among them, receptor-mediated endocytosis is a commonly pursued mechanism for targeted delivery or imaging applications.

In general, NPs are shown to be taken up by macropinocytosis and phagocytosis. Both of these endocytic processes require membrane ruffling with actin polymerization, leading to enclosure of a particle. It has been shown that the geometry of the particle modulates the phagocytic potential, and the flexibility of the particle surface dictates the extent of particle internalization (Champion and Mitragotri 2009). In addition to endocytosis, some positively charged supramolecular polymers, such as dendrimers, induce the formation of transient cell holes.

Because a plethora of diverse nanostructures are used for a wide variety of biomedical applications, no single mechanism of internalization is considered to be the most prominent. However, manipulation of the physicochemical properties of nanomaterials—such as size, geometry, radius of curvature, charge, and surface functionalization—were shown to influence the uptake pathway and intracellular fate of NPs (Herd 2002; Braeckmans 2010; Jimenez de Aberasturi 2012). For instance, silica nanoparticles of different geometries were shown to interact with cell surface guanosine triphosphatases differently and smaller particles have shown a reduced rate of uptake.

Experimental data in corroboration with mathematical modeling have confirmed that membrane invaginations are due to elastic deformation, which is dependent on the critical nanoparticle radius to provide minimum possible free energy; when the particle radius is above this critical value, the cell membrane is unable to invaginate the particle (Lee 2011). Although spherical particles of <200 nm are taken up by clathrin-mediated endocytosis, rod-like particles are rapidly taken up by clathrin-mediated endocytosis and macropinocytosis. Rod-like particles were oriented in the perinuclear region unlike spherical particles, indicating that the intracellular fate of the nanoparticles is dependent on their geometry (Yoo 2010). Arnida et al. (2010, 2011) have reported that gold nanorods were taken up to a lesser extent by the liver; they also had longer half-lives and higher tumor localization compared with their spherical counterparts. PEGylated gold nanorods were also taken up to a lesser extent by macrophages, indicating their low immunogenic potential.

8.5.1.2 Biomedical applications and cytotoxicity

High surface-to-volume ratio, broad optical properties, and easy surface functionalization have made metallic NPs uniquely suitable for cancer imaging and photothermal therapy (Sperling 2008; Sau 2010). Quantum dot clearance was reported to occur rapidly via kidneys if the hydrodynamic radius is below 5.5 nm, whereas larger nanoparticles showed long circulation times (Choi 2007).

Gold nanocrystals are one of the most widely studied nanomaterials. They recently entered into clinical trials for TNF-α delivery (Cormode 2013). Gold nanorods and nanoshells absorb light very strongly and heat is emitted when they absorb light; hence, they have been proposed for photothermal ablation therapy. Gold nanoparticles are approved by the U.S. Food and Drug Administration for use in diagnostic devices for various diseases and therapies, including radiosensitizers for radiotherapy, contrast agents in fluorescence imaging, surface-enhanced Raman spectroscopy, computed tomography, and related spectroscopies such as spectral CT (Qian 2008). Iron oxide nanoparticles have applications in MRI and as contrast agents for magnetic particle imaging agents and hyperthermia. Magnetic nanoparticles have been used for targeted drug delivery by employing external magnetic fields for site-specific localization and alternating magnetic field-induced hyperthermia for local cell death.

A number of tetrahedral gold(I) compounds were reported to exhibit cytotoxicity to a variety of tumor cell lines; however, they have also shown toxicity to normal organs such as the heart, liver, and lungs in animal models by virtue of their mitochondrotoxicity. Gold(III) compounds with imine donors were found to be more potent than cisplatin in cisplatin-resistant cell lines. Some gold(III) complexes of N-confused porphyrins (Chmielewski 2005) and corroles (Gross 1999) were suggested to display phototoxicity (Rabinovich 2011).

The sustained release of gold complexes from drug-carrying capsules could be a useful approach to decreasing their toxicity towards normal cells.

Zhang et al. (2012) prepared a self-assembled supramolecular polymer complexing Au(III) with 2,6-diphenylpyridine and 2,4-diamino-6-(4-pyridyl)-1,3,5-triazine), resulting in concentration-dependent formation of nanofibrillar networks. The gold complex displayed sustained cytotoxicity and selective cytotoxicity toward cancerous cells and was capable of encapsulating other cytotoxic agents.

Although many gold(III) organic complexes have shown good anticancer activity, poor stability in solution is still a problem. Gold(III) porphyrins were found to be stable in dimethyl sulfoxide and physiological buffers, and no loss of metal ions was found in phosphate-buffered saline (even in the presence of glutathione). The chlorido(tetraphenyl porphyrin)aurate(III) ion complex was highly cytotoxic to a variety of cancer cell lines, including cisplatin-resistant and multidrug-resistant cancer cell lines. Interestingly, the gold complex was more than 10-fold more toxic to nasopharyngeal cancer cells as compared to normal peripheral blood mononuclear cells, with >80% reduction in tumor volume in vivo (Sun 2007). In addition, the research group also prepared cyclometalated gold and platinum complexes that have shown potent anticancer activity.

Cisplatin interacts with DNA to trigger apoptosis, but it is not cancer-cell specific. Several transition-metal complexes are under investigation to improve the cytotoxicity and selectivity of cisplatin toward cancer cells. Similarly, terpyridine-bis(glyco-arylacetamide)platinate(II) complexes were found to be 100 times more potent than cisplatin. The half maximal inhibitory concentration (IC_{50}) values against keratin-based epidermal carcinoma (HeLa KB-3-1) and its multidrug-resistant strain (KB-V-1) were 1.7 and 1.0 μM, which are approximately 10 and 40 times more potent than cisplatin, respectively.

Chen et al. (2012) reported synthesis of a gold-liriodenine complex (liriodenine is an oxoaporphine alkaloid with anticancer activity that is isolated from *Zanthoxylum nitidum*). The resultant complexes exhibited an IC_{50} of <15 μM; they also were found to induce S-phase cell cycle arrest, inhibit topoisomerase I, and bind to DNA by intercalation.

Among the metal complexes, organogold complexes have emerged as the most widely investigated agents since the serendipitous discovery of cisplatin as an anticancer agent. Gold N-heterocyclic carbene (NHC) complexes (Lin 2005) have been investigated as alternatives to gold tertiary phosphane compounds (Werner 2004). Au(NHC)L, where L is a two-electron donor, has shown promising apoptosis activity (Weaver 2011). These complexes were shown to induce apoptosis by increasing mitochondrial membrane permeabilization, which is directly dependent on the lipophilicity of the complexes. Furthermore, lipophilic cationic gold(I) complexes of NHC were found to specifically target cancer cells. Au-NHC complexes were reported to induce cancer cell-specific cell death through mitochondrial apoptotic pathway and inhibition of thioredoxin reducatse activity (Weaver 2011).

Arsenijevic et al. (2012) reported the synthesis of three new gold(III) complexes: dichlorido(ethylendiamine)aurate(III) ion, dichlorido(2,2'-bipyridyl)-aurate(III) ion [Au(bipy)Cl$_2$], and dichloride(1,2-diaminocyclohexane)aurate(III) ion, with guanosine 5'-monophosphate (a fragment of DNA, and the target for gold complexes is DNA). Among these three complexes, Au(bipy)Cl$_2$ displayed better apoptotic activity in A549 cells (lung epithelial carcinoma) than the others, and it was comparable to cisplatin.

Iron complexes are important intermediates in many enzymatic processes in vivo. Several iron complexes have been used in various biomedical processes. Among all the reported complexes, iron-bleomycin is the most notable example; it causes oxidative DNA damage in the presence of oxygen and hydrogen peroxide. Sun et al. (2007) developed iron complexes containing a pentadentate pyridyl ligand; these complexes are stable under physiological conditions, induce S-phase cell cycle arrest, and generate reactive oxygen species.

Superparamagnetic nanoparticles, by virtue of their reversible magnetization and hyperthermic potential, are considered to be powerful contrast agents for MRI, hyperthermic cancer treatment, spatially controllable drug release, and targeting applications. Several attempts are underway to optimize their magnetic and hyperthermic characteristics without compromising their biocompatibility and pharmacokinetics/toxicity in vivo.

Core-shell iron oxide nanostructures have been shown to exhibit high hyperthermic potency. The shape and size of the iron oxide nanoparticles were shown to have a significant impact on hyperthermic power (measured by the specific absorption rate) and MRI relaxivities (Lee 2011). Moreover, their architecture, such as the ability to orient as linear assemblies and isotropic clusters, may profoundly influence their magnetic behavior (Lartigue 2012). The same research group synthesized highly crystalline multicore iron oxide nanoparticles by a polyol process, demonstrating that these multicore NPs showed better magnetic properties and T1 and T2 relaxivities for MRI contrast generation and hyperthermia (Lartigue 2012). Moreover, coating the iron oxide nanostructures with citric acid rendered them nontoxic, highly stable in biologic media, and able to be rapidly internalized by cancer cells for MRI detection.

Liu et al. (2011) prepared silver-heterocyclic carbine complexes with 4,5-diarylimidazolium halides and tested cytotoxicity against breast and colon carcinoma cells. The silver oxide complexes were more active against the MD Anderson metastatic breast cell line 231 than the Michigan Cancer Foundation cancer cell line 7, with activity that was comparable to cisplatin. The complexes were only marginally interactive with the DNA, estrogen receptors, and the cyclooxygenase enzymes, indicating that these are not the targets for their mechanism of action. The complexes have also shown significant bacterial growth inhibitory activity.

The properties of metal-containing supramolecular polymers were shown to change with alteration of the metal atoms and use of other ligands. Organic complexes of gold, platinum, and copper are known to self-assemble into polymeric nanostructures that have potential applications in biomedical sciences and medicine. A classic example of metal-based organic nanostructures is the complexation of gadolinium (Gd) with tetraaza macrocyclic moiety attached to peptides for MRI applications (Bull 2005a, 2005b). Despite the great potential of organic gold complexes in clinical applications, toxicity remains unresolved. Chloro(triethylphosphine)aurate(I) is an organogold compound that has potential therapeutic activity in rheumatoid arthritis in animal models but has liver mitochondrotoxicity (Agusti 2008).

Stimuli-responsive polymers have gained interest for potential biomedical applications because their reversible size, shape, and physicochemical properties enable them to release entrapped materials in response to an external stimulus. Kumpfer and Rowan (2011) synthesized metallosupramolecular polymers using metal ligand-coordination bonds. The materials had thermo-, photo-, and chemoresponsive shape-memory properties. Any stimulus that causes decomplexation and exchange of metal ligand complexes may be used to revert back to the original material shape and properties. Although several potential biomedical applications, especially pulsatile delivery of pharmaceuticals (e.g., hormones) may be envisaged using this technology, no studies on their application and safety are reported.

8.5.1.3 Toxicity and safety of nanomaterials

In general, it is well known that the biofate of NPs depends on their size, shape, and surface chemistry. Hydrophilic NPs or supramolecular polymers (50–100 nm) do not significantly activate the reticuloendothelial system, but they are large enough to avoid renal filtration. Metal NPs, such as silver, gold, and platinum (Conde 2012), have been shown to induce DNA damage and oxidative stress (Limbach 2007). Gold NPs are generally considered to be benign, but their toxicity is size dependent. AuNPs that are 1–2 nm in size cause cell type-dependent cytotoxicity, with IC_{50} in the high micromolar range; however, the IC_{50} of 15-nm particles was 50-fold higher than the smaller particles (Pan 2007).

Research shows diverse observations in terms of immunogenicity of AuNPs. Yen et al. (2009) showed that small AuNPs inhibited macrophage growth and upregulated proinflammatory genes IL-1, IL-6, and TNF-α. The effect of the shape of AuNPs on in vivo toxicity demonstrated that rod-shaped NPs were the most toxic, whereas Au spheres had the best biocompatibility. Traces of cetyltrimethyl ammonium bromide used in the synthesis of Au rods was suggested to be the reason for their toxicity, although this remains to be confirmed.

Similarly, there are conflicting reports on toxicity of gold versus silver NPs. Silver NPs are generally considered to be more toxic than gold NPs. However, Yen et al. (2009) demonstrated the low toxicity of AgNPs as compared to

AuNPs, which was attributed to the difference in their surface charges (Braydich-Stolle 2005). Platinum NPs were shown to enter the cells by diffusion and cause DNA damage and apoptosis. In an interesting comparative toxicity study, silver NPs were more toxic than platinum NPs, whereas AuNPs were nontoxic (Asharani 2011). It should be noted that these reports focused on the toxicity of native metallic NPs. Surface functionalization using natural polymers, targeting ligands, or other biocompatible materials, such as amino acids, may not only change their toxicity profile but also their biodisposition. In this context, developing metal-organic frameworks and metal-supramolecular polymers have great potential for biomedical applications in near future. **Figure 8.12** shows that both nanoframeworks and nanoparticles act 10-fold quicker at higher doses; however even at one-tenth of the concentration, an IC_{50} of approximately 0.02 ppm is observed.

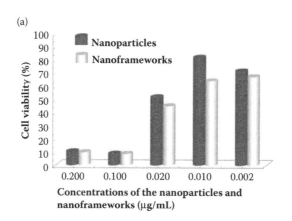

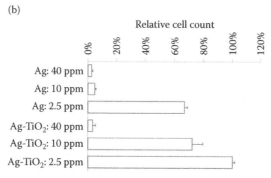

Figure 8.12 *Efficacy of nanomaterials. (a) Toxicity analysis of magnetic Fe_3O_4 nanoparticle and iron organo-frameworks in human blood cells. (b) Toxicity analysis of Ag and Ag-TiO_2 nanoparticles on NIH 3T3 cell lines. (Courtesy of K. Chamakura and R. Perez-Ballestero.)*

8.5.2 Mechanistic study

Nanoparticles may be engineered to target specific stages of cell development, such as those that may also be targeted during interphase. Contact inhibition at either the interphase growth cycle (G_0), first gap (G_1), synthesis (S), and second gap (G_2), are necessary processes in preparing the cell for mitosis (M). The developed resistance to certain drugs such as adriamycin (ARDR) by cells may be circumvented by coadministering other drugs such as verpamil; however, such circumvention may not be a universal phenomenon (Gibby, 1987), necessitating other approaches as discussed below.

The simplified schematic in **Figure 8.13** demonstrates eukaryotic cell interactions with an injected nanoparticle. This injection may be nonspecific or targeted using tumor-specific ligands on the nanoparticles, which bind to cell-surface receptors and trigger the internalization process into the cell via the endosome. Within the cell, a trigger may initiate the action of nanoparticle or release of drug, such as lowering of the pH, where the drug affects the cytoplasm. Intracellular trafficking will distribute the drug to the organelles such as DNA, nucleus, endoplasmic reticulum, mitochondria, or cytoskeleton.

Common biochemical modes (Tatum 1950) of intervention are the enhancement of recombinase A [recA, protein Rad51 in eukaryotes]-dependent recombational repair (Howard-Flanders 1966), scavenging and neutralization of reactive oxygen species (Doroshow 1980), inhibition of nonhomologous DNA end-joining, and lessened repressor protein LexA inhibition towards "save our souls" repair (Radman 1975). The targeting or general receptors, such as glycoprotein P or receptors for folic acid, may be one

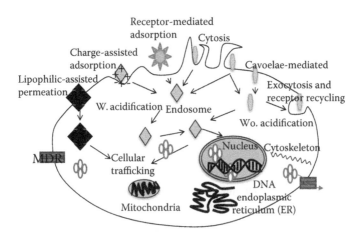

Figure 8.13 *Schematic demonstration of eukaryotic cells interacting with various of nanoparticles, which are injected through adsorption, permeation, and receptor-mediated endocytosis.*

broad method, specific receptors through the use of specific antibodies, or via receptor-mediated endocytosis. One common approach is for nanoparticle encapsulation using organic frameworks, such as linear coblock polymers (Duncan 2003), amphiphilic block copolymers forming micelles (Oerlemans 2010), hyperbranched synthetic dendrimers (Duncan 2005), self-assembling lipid-based liposomes (Kostarelos 2007), multivalent cage-like structures (Douglas 2006), and carbon-based nanotubes (Liu 2008) or C60 (Chaudhuri 2009).

In addition to interacting with tumor cell metalloproteinases, glucose turnover (Pelicano, 2006) may be a mechanism for therapy. Nanoparticles may be used for precise delivery, control, and apoptosis by a number of mechanisms, such as cell-surface antigens/receptor-based targeting using folate (Sudimack, 2000), which is often expressed in tumor cells to a greater degree than nontumor cells (Leamon, 2001), or through the use of heparin-folate-drug triconjugate for drug-resistant cells (Cho, 2008). Polymeric oligonucletoides, which have affinity for specific antigens, may also be utilized in drug delivery (Farokhzad, 2004) and has been shown to be effective in animal models (Farokhzad, 2006) in addition to encapsulating drug-loaded nanoparticles (Nagase, 1999) through direct polymer-drug conjugates (Lambert, 2005). Transporter proteins, such as transferrin, may also be conjugated with polymers (Dautry-Varsat 1983) or liposome-loaded drugs (Goldstein 1985), which have been demonstrated to be effective against certain cells, such as MCF-7 (Sahoo 2005). Lastly, carbohydrate-binding lectins may also be conjugated using a similar approach to transferrin for lectin targeting at the cell surface (Lehr 2000) or nanoparticle-carbohydrate tagging, which bind to the lectin in the reverse sense to lectin targeting (Minko 2004); this approach has been applied in metastatic liver cancer models (Duncan 2006).

Our research group has been working on cisplatin-polymer complexes to increase cisplatin uptake in cisplatin-resistant cells. Cisplatin accumulates in the cell by passive diffusion or facilitated transport. Many of the cisplatin-resistant cells show a decreased platinum accumulation, most likely due to decreased uptake rather than enhanced efflux (Johnson 2001). Cisplatin forms stable complexes with both $-NH_2$ and $-COOH$ terminal poly(amido) amine (PAMAM) dendrimers (Yellepeddi 2011). The therapeutic efficacy of dendrimer-cisplatin complexes was evaluated in a Sloan-Kettering Ovarian HER2 3+ Cancer (SKOV-3) xenografted murine ovarian cancer model (**Figure 8.14**). Treatment with intraperitoneal $PAMAMG_4$ NH_2-cisplatin complexes resulted in 41.93% and 25.8% increases in lifespan when compared with treatment with cisplatin alone. Furthermore, a strong correlation was seen between the survival data and the cellular uptake of platinum ($R_2 = 0.912$), as well as survival data and DNA adduct formation in SKOV-3 cells ($R_2 = 0.904$). Results suggest that G_4 NH_2 dendrimers exhibit superior efficacy through enhanced cellular uptake and improved DNA adduct formation (Yellepeddi 2011).

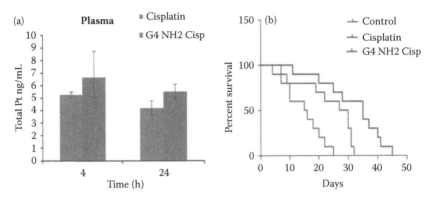

Figure 8.14 *Total Pt concentration-time profile in plasma (a) and survival studies in SKOV xenografted athymic nude mice (b). (From Yellepeddi, V.K. 2012. In vivo efficacy of PAMAM-dendrimer-cisplatin complexes in SKOV-3 xenografted Balb/c nude mice.* J Biotech Biomaterials *S13:003. With permission.)*

8.6 Comparison with our published data

To put the results from **Figure 8.12** into perspective, one of the lowest effective doses reported using diheteroaryl-substituted pyrazole against NIH 3T3 cells was 0.089 μM, which corresponds to [(0.089 μM/304)/1000], or 0.034 ppm (Sawyer 2003). A study by Yu et al. (2004) on the dose-curve of microtubule-disrupting agents in various human breast cancer cell lines used mitomycin C as an anticancer agent. The authors found the lowest dose to be 8.6 mg/mL against the MCF-7 cell line, which corresponds to an equivalent concentration of 8,600 ppm (mg/mL × 1,000 = ppm), noting that our effectiveness was lower (1 ppm = 1 mg/L = 1 μg/mL = 1,000 μg/L). In moles per liter, 1 ppm of mitomycin C would correspond to an equivalent concentration of 3 μM (1/MW/1,000= 1 ppm, molarity × FW × 1,000 = ppm) and 120 nM for doxorubicin, corresponding to a concentration of [120 × 10^{-9} × 580 × 1,000] = 0.07 ppm, also against MCF-7 cell line (Yu, 2004). It can be seen that the unoptimized nanosystems are competitive in terms of concentration in the sub-ppm threshold, whereas slightly large silver and ultrafine silver nanoparticles yield an IC_{50} of approximately 3.5 ppm (~32 μM), or 18.5 μM of $AgTiO_2$.

8.7 Future trends

A review of the literature indicates that drug development and therapy can be bracketed into a number of distinct phases (Vickers 2002). The early phase (the 1800s) was devoted to isolation, purification, characterization, and cataloging of bioactive pharmacopoeia. The second phase may be considered from the early 1900s, when it was recognized that the immune system (now known as P-glycoprotein and related systems) eliminates active drugs by filtering or receptor-mediated endocytosis, giving rise to extensive research in the science of encapsulation (Panyam 2002). The third phase, from the mid-1950s until the present, has focused on developing nucleotide-based therapies (Patil 2005); the use of metals, such as platinum (Bruijnincx 2008); ultrasound (Pitt 2004) and radiation (Tepper 1976); natural products (Ulrich 2006); and intercalating ligands (Denny 1989), which promote apoptosis (Hickman 1992).

Since the 1980s, cells have been modified using DNA viruses as vectors, direct gene transfer (Niidome, 2002), or stem cell-based therapies (Aggarwal, 2005) to treat cancer cells. In the last decade, these "traditional" techniques have been supplemented by material science-based treatments, which have been extensively applied to cancer therapy. Here, nanoparticles have been fabricated with specific shapes and dimensions and coated with water-soluble polymers (Liu 2008), micelles (Sutton 2007), dendrimers (Majoros 2006), and carbon nanotubes (Bhirde 2009). To this, nanocrystals have recently been applied to exploit the optical, electrical (Lal 2008), or magnetic properties (Kircher 2003) of these systems, such as induction of magnetic-induced hyperthermia (Jordan 1999) or photodynamic therapy (Cheng 2008), as well as to tag the surface with epitopes or coat/load with drugs.

A new addition to nanoparticles and nanorods are metal-organic frameworks, which allow engineering controls such as encapsulation, surface modification, and drug loading inside the MOF due to its ultrahigh surface area, extreme stability, and real-time tracking and imaging, thus supplementing earlier strategies (Matsumura 1986) with minimal side effective (**Figure 8.13**). The reviewed trends and our own experiences suggest that cancer nanotheranostics discoveries in the laboratory environment are being rapidly transferred to translational medicine in cancer diagnosis, therapy, and monitoring. Here, the developed systems ideally exhibit minimal body toxicity to noncancer cells, long-term activity with low clearance from the target cells/organs, and a broad spectrum of activity, particularly against drug-tolerant cancer cells. The future holds much promise. Hopefully the new generation of nanomaterials, including MOFs, will deliver on their potential as nanotheranostic agents of change.

Acknowledgments

The authors thank the College of Arts and Sciences at Texas A&M University–Kingsville for funding (Dr. Bashir: 160336-00002; Dr. Liu: 160310-00004) and student support. The Microscopy and Imaging Center at Texas A&M University and the Department of Chemistry at Texas A&M University–Kingsville are also acknowledged for their technical support and nanostructure characterization. The Welch Foundation (AC-006) is duly acknowledged for facilitating the research and disseminating the results.

Authors' contributions

S. Bashir conducted the experimental design and analyzed data (XRD and zetapotential). S. Palakurthi conceived the nanocancer research idea and supervised the biological experimental work. H.-C. Zhou originated the use and applications for gas storage and allowed the use of his laboratory for MOF fabrication for therapy by J. Liu. J. Liu also conceived the nanotechnological research idea and prepared nanomaterials (her students, the Welch scholars, are currently working on collecting more data based on the previous results), and proposed the scope, aims, and thrust of the current book chapter as invited by the editor. All authors collectively wrote the first draft and revised manuscript. J. Liu supervised the submission of this book chapter.

References

Aggarwal, S., et al. 2005. Human mesenchymal stem cells modulate allogeneic immune cell responses. *Blood* 105:1815–1822.

Agusti, G.M. 2008. Spin-crossover behavior in cyanide-bridged iron(II)-gold(I) bimetallic 2D Hofmann-like metal-organic frameworks. *Inorg Chem* 47:2552–2561.

Andersen, J.K. 2004. Oxidative stress in neurodegeneration: Cause or consequence? *Nat Rev Neurosci* 5:S18–S25.

Arbab, A.S. 2003. Characterization of biophysical and metabolic properties of cells labeled with superparamagnetic iron oxide nanoparticles and transfection agent for cellular MR imaging 1. *Radiology* 22:838–846.

Arnida, J.-A.M. 2011. Geometry and surface characteristics of gold nanoparticles influence their biodistribution and uptake by macrophages. *Eur J Pharm Biopharm* 77:417–423.

Arnida, M.A. 2010. Cellular uptake and toxicity of gold nanoparticles in prostate cancer cells: A comparative study of rods and spheres. *J Appl Toxicol* 30:212–217.

Arsenijevic, M.M. 2012. Cytotoxicity of gold (III) complexes on A549 human lung carcinoma epithelial cell line. *Medicinal Chemistry* 8:2–8.

Asharani, P.V. 2011. Comparison of the toxicity of silver, gold and platinum nanoparticles in developing zebrafish embryos. *Nanotoxicology* 5:43–54.

Babu, K. 2008. Grafting of poly (methyl methacrylate) brushes from magnetite nanoparticles using a phosphonic acid based initiator by ambient temperature atom transfer radical polymerization (ATATRP). *Nanoscale Res Lett* 3:109–117.

Barth, J.V. 2007. Molecular architectonic on metal surfaces. *Annu Rev Phys Chem* 58:375–407.

Bentzen, S.M. 2005. Theragnostic imaging for radiation oncology: Dose-painting by numbers. *Lancet Oncol* 6:112–117.

Bhirde, A.A. 2009. Targeted killing of cancer cells in vivo and in vitro with EGF-directed carbon nanotube-based drug delivery. *ACS Nano* 3:307–316.

Borm, P.J. 2006. The potential risks of nanomaterials: A review carried out for ECETOC. *Particle Fibre Toxicol* 31:11-1-11.35.

Bourgeat-Lami, E. 2002. Organic-inorganic nanostructured colloids. *J Nanosci Nanotechnol* 2:1–24.

Braeckmans, K.B. 2010. Advanced fluorescence microscopy methods illuminate the transfection pathway of nucleic acid nanoparticles. *J Control Release* 148:69–74.

Braydich-Stolle, L.S. 2005. In vitro cytotoxicity of nanoparticles in mammalian germline stem cells. *Toxicol Sci* 88:412–419.

Bridot, J.L. 2007. Hybrid gadolinium oxide nanoparticles: multimodal contrast agents for in vivo imaging. *J Am Chem Soc* 129:5076–5084.

Bruijnincx, P.C. 2008. New trends for metal complexes with anticancer activity. *Curr Opin Chem Biol* 12:197–220.

Bull, S.R. 2005. Magnetic resonance imaging of self-assembled biomaterial scaffolds. *Bioconjug Chem* 16:1343–1348.

Bull, S.R. 2005. Self-assembled peptide amphiphile nanofibers conjugated to MRI contrast agents. *Nano Lett* 5:1–4.

Capek, I. 2004. Preparation of metal nanoparticles in water-in-oil (w/o) microemulsions. *Adv Colloid Interface Sci* 110:49–74.

Carlson, C.H.-S. 2008. Unique cellular interaction of silver nanoparticles: size-dependent generation of reactive oxygen species. *J Phys Chem B* 112:13608–13619.

Chamakura, K.P.-B. 2011. Comparison of bactericidal activities of silver nanoparticles with common chemical disinfectants. *Colloids Surf B Biointerface* 84:88–96.

Chaudhuri, P.P. 2009. Fullerenol—Cytotoxic conjugates for cancer chemotherapy. *ACS Nano* 3:2505–2514.

Chen, X.M. 2007. Solvothermal in situ metal/ligand reactions: A new bridge between coordination chemistry and organic synthetic chemistry. *Acc Chem Res* 40:162–170.

Chen, Z.F. 2012. Synthesis, characterization, and in vitro antitumor properties of gold(III) compounds with the traditional Chinese medicine (TCM) active ingredient liriodenine. *J Biol Inorg Chem* 17:247–261.

Cheng, Y.C. 2008. Highly efficient drug delivery with gold nanoparticle vectors for in vivo photodynamic therapy of cancer. *J Am Chem Soc* 130:10643–10647.

Chmielewski, P.J.-G. 2005. Core modified porphyrins—A macrocyclic platform for organometallic chemistry. *Coordination Chem Rev* 249:2510–2533.

Cho, K.W. 2008. Therapeutic nanoparticles for drug delivery in cancer. *Clin Cancer Res* 14:1310–1316.

Choi, H.S. 2007. Renal clearance of quantum dots. *Nat Biotechnol* 25:1165–1170.

Conde, J.G. 2012. Noble metal nanoparticles applications in cancer. *J Drug Deliv* 751075: 1–12.

Cormode, D.P.-G. 2013. Inorganic nanocrystals as contrast agents in MRI: Synthesis, coating and introduction of multifunctionality. *NMR Biomed* 26:766–780.

Dautry-Varsat, A.C. 1983. pH and the recycling of transferrin during receptor-mediated endocytosis. *Proc Natl Acad Sci USA* 80:2258–2262.

Denny, W.A. 1989. DNA-intercalating ligands as anti-cancer drugs: prospects for future design. *Anticancer Drug Des* 4:241–263.

Doroshow, J.H. 1980. Enzymatic defenses of the mouse heart against reactive oxygen metabolites: Alterations produced by doxorubicin. *J Clin Invest* 65:128–135.

Douglas, T. 2006. Viruses: Making friends with old foes. *Science* 312:873–875.

Drake, J.W. 1976. The biochemistry of mutagenesis. *Ann Rev Biochem* 45:11–37.

Dretler, S.P. 2005. Laser lithotripsy: A review of 20 years of research and clinical applications. *Lasers Surg Med* 8:341–356.

Duncan, R. 2005. Dendrimer biocompatibility and toxicity. *Adv Drug Deliv Rev* 57: 2215–2237.

Duncan, R. 2003. The dawning era of polymer therapeutic. *Nat Rev Drug Discov* 2:347–360.

Duncan, R. 2006. Polymer conjugates as anticancer nanomedicines. *Nat Rev Cancer* 6:688–701.

Eddaoudi, M.M. 2001. Modular chemistry: secondary building units as a basis for the design of highly porous and robust metal-organic carboxylate frameworks. *Acc Chem Res* 34:319–330.

Ercal, N.G.-O.-B. 2001. Toxic metals and oxidative stress part I: Mechanisms involved in metal-induced oxidative damage. *Curr Topics Med Chem* 1:529–539.

Evan, G.I. 2001. Proliferation, cell cycle and apoptosis in cancer. *Nature* 411:342–348.

Farokhzad, O.C. 2004. Nanoparticle-aptamer bioconjugates: A new approach for targeting prostate cancer cells. *Cancer Res* 64:7668–7672.

Farokhzad, O.C. 2006. Targeted nanoparticle-aptamer bioconjugates for cancer chemotherapy in vivo. *Proc Natl Acad Sci USA* 103:6315–6320.

Farrugia, L.J. 1999. WinGX program features. *J Appl Cryst* 32:837–838.

FEI Company. 2002. *Tecnai G2 F20 Cryo. Advanced TEM for Beam Sensitive Samples.* Hillsboro, OR: FEI Company.

Ferrucci, J.T. 1990. Iron oxide-enhanced MR imaging of the liver and spleen: Review of the first 5 years. *Am J Roentgenol* 155:943–950.

Foldbjerg, R.O. 2009. PVP-coated silver nanoparticles and silver ions induce reactive oxygen species, apoptosis and necrosis in THP-1 monocytes. *Toxicol Lett* 190:156–162.

Fowler, P.A. 2002. Environmental chemical effects on testicular function. *Reprod Med Rev* 10:77–100.

Fujishima, A. 1972. Photolysis-decomposition of water at the surface of an irradiated semiconductor. *Nature* 238:37–38.

Gao, J.G. 2009. Multifunctional magnetic nanoparticles: Design, synthesis, and biomedical applications. *Acc Chem Research* 42:1097–1107.

Gibby, E.M. 1987. Selective interactions of verapamil with anthraquinones in adriamycin-sensitive and -resistant murine and human tumour cell lines in vitro. *Cancer Chemother Pharmacol* 20:5–7.

Ginsburg, G.S. 2009. Genomic and personalized medicine: Foundations and applications. *Transl Res* 154:277–287.

Goldstein, J.L. 1985. Receptor-mediated endocytosis: Concepts emerging from the LDL receptor system. *Ann Rev Cell Biol* 1:1–39.

Goya, G.F. 2008. Magnetic nanoparticles for cancer therapy. *Curr Nanosci* 4:1–16.

Gratias, D.P. 1979. Crystallographic description of coincidence-site lattice interfaces in homogeneous crystal. *Acta Crystallogr A* 35:885–894.

Grégoire, V.H. 2008. PET-based treatment planning in radiotherapy: A new standard? *J Nucl Med* 48:68S–77S.

Gross, Z.G. 1999. The first direct synthesis of corroles from pyrrole. *Angew Chem Int Ed Engl* 38:1427–1429.

Gupte, A. 2009. Elevated copper and oxidative stress in cancer cells as a target for cancer treatment. *Cancer Treat Rev* 35:32–46.

Hamilton, A.J. 1999. A species of small antisense RNA in posttranscriptional gene silencing in plants. *Science* 286:950–952.

Han, G.G. 2007. Functionalized gold nanoparticles for drug delivery. *Nanomedicine* 2:113–123.

Heinlaan, M.I. 2008. Toxicity of nanosized and bulk ZnO, CuO and TiO 2 to bacteria Vibrio fischeri and crustaceans Daphnia magna and Thamnocephalus platyurus. *Chemosphere* 71:1308–1316.

Herd, H. 2002. Screening for diabetic retinopathy: Developing a standardized photographic protocol for St. John's Hospital. *J Audiov Media Med* 25:28–33.

Hickman, J.A. 1992. Apoptosis induced by anticancer drugs. *Cancer Metastasis Rev* 11:121–139.

Horisberger, M. 1981. Colloidal gold: A cytochemical marker for light and fluorescent microscopy and for transmission and scanning electron microscopy. *Scan Electron Microsc* 2:9–31.

Howard-Flanders, P.B. 1966. Three loci in Escherichia coli K-12 that control the excision of pyrimidine dimers and certain other mutagen products from DNA. *Genetics* 53:1119–1136.

Huang, S.M. 2009. Application of pharmacogenetics and pharmacogenomics in drug development and regulatory review. In *Genomics: Fundamentals and Applications*, ed. S. Choudhuri, 357–378. New York, NY: Informa Healthcare.

Huang, X. J.-S.-S. 2007. Gold nanoparticles: Interesting optical properties and recent applications in cancer diagnostics and therapy. *Nanomedicine* 2:681–693.

James, S.L. 2003. Metal-organic frameworks. *Chem Soc Rev* 32:276–288.

Janiak, C. 2003. Functional organic analogues of zeolites based on metal–organic coordination frameworks. *Angew Chem Int Ed Engl* 36:1431–1434.

Janib, S.M. 2010. Imaging and drug delivery using theranostic nanoparticles. *Adv Drug Deliv Rev* 62:1052–1063.

Jimenez de Aberasturi, D.M.-P.-M. 2012. Optical sensing of small ions with colloidal nanoparticles. *Chem Materials* 24:738–745.

Johnson S.W. 2001. Cisplatin and its analogues. In *Cancer: Principles and Practice of Oncology*, 6th ed., ed. H. Vincent T. DeVita, 376–388. Philadelphia, PA: Lippincott Williams and Wilkins.

Johnston, H.J. 2010. A review of the in vivo and in vitro toxicity of silver and gold particulates: Particle attributes and biological mechanisms responsible for the observed toxicity. *Crit Rev Toxicol* 40:328–346.

Jordan, A.S. 1999. Magnetic fluid hyperthermia (MFH): Cancer treatment with AC magnetic field induced excitation of biocompatible superparamagnetic nanoparticles. *J Magnetism Magnetic Materials* 201:413–419.

Kawata, K.O. 2009. In vitro toxicity of silver nanoparticles at noncytotoxic doses to HepG2 human hepatoma cells. *Environ Sci Technol* 43:6046–6051.

Kircher, M.F. 2003. A multimodal nanoparticle for preoperative magnetic resonance imaging and intraoperative optical brain tumor delineation. *Cancer Res* 63:8122–8125.

Kohen, R. Invited review: Oxidation of biological systems: oxidative stress phenomena, antioxidants, redox reactions, and methods for their quantification. *Toxicol Pathol* 30:620–650.

Koizumi, N.L. 2008. A framework of the non-invasive ultrasound theragnostic system. In *Lecture Notes in Computer Science*, ed. T.S. Dohi, 231–240). Tokyo: Springer.

Kooi, M.E. 2003. Accumulation of ultrasmall superparamagnetic particles of iron oxide in human atherosclerotic plaques can be detected by in vivo magnetic resonance imaging. *Circulation* 107:2453–2458.

Kostarelos, K. 2007. Liposome–nanoparticle hybrids for multimodal diagnostic and therapeutic applications. *Nanomedicine* 2:85–98.

Krishnan, K.M. 2006. Nanomagnetism and spin electronics: Materials, microstructure and novel properties. *J Mater Sci* 41:793–815.

Kumpfer, J.R. 2011. Thermo-, photo-, and chemo-responsive shape-memory properties from photo-cross-linked metallo-supramolecular polymers. *J Am Chem Soc* 133: 12866–12874.

Kuppler, R.J. 2009. Potential applications of metal-organic frameworks. *Coordin Chem Rev* 253:3042–3066.

Kurmoo, M. 2009. Magnetic metal–organic frameworks. *Chem Soc Rev* 38:1353–1379.

Lal, S.C. 2008. Nanoshell-enabled photothermal cancer therapy: Impending clinical impact. *Acc Chem Res* 41:1842–1851.

Lambert, J.M. 2005. Drug-conjugated monoclonal antibodies for the treatment of cancer. *Curr Opin Pharmacol* 5:543–549.

Lartigue, L.P. 2012. Cooperative organization in iron oxide multi-core nanoparticles potentiates their efficiency as heating mediators and MRI contrast agents. *ACS Nano* 6:10935–10949.

Leamon, C.P. 2001. Folate-mediated targeting: From diagnostics to drug and gene delivery. *Drug Discovery Today* 6:44–51.

Lee, N.K. 2011. Magnetosome-like ferrimagnetic iron oxide nanocubes for highly sensitive MRI of single cells and transplanted pancreatic islets. *Proc Natl Acad Sci* 108:2662–2667.

Lee, Y.E. 2009. Nanoparticle PEBBLE sensors in live cells and in vivo. *Ann Rev Anal Chem* 2:57–76.

Lehr, C.M. 2000. Lectin-mediated drug delivery: The second generation of bioadhesives. *J Control Release* 65:19–29.

Li, C.L., Z. Li, X. Du, C. Du, and J. Liu. 2012. Structural characterization and magnetic property of iron-platinum nanoparticles fabricated by pulse electrodeposition. *Rare Metals* 31:31–34.

Li, C.T. 2009. Effects of in utero exposure to nanoparticle-rich diesel exhaust on testicular function in immature male rats. *Toxicol Lett* 185:1–8.

Li, J.R. 2009. Interconversion between molecular polyhedra and metal: Organic frameworks. *J Am Chem Soc* 131:6368–6369.

Li, J.R. 2010. Bridging-ligand-substitution strategy for the preparation of metal–organic polyhedra. *Nat Chem* 2:893–898.

Limbach, L.K. 2007. Exposure of engineered nanoparticles to human lung epithelial cells: Influence of chemical composition and catalytic activity on oxidative stress. *Environ Sci Technol* 41:4158–4163.

Lin, I.J. 2005. Review of gold (I) N-heterocyclic carbenes. *Can J Chem* 83:812–825.

Liong, M.L. 2008. Multifunctional inorganic nanoparticles for imaging, targeting, and drug delivery. *ACS Nano* 2:889–896.

Liu, J.C.-B. 2012. Historical overview of the first two waves of bactericidal agents and development of the third wave of potent disinfectants. In *Nanomaterials in Biomedicine (ACS Symposium Series)* ed. R. Nagarajan, 129–154. Washington, DC, American Chemical Society.

Liu, W.K. 2011. Synthesis and biological studies of silver N-heterocyclic carbene complexes derived from 4,5-diarylimidazole. *Eur J Med Chem* 46:5927–5934.

Liu, Y. M. 2007. Nanomedicine for drug delivery and imaging: A promising avenue for cancer therapy and diagnosis using targeted functional nanoparticles. *Intl J Cancer* 120:2527–2537.

Liu, Z.C. 2008. Drug delivery with carbon nanotubes for in vivo cancer treatment. *Cancer Res* 68:6652–6660.

Liu, Z.R. 2008. PEGylated nanographene oxide for delivery of water-insoluble cancer drugs. *J Am Chem Soc* 130:10876–10877.

Longmire, M.C. 2008. Clearance properties of nano-sized particles and molecules as imaging agents: Considerations and caveats. *Nanomedicine* 3:703–717.

Lu, C.W. 2007. Bifunctional magnetic silica nanoparticles for highly efficient human stem cell labeling. *Nano Letters* 7:149–154.

Macgillivray, L.R. 2008. Supramolecular control of reactivity in the solid state: From templates to ladderanes to metal-organic frameworks. *Acc Chem Res* 41:280–291.

Maeda, H. 2001. The enhanced permeability and retention (EPR) effect in tumor vasculature: The key role of tumor-selective macromolecular drug targeting. *Adv Enzyme Regul* 41:189–207.

Majoros, I.J. 2006. PAMAM dendrimer-based multifunctional conjugate for cancer therapy: Synthesis, characterization, and functionality. *Biomacromolecules* 7:572–579.

Marshall, A. 1997. Genset–Abbott deal heralds pharmacogenomics era. *Nat Biotechnol* 15:829–830.

Matsumura, S.W. 2008. Radiation-induced CXCL16 release by breast cancer cells attracts effector T cells. *J Immunol* 181:3099–3107.

Matsumura, Y. 1986. A new concept for macromolecular therapeutics in cancer chemotherapy: Mechanism of tumoritropic accumulation of proteins and the antitumor agent smancs. *Cancer Res* 46:6387–6392.

Matsumura, Y.M. 2004. Resistance of CD1d–/– mice to ultraviolet-induced skin cancer is associated with increased apoptosis. *Am J Pathol* 165:879–887.

McConville, J.H. 2010. New advances in the zeta potential measurement. Boston, MA: American Chemical Society.

McKeown, C.M. 1988. Sinusoidal lining cell damage: The critical injury in cold preservation of liver allografts in the rat. *Transplantation* 46:178–190.

Mead, A. 1972. Clinical and pharmacological studies with 5-hydroxy-2-formylpyridine thiosemicarbazone. *Cancer Res* 32:1455–1462.

Medina, C.S.-M. 2009. Nanoparticles: Pharmacological and toxicological significance. *Brit J Pharmacol* 150:552–558.

Medina-Ramirez, I.B. 2009. Green synthesis and characterization of polymer-stabilized silver nanoparticles. *Colloids Surf B* 73:185–191.

Medina-Ramirez, I.G.-G. 2012. Application of nanometals fabricated using green synthesis in cancer diagnosis and therapy. In *Green Chemistry—Environmentally Benign Approaches*, ed. M. Kidwai, 33–61. Rijeka, Croatia: InTech.

Mijatovic, D.E. 2005. Technologies for nanofluidic systems: Top-down vs. bottom-up—A review. *Lab on a Chip* 5:492–500.

Minko, T. 2004. Drug targeting to the colon with lectins and neoglycoconjugates. *Adv Drug Deliv Rev* 56:491–509.

Misra, R.A. 2010. Cancer nanotechnology: Application of nanotechnology in cancer therapy. *Drug Discov Today* 15:842–850.

Mori, H.S. 2002. Hybrid nanoparticles with hyperbranched polymer shells via self-condensing atom transfer radical polymerization from silica surfaces. *Langmuir* 18:3682–3693.

Na, H.B. 2009. Inorganic nanoparticles for MRI contrast agents. *Adv Mat* 21:2133–2148.

Nagase, H. 1999. Matrix metalloproteinases. *J Biol Chem* 274:21491–21494.

Naranjo, C.A. 1997. Using fuzzy logic to predict response to citalopram in alcohol dependence. *Clin Pharmacol Therapeutics* 62:209–224.

Néel, L. 1971. Magnetism and the local molecular field. *Science* 174:985–992.

Ni, Z.Y. 2005. Porous metal-organic truncated octahedron constructed from paddle-wheel squares and terthiophene links. *J Am Chem Soc* 127:12752–12753.

Niederberger, M. 2007. Nonaqueous sol-gel routes to metal oxide nanoparticles. *Acc Chem Res* 40:793–800.

Niidome, T. 2002. Gene therapy progress and prospects: Nonviral vectors. *Gene Ther* 9:1647–1652.

Oerlemans, C.B. 2010. Polymeric micelles in anticancer therapy: Targeting, imaging and triggered release. *Pharm Res* 27:2569–2589.

Ogden, S.G. 2008. *Silane functionalisation of iron oxide nanoparticles.* Melbourne, Australia: SPIE.

Ozdemir, V.W.-J. 2006. Shifting emphasis from pharmacogenomics to theragnostics. *Nature* 200:942–946.

Pan, Y.S.-D. 2007. Size-dependent cytotoxicity of gold nanoparticles. *Small* 3:1941–1949.

Panyam, J.Z. 2002. Rapid endo-lysosomal escape of poly (DL-lactide-co-glycolide) nanoparticles: Implications for drug and gene delivery. *FASEB* 16:1217–1226.

Park, K. 2007. Nanotechnology: What it can do for drug delivery. *J Control Rel* 120:1–3.

Patil, S.D. 2005. DNA-based therapeutics and DNA delivery systems: A comprehensive review. *AAPS J* 7:E61–E77.

Peer, D.K. 2007. Nanocarriers as an emerging platform for cancer therapy. *Nature Nanotechnol* 2:751–760.

Pelicano, H.M. 2006. Glycolysis inhibition for anticancer treatment. *Oncogene* 25:4633–4646.

Pene, F.C. 2009. Toward theragnostics. *Crit Care Med* 37:S50–S58.

Perry Iv, J.J. 2009. Design and synthesis of metal–organic frameworks using metal–organic polyhedra as supermolecular building blocks. *Chem Soc Rev* 38:1400–1417.

Pitt, W.G. 2004. Ultrasonic drug delivery—A general review. *Expert Opin Drug Deliv* 1:37–56.

Qian, X. X.-G. 2008. In vivo tumor targeting and spectroscopic detection with surface-enhanced Raman nanoparticle tags. *Nat Biotechnol* 26:83–90.

Qiu, S. 2009. Molecular engineering for synthesizing novel structures of metal–organic frameworks with multifunctional properties. *Coordin Chem Rev* 253:2891–2911.

Qiu, Y.L. 2010. Surface chemistry and aspect ratio mediated cellular uptake of Au nanorods. *Biomaterials* 31:7606–7619.

Rabinovich, E.I. 2011. Gold(I) and gold(III) corroles. *Chemistry* 17:12294–12301.

Radman, M. 1975. Phenomenology of an inducible mutagenic DNA repair pathway in Escherichia coli: SOS repair pichulein hypothesis. *Basic Life Sci* 5:355–367.

Rhee, Y.S. 2011. Nanopharmaceuticals I: Nanocarrier systems in drug delivery. *Intl J Nanotechnol* 8:84–114.

Rothen-Rutishauser, B.G. 2009. Direct combination of nanoparticle fabrication and exposure to lung cell cultures in a closed setup as a method to simulate accidental nanoparticle exposure of humans. *Environ Sci Technol* 43:2634–2640.

Sahoo, S.K. 2005. Characterization of porous PLGA/PLA microparticles as a scaffold for three dimensional growth of breast cancer cells. *Biomacromolecules* 6:1132–1139.

Sau, T.K. 2010. Properties and applications of colloidal nonspherical noble metal nanoparticles. *Adv Mater* 22:1805–1825.

Sawyer, J.S. 2003. Synthesis and activity of new aryl- and heteroaryl-substituted pyrazole inhibitors of the transforming growth factor-β type I receptor kinase domain. *J Med Chem* 46:3953–3956.

Schrand, A.M. 2010. Metal-based nanoparticles and their toxicity assessment. *Wiley Interdiscip Rev Nanomed Nanobiotechnol* 2:544–568.

Sevick-Muraca, E.M. 2012. Translation of near-infrared fluorescence imaging technologies: Emerging clinical applications. *Ann Rev Med* 63:217–231.

Sheldrick, G.M. 2002. *Shelxtl-plus software package.* Madison, WI: Bruker Analytical X-ray Division.

Shukla, R.B. 2005. Biocompatibility of gold nanoparticles and their endocytotic fate inside the cellular compartment: A microscopic overview. *Langmuir* 21:10644–10654.

Shulaev, V. 2006. Metabolic and proteomic markers for oxidative stress. New tools for reactive oxygen species research. *Plant Physiol* 141:367–372.

Smith, D. 1989. Materials science: Metals, ceramics, and semiconductors. In *High-Resolution Transmission Electron Microscopy and Associated Techniques*, ed, P.C. Buseck, 477–518. Oxford: Oxford University Press.

Spence, J. 1989. Techniques closely related to high-resolution electron microscopy. *High-Resolution Transmission Electron Microscopy and Associated Techniques*, ed. P.C. Buseck, 190–243. Oxford: Oxford University Press.

Sperling, R.A. 2008. Biological applications of gold nanoparticles. *Chem Soc Rev* 37:1896–1908.

Steel, P.J. 2005. Ligand design in multimetallic architectures: Six lessons learned. *Acc Chem Res* 38:243–250.

Su, C.Y. 2004. Exceptionally stable, hollow tubular metal-organic architectures: Synthesis, characterization, and solid-state transformation study. *J Am Chem Soc* 126:3576–3586.

Sudimack, J. 2000. Targeted drug delivery via the folate receptor. *Adv Drug Delivery Rev* 41:147–162.

Sun, C.D. 2010. PEG-mediated synthesis of highly dispersive multifunctional superparamagnetic nanoparticles: Their physicochemical properties and function in vivo. *ACS Nano* 4:2402–2410.

Sun, D.C. 2006. Construction of open metal–organic frameworks based on predesigned carboxylate isomers: From achiral to chiral nets. *Chemistry* 12:3768–3776.

Sun, D.K. 2006. Stability and porosity enhancement through concurrent ligand extension and secondary building unit stabilization. *Inorganic Chem* 45:7566–7568.

Sun, R.W. 2007. Some uses of transition metal complexes as anti-cancer and anti-HIV agents. *Dalton Trans* 43:4884–4892.

Sutton, D.N. 2007. Functionalized micellar systems for cancer targeted drug delivery. *Pharm Res* 24:1029–1046.

Tatum, E.L. 1950. Genetics of microorganisms. *Ann Rev Microbiol* 4:129–150.

Tepper, J.N. 1976. Carcinoma of the pancreas: Review of MGH experience from 1963 to 1973—Analysis of surgical failure and implications for radiation therapy. *Cancer* 37:1519–1524.

Thompson, L.H. 2001. Homologous recombinational repair of DNA ensures mammalian chromosome stability. *Mutat Res* 477:131–153.

Traversa, E.D. 2001. Photoelectrochemical properties of sol-gel processed Ag-TiO2 nanocomposite thin films. *J Sol-Gel Sci Technol* 22:115–123.

Ulrich, C.M. 2006. Non-steroidal anti-inflammatory drugs for cancer prevention: Promise, perils and pharmacogenetics. *Nat Rev Cancer* 6:130–140.

Vayssieres, L. 2003. Growth of arrayed nanorods and nanowires of ZnO from aqueous solutions. *Adv Mat* 15:464–466.

Vayssieres, L. 2004. On the design of advanced metal oxide nanomaterials. *Intl J Nanotechnol* 1:1–41.

Veiseh, O.G. 2010. Design and fabrication of magnetic nanoparticles for targeted drug delivery and imaging. *Adv Drug Deliv Rev* 62:284–304.

Vickers, A. 2002. Botanical medicines for the treatment of cancer: Rationale, overview of current data, and methodological considerations for phase I and II trials. *Cancer Invest* 20:1069–1079.

Wagner, A.J. 2007. Cellular interaction of different forms of aluminum nanoparticles in rat alveolar macrophages. *J Phys Chem B* 111:7353–7359.

Wall Street Transcript. 2004. *CEO Interview: John Funkhouser–Pharmanetics Inc.* Available from http://www.twst.com/interview/3155.

Wang, L.S. 2012. Nanotheranostics—A review of recent publications. *Intl J Nanomed* 7:4679–4695.

Wang, M. 2010. Targeting nanoparticles to cancer. *Pharmacol Res* 62: 90–99.

Wang, X.S. 2008. Metal-organic frameworks based on double-bond-coupled di-isophthalate linkers with high hydrogen and methane uptakes. *Chem Mat* 20:3145–3152.

Weaver, J.G. 2011. Cytotoxicity of gold (I) N-heterocyclic carbene complexes assessed by using human tumor cell lines. *Chemistry* 17:6620–6624.

Weissleder, R.A. 1989. Superparamagnetic iron oxide: Pharmacokinetics and toxicity. *Am J Roentgenol* 152:167–173.

Werner, H. 2004. The way into the bridge: A new bonding mode of tertiary phosphanes, arsanes, and stibanes. *Angew Chem Int Ed Engl* 43:938–954.

Wildgoose, G.G. 2005. Metal nanoparticles and related materials supported on carbon nanotubes: Methods and applications. *Small* 2:182–193.

Xia, Y. 2008. Nanomaterials at work in biomedical research. *Nat Mat* 7:758–760.

Xie, J.L. 2010. Nanoparticle-based theranostic agents. *Adv Drug Deliv Rev* 62:1064–1079.

Xie, J.X. 2007. Controlled PEGylation of monodisperse fe3o4 nanoparticles for reduced non-specific uptake by macrophage cells. *Adv Mat* 19:3163–3166.

Yaghi, O.M. 2003. Reticular synthesis and the design of new materials. *Nature* 423:705–714.

Yan, H.H. 2003. Morphogenesis of one-dimensional ZnO nano-and microcrystals. *Adv Mat* 15:402–405.

Ye, B.H. 2005. Metal-organic molecular architectures with 2, 2′-bipyridyl-like and carboxylate ligands. *Coordin Chem Rev* 249:545–565.

Yellepeddi, V.K. 2011. Biotinylated PAMAM dendrimers for intracellular delivery of cisplatin to ovarian cancer: Role of SMVT. *Anticancer Res* 31:897–906.

Yen, H.J. 2009. Cytotoxicity and immunological response of gold and silver nanoparticles of different sizes. *Small* 5:1553–1561.

Yoo, J.W. 2010. Endocytosis and intracellular distribution of PLGA particles in endothelial cells: Effect of particle geometry. *Macromol Rapid Commun* 31:142–148.

Yu, J.L. 2004. Precipitous dose-response curves for the anticancer. *Med Hypoth Res* 1:267–274.

Yuan, F.D. 1995. Vascular permeability in a human tumor xenograft: Molecular size dependence and cutoff size. *Cancer Res* 55:3752–3756.

Zeng, H.L. 2004. Bimagnetic core/shell FePt/Fe3O4 nanoparticles. *Nano Lett* 4:187–190.

Zhang, B.L. 2010. Development of silver–zein composites as a promising antimicrobial agent. *Biomacromolecules* 11:2366–2375.

Zhang, J.J. 2012. Organogold(III) supramolecular polymers for anticancer treatment. *Angew Chem Int Ed Engl* 51:4882–4886.

Zhang, S. 2003. Fabrication of novel biomaterials through molecular self-assembly. *Nat Biotechnol* 21:1171–1178.

Zhao, D.T. 2010. tuning the topology and functionality of metal-organic frameworks by ligand design. *Acc Chem Res* 44:123–133.

Zhao, D.Y. 2009. Stabilization of metal-organic frameworks with high surface areas by the incorporation of mesocavities with microwindows. *J Am Chem Soc* 131:9186–9188.

Zhou, H.C. 2012. Introduction to metal-organic frameworks. *Chem Rev* 112:673–674.

Zhuang, W.Y. 2012. Highly potent bactericidal activity of porous metal-organic frameworks. *Adv Healthc Mat* 1:225–238.

9

Application of Metals in Traditional Chinese Medicine

Bhaskar Mazumder, Shufeng Zhou,
Jiayou Wang, and Yashwant Pathak

Contents

9.1   Introduction

The theoretical view of traditional Chinese medicine (TCM) is based on experience and is mainly guided by a holistic concept. The integrity of the human body and the relationship of the body with the environment are emphasized in TCM. From this perspective, TCM practitioners believe that no single body part or symptom can be understood apart from its relationship to the whole. In contrast to the Western style of medicine and treatment, which searches to uncover a distinct entity or causative factor for a particular ailment, TCM looks at patterns of disorderliness, which include signs and symptoms as well as a patient's emotional and psychological behavior. In TCM, humans are viewed both as a reflection of and as an integral part of nature, and health results from maintaining harmony and balance within the body and nature. The backbone of TCM is generally composed of a few key components, as discussed in the following sections.

9.2 Yin-yang theory

The yin-yang theory describes how things function in relationship to each other and with respect to the universe. It encompasses the concept of two opposing, yet complementary, forces that shape the world and all its life. It describes naturally occurring phenomena that are grouped in pairs of opposites, such as heaven and earth, sun and moon, night and day, winter and summer, male and female, up and down, inside and outside, and movement and stasis. Yin and yang are natural complements in the sense that they counterbalance each other. The famous symbol of yin and yang (**Figure 9.1**) depicts the yin to be the dark side flowing into the yang, which is designated as the light side, and vice versa. The dots within each side symbolize that there is always a bit of yin within yang and a bit of yang within yin. All physiological functions of the body, as well as the signs and symptoms of disease, can be differentiated on the basis of yin and yang characteristics. The yin-yang theory finds various applications in the Chinese system of medicine. This concept can be applied for several analyses, as described in the following sections.

Figure 9.1 *The famous symbol of yin and yang.*

9.2.1 Anatomical analysis

Each organ in a human being has a yin and a yang feature. The two primary elements of the human body, blood and qi (pronounced "chee"), may also be thus categorized, with blood being yin and qi being yang. The liver, heart, spleen, lung, and kidney are yin; the gallbladder, stomach, intestines, bladder, and triple burner are yang (**Figure 9.2**). There is no organ in Western medicine that corresponds to San Jiao (triple burner). The triple burner is said to occupy the thoracic and abdominopelvic cavities. It is related to the fire element of the Chinese five elements.

9.2.2 Physiological analysis

Yin and yang provide a general method to analyze the functioning of the human body with respect to four types of movement: upbearing, downbearing, issue, and entry. These movements serve to explain the interactions between blood and qi, as well as the organs and channels.

9.2.3 Analysis of pathologic conditions

When yin prevails, there is cold; when yang prevails, there is heat. When yang is deficient, there is cold; when yin is deficient, there is heat.

9.2.4 Concept of qi

Qi signifies a vital energy or life force that circulates in the body through a system of pathways called meridians. The main functions of meridians are

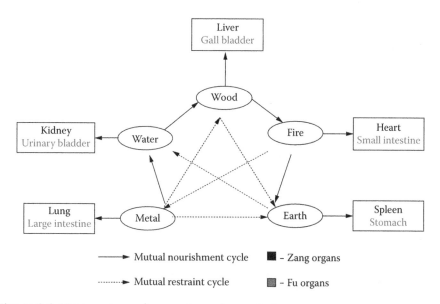

Figure 9.2 Human organs have a yin and yang feature.

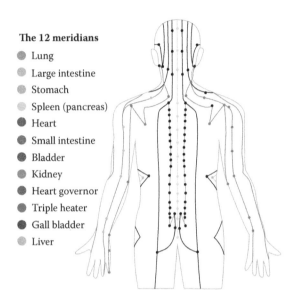

The 12 meridians

- Lung
- Large intestine
- Stomach
- Spleen (pancreas)
- Heart
- Small intestine
- Bladder
- Kidney
- Heart governor
- Triple heater
- Gall bladder
- Liver

Figure 9.3 *The body's energy pathways, commonly known as meridians.*

to connect the internal organs with the exterior of the body, to harmonize the yin and yang principles within the body's organs and five substances, to distribute qi within the body, and to protect the body against external imbalances. For yin and yang to be balanced and for the body to be healthy, qi must be balanced and flowing freely. When there is too little or too much qi in one of the body's energy pathways, called meridians (**Figure 9.3**), or when the flow of qi is blocked, it causes illness. The ultimate goal of TCM treatment is to balance the yin and yang by promoting the natural flow of qi.

9.2.5   Five phases (elements) theory

To understand this theory, it helps to explain how the body works as these five elements—fire, earth, metal, water, and wood—correspond to particular organs and tissues in the body. Along with the law of yin-yang, ancient Taoist scholars noticed a pattern of expression in nature and named them the five elements (or five phases). These elements or energies were represented as wood, fire, earth, metal, and water and were understood to be the prime building blocks from which all material substances in the phenomenal world are made. There prevails a dynamic balance and relationship among the elements through some cycles, such that if the balance is disturbed or destroyed, pathophysiological changes may occur.

The five phases have a flow in which they move toward each other, which is called the generating/sheng cycle: water generates wood, wood generates fire, fire generates earth, earth generates metal, and metal generates water. Meanwhile, to preserve the balance, the five phases also have a system of checks and balances to ensure that no phase is too long or too strong, which

Table 9.1 Five-Element Theory in Traditional Chinese Medicine

Indicator	1	2	3	4	5
Tastes	Sour	Bitter	Sweet	Pungent	Salty
Colors	Blue (dark)	Red	Yellow	White	Black
Evils	Wind	Heat	Damp	Dry	Cold
Organs	Liver	Heart	Spleen	Lung	Kidney
Emotions	Anger	Joy	Worry	Grief	Fear
Senses	Eyes	Tongue	Mouth	Nose	Ear
Expressions	Shouting	Laughing	Singing	Wailing	Groaning
Orientations	East	South	Middle	West	North
Elements	Wood	Fire	Earth	Gold	Water

Each element has a correlative connection. For example, impaired function of the liver may produce problems in the eyes; excessive joy is harmful to the "heart," which includes the functions of the brain; also, herbs with a black color would be likely to affect the function of the kidney.

is termed the controlling/ke cycle: water controls fire, fire controls metal, metal controls wood, wood controls earth, and earth controls water. If any phase is extremely strong, it can actually turn around and put down the phase that normally controls it, which is called the insulting cycle (**Table 9.1**).

Proper formation, maintenance, and circulation of these energies are essential for health. A traditional Chinese medicine practitioner takes into account these concepts while treating patients. Drugs or herbs are used to correct these imbalances of yin and yang in the human body.

9.3   Are Chinese herbal remedies really safe?

Chinese herbal medicine, one of the branches of traditional Chinese medicine, is a highly complex system of diagnosis and treatment that uses medicinal herbs. Herbs range from nontoxic and rejuvenating, used to support the body's healing system, to highly toxic ones, used to treat complicated ailments.

Contamination may be a problem with Chinese herbal remedies. Chinese herbal pills have been found to contain undeclared mefenamic acid and diazepam, with adverse effects of massive gastrointestinal bleeding and acute interstitial nephritis followed by acute renal failure. Another Chinese herbal remedy was shown to contain toxic amounts of arsenic and mercury. Similarly, there have been cases of lead, mercury, thallium, arsenic, and cadmium poisoning due to the use of contaminated Chinese herbal medicines (Ernst 1998). Excessive levels of nonessential toxic elements, such as lead, cadmium, mercury, and aluminum, can have a "deranging" effect on trace element balances in the body's cells, as shown in **Table 9.2**.

Table 9.2 Effect of Toxic Minerals on Body Organs and Tissues

Toxic Metals	Organs Affected
Lead	Bone, liver, kidney, pancreas, heart, brain, nervous system
Cadmium	Renal cortex of the kidney, heart, blood vessels to the brain, appetite and smell center of the brain, development of cancer
Aluminum	Stomach, bones, brain
Arsenic	Cells (cellular metabolism)
Mercury	Nervous system, appetite and pain centers of the brain, immune system, cell membranes

Source: Data from http://www.traceminerals.com/research/humanhealth.

9.4 Health benefits of some important minerals and trace minerals

Minerals in trace amounts may play a significant role against a variety of degenerative diseases and processes induced by heavy metal toxicity. They may also prevent and reduce injury from environmental pollutants and enhance the ability to work and learn. They can also protect the body from the effects of toxic minerals. Numerous minerals, when in proper balance with one another, may perform important biochemical functions that are especially important for age-related health problems. As stated in **Table 9.3**, these minerals and trace minerals may be most beneficial if they are in balance with other elements they interact with in the body.

9.5 Stone drugs as natural mineral supplements in traditional Chinese medicine

Mineral drugs directly supply minor and trace elements to the body. The value of a mineral drug is determined by its structure, biochemical potential, and biophysical potential of the elements that it contains. The curative effect is not only a biochemical effect but also a biophysical one. Minerals as drugs have played important roles in the development and prosperity of traditional Chinese medicine since its inception.

The majority of Chinese stone drugs are naturally occurring minerals, such as pyrite (pyritum) or cinnabar (cinnabaris). Mirabilite (mirabilitum depuratum) and calomel (calomelas) are examples of processed minerals. Apart from this, a few stone drugs consist of animal fossils and are popularly known as dragon's teeth (dens draconis) or dragon's bones (os draconis).

9.6 History of Chinese mineral drugs

Application of Chinese stone drugs has gradually developed over a history of more than 2,000 years. Excavation conducted in Mawangdui, Changsha,

Table 9.3 The Health Benefits of Some Important Minerals and Trace Minerals

Minerals and Trace Minerals	Physiological Roles
Calcium	• Essential for developing and maintaining healthy bones and teeth • Assists in blood clotting, muscle contraction, nerve transmission, oxygen transport, cellular secretion of fluids, and enzyme activity • Reduces risk of osteoporosis
Chromium	• Aids in glucose metabolism • Helps to regulate blood sugar
Cobalt	• Promotes the formulation of red blood cells • Serves as a component of vitamin B12
Copper	• Essential to normal RBC formation and connective tissue formation • Acts as a catalyst to store and release iron to assist hemoglobin formation • Regulates and coordinates central nervous system function
Iodine	• Promotes thyroid hormone production
Iron	• Helps in transport of oxygen throughout the body • Helps in maintaining brain function
Magnesium	• Helps in nerve and muscle function • Helps to maintain the integrity of cell membranes and stabilizes the cell electrically
Manganese	• Supports brain function and reproduction • Regulates blood sugar
Molybdenum	• Regulates normal growth and development • Key component in many enzyme systems
Phosphorous	• In conjunction with calcium, helps to maintain strong bones and teeth • Important role in cell membrane integrity and intercellular communication
Potassium	• Regulates proper heartbeat, maintains fluid balance, and helps to maintain muscle contraction and function
Selenium	• Important component of a key antioxidant enzyme that is necessary for normal growth and development • Role in detoxification of heavy metals
Zinc	• Assists in digestion, metabolism, reproduction, and wound healing • Plays an important role in immune response • Is an important antioxidant

Source: Data from http://www.traceminerals.com/research/humanhealth.

China in 1973 provided information about 242 types of medications, out of which 20 were stone drugs. Several artifacts that were excavated from various Chinese dynasties (Qin Dynasty, 221–207 B.C.; West Han Dynasty, 206 B.C.–220 A.D.) also substantiate the claim that stone drugs were used in Chinese traditional medicine. Shen Nong's *Herbal Classic*, the popular classic book from these dynasties, lists 365 medicines, of which 46 were derived from minerals. This book has also provided details of the source, properties, and effects of each stone drug and described details of the chemical reactions. Shen Nong's *Herbal Classic* as compiled by Tao Hong Jing

(452–536 A.D.) included details of 365 more medicines that were not mentioned in Shen Nong's original publication, of which 32 were stone drugs. Tao Hon Jing also emphasized the necessity of breaking and grinding minerals before their use.

During the Tang Dynasty (618–907 AD), Su Jing and coauthors compiled the first officially sanctioned Chinese pharmacopeia. It was published in 659 A.D. and was known as *The Newly- Compiled Materia Medica*. The pharmacopoeia provided information about 844 types of traditional medications, 83 of which were stone drugs. A detailed review of all these compilations shows that 104 types of stone drugs were being used by Chinese physicians during the Tang Dynasty.

In the year 1590, at the time of the Ming Dynasty (1368–1644 A.D.), the great medical scientist Li Shizhen compiled the *Compendium of Materia Medica*, which contained approximately 1,892 medicinal substances, of which 355 (19%) were stone drugs. This publication provided a comprehensive description of each mineral's name and sources, method of collection, and physiochemical nature. In addition, Li Shizhen gave detailed information on the processing, chemical characteristics, taste, toxicity, application, and dose of each stone drug.

Success stories attached with Chinese stone drugs were also associated with some failures in establishing use of these drugs for several reasons. During the twentieth century, epidemiological research generally stressed the adverse effects of heavy metals on human health. As a consequence, many inorganic medications were looked upon with suspicion. Recently, however, a revival of stone drug use seems to be occurring once again in China. There are many reasons for this. Medical geographers, for instance, have identified numerous endemic illnesses that appear to involve mineral imbalances and may be prevented or cured through the use of mineral drugs (Tan 1989). In addition, advancements in physicochemical analysis now can ensure greater stone drug purity, thus reducing the chance of accidental adverse side effects. In addition, scientific evidence of their value is growing. Ethnopharmacological experiments continue to be conducted with animals and in clinics and laboratories to scientifically validate the efficacy of numerous mineral medicines that have been used traditionally by Chinese physicians.

9.7   Chinese stone drugs for the prevention and treatment of diseases

Over the past 2,000 years or so, Chinese traditional medicine practitioners have identified approximately 350 mineral drugs with medicinal potential; approximately 60 of these are still commonly used. These mineral drugs are generally given as concoctions with herbal preparations to treat specific diseases.

9.7.1 Endemic diseases

Goiter and fluorosis, which are associated with bulk and trace element deficiency, are quite common in China. Because stone drugs are excellent sources of many of the bulk and trace elements involved in such endemic diseases, it is not surprising that they have been of value in prevention and treatment in the traditional system of medicine.

Chinese physicians in the Tang Dynasty were the first to successfully treat patients with goiter by using the iodine-rich thyroid gland of animals such as sheep and pigs (Temple 1986). Seaweed (a genus of brown macroalgae), which has high iodine content, traditionally has been used in Chinese medicine to prevent and treat goiter. In addition, some goiter-related formulae include the stone drug pyrite (pyritum), which consists predominantly of iron disulphide (FeS_2).

Fluorosis is endemic in many regions of China, where it is caused mostly by ingestion of excessive amounts of fluorine from drinking water. It is associated with a mottled discoloration of the enamel of the teeth and can lead to combined production of abnormally dense bone and impaired bone mineralization. Borax (sodium tetraborate hexahydrate), which is obtained from dry salt lakes, has been used for centuries in Chinese traditional medicine. In an experiment conducted by Fan et al. (1987), it was suggested that borax may be a very effective treatment for endemic fluorosis in humans.

9.7.2 Malignant neoplasms

Mineral drugs are also effective against cancer. Jiangshi, a ginger-like stone that is available naturally in north and northwestern China (Shaanxi, Gansu, Hebei, and Shagdong provinces), was found to be effective against malignant neoplasms, especially against esophageal cancer. However, the composition of Jiangshi differs depending upon the geographical availability (**Tables 9.4 and 9.5**).

9.7.3 Infectious diseases

Minerals are widely used against infectious diseases in traditional Chinese medicine. There are several minerals that are effective against infectious disease. Aluminum potassium sulfate ($KAl(SO_4)_2·12H_2O$), commonly known as Alum (alumen), has been found to be effective against *Staphylococcus aureus*, *Bacillus proteus*, *Bacillus coli*, *Bacillus dysenteriae*, *Bacillus diphtheria*, and *Streptococcus viridans*.

Gypsum is used as a major component of several Chinese medicines (e.g., Baihutang) against Japanese B encephalitis (Yuan 1956). Borax is used as a bacteriostatic and antifungal agent in traditional Chinese medicine. Calamine has a history of use in acute conjunctivitis and mirabilite is used for common surgical infections.

Table 9.4 Major Constituents of Three Varieties of Jiangshi: Origin and Composition

Element	Zhangxian Gansu	Shexian Hebei	Xingtai Hebei
Calcium	29.22	27.04	23.21
Silicon	8.31	10.07	13.22
Aluminum	2.04	2.28	2.99
Iron	0.99	1.11	1.44
Magnesium	0.42	0.73	0.59
Potassium	0.65	0.62	0.91
Sodium	0.42	0.54	0.60
Barium	0.18	0.17	0.15
Titanium	0.12	0.15	0.19
Manganese	0.022	0.021	0.025
Phosphorus	0.017	0.009	0.027
Strontium	0.019	0.020	0.024
Fluorine	0.06	0.07	0.06
Sulfur	0.008	0.008	0.008

Source: Wang, X.Y., et al. (1983). Study of Jiangshi composition. *Bulletin of Chinese Materia Medica* 8:28–30.
Percentage values are by weight.

9.8 The processing of mineral drugs

Mineral drugs in TCM are mostly used in combination with herbal preparations. They are treated differently depending on the nature of the disease and the mode of application of the metal drugs. For example, raw untreated fossil bones (os dracunic) are recommended as a tranquilizer, whereas calcined fossils (strongly heated) are recommended as anti-inflammatory analgesics (Yu et al. 1995).

Mineral drugs that are used in a wide variety of Chinese medicines are processed mainly in four ways: powder refining, recrystallization, calcining, and tempering (Zhang 1959) (**Table 9.6**).

9.8.1 Powder refining

Powder refining is done to convert the crude mineral drugs into fine particles. Mineral drugs are ground into powder in the presence of water, and the mass is allowed to settle down freely into a pot containing clean water. The larger particles, due to the higher gravitational force, settle quickly to the bottom and the finer particles remains dispersed/suspended in the water. The supernatant water containing the finer particles are decanted. The process

Table 9.5 The Minor Constituents of Three Varieties of Jiangshi: Origin and Composition (parts per million)

Element	Zhangxian Gansu	Shexian Hebei	Xingtai Hebei
Zinc	60	60	60
Cobalt	17	16	15
Nickel	16.69	16.75	14.85
Iodine	15	15	12
Vanadium	8.78	8.05	13.17
Chromium	3.42	6.84	10.26
Copper	3	1	1.1
Tin	1	2	2
Selenium	0.01	0.04	0.02
Molybdenum	0.2	0.35	0.95
Boron	1.15	8.07	1.04
Bromine	51	1	45
Tungsten (wolfram)	0.18	0.71	0.73
Lead	0.9	1	1.1
Scandium	2.09	2.22	2.28
Beryllium	0.72	0.72	0.72
Silver	4.6	4.5	3.8
Tellurium	0.008	0.003	0.008
Arsenic	2	2	2
Uranium	0.55	1	0.7
Thorium	0.85	0.5	1.1

Source: Wang, X.Y., et al. (1983). Study of Jiangshi composition. *Bulletin of Chinese Materia Medica* 8:28–30.

is repeated several times to obtain the complete conversion of crude drugs into the desired range of particle size. Because heavy metals, due to their higher density, sediment rapidly in every step and are discarded at the end, the purity of the mineral drugs is also increased remarkably in this process.

9.8.2 Recrystallization

The objective of recrystallization is to purify the soluble mineral drugs and recrystallize them. The soluble mineral drugs are ground and then dissolved in water until a saturated solution is formed. Crystallization is then carried out either by adiabatically cooling with or without gentle agitation or by a solvent evaporation method (heating). The latter method is applied for those compounds with a solubility profile that is not remarkably varied with

Table 9.6 Some Mineral Drugs Used in Traditional Chinese Medicine

Minerals	Diseases Treated
Alum (Alumen)	Bacteriostatic, inhibits *Candida albicans*, treatment of proctoptoma
Borax	Bacteriostatic, antifungal
Calamine	Acute conjunctivitis
Calcined alum with honey	Treatment of gastroduodenal ulcer
Cinnabar	Treatment of angina and insomnia
Chalcanthite	Antiemetic, treatment of mouth ulcer
Gypsum	Antipyretic, treatment of Japanese B encephalitis
Minium	Treatment of skin disease, epilepsy, and depression
Mirabilite	Treatment of common surgical infections, digestive system disorders, diuretic
Os draconic (dragon's bone)	Nervous system disorder
Powdered alum	Antiepileptic
Pyrites	Treatment of bone fractures, promotes bone growth
Red ochre (hematite)	Treatment of hematemesis, hemoptysis

temperature. The pure crystals are collected by filtering and dried before their use.

9.8.3 Calcining

Calcining is employed to increase the efficacy of hydrated mineral drugs by removing their water of crystallization. Most common among these types of chemicals are alum [$KAl(SO_4) \cdot 12H_2O$] and gypsum [$CaSO_4 \cdot 2H_2O$]. Calcining is done either by excessive heating or by prolonged atmospheric exposure. In the case of excessive heating, care should be taken regarding the thermal stability/decomposition of the mineral. Recrystallization by atmospheric exposure is time consuming and depends upon atmospheric conditions, such as humidity and temperature (**Table 9.7**).

9.8.4 Tempering

Tempering is another method adopted in TCM to increase or modify the chemical composition and efficacy of mineral drugs. Here, the mineral drugs are heated to a red-hot condition and treated under cold water or vinegar. Depending upon the nature of the minerals under treatment, three types of reactions are mainly taking place in the tempering of minerals: oxidation, decomposition, and salt formation.

Sulfide minerals—such as acanthite (Ag_2S), chalcocite (Cu_2S), bornite (Cu_5FeS_4), galena (PbS), sphalerite (ZnS), chalcopyrite ($CuFeS_2$), millerite (NiS), pentlandite [$(Fe,Ni)_9S_8$], covellite (CuS), cinnabar (HgS), realgar (AsS), orpiment (As_2S_3), stibnite (Sb_2S_3), pyrite (FeS_2), and molybdenite (MoS_2)—are subjected

Table 9.7 Typical Presentation of the Most Commonly Encountered Metals and Their Treatment

Metal	Acute Conditions	Chronic Conditions	Toxic Concentration	Treatment
Arsenic	Nausea, vomiting, "rice water" diarrhea, encephalopathy, MODS, LoQTS, painful neuropathy	Diabetes, hypopigmentation/ hyperkeratosis, skin cancer, lung cancer, bladder cancer, encephalopathy	24-h urine: $\geq$50 µg/L urine, or 100 µg/g creatinine	BAL (acute, symptomatic), succimer, DMPS (Europe)
Bismuth	Renal failure, acute tubular necrosis	Diffuse myoclonic encephalopathy	No clear reference standard	No accepted chelation regimen; contact a medical toxicologist regarding treatment plan
Cadmium	Pneumonitis (oxide fumes)	Proteinuria, lung cancer, osteomalacia	Proteinuria and/or $\geq$15 µg/g creatinine	No accepted chelation regimen; contact a medical toxicologist regarding treatment plan.
Chromium	GI hemorrhage, hemolysis, acute renal failure (Cr^{6+} ingestion)	Pulmonary fibrosis, lung cancer (inhalation)	No clear reference standard	NAC (experimental)
Cobalt	Dilated cardiomyopathy	Pneumoconiosis (inhaled), goiter	Normal excretion: 0.1–1.2 µg/L (serum) or 0.1–2.2 µg/L (urine)	NAC, $CaNa_2$ EDTA
Copper	Blue vomitus, GI irritation/ hemorrhage, hemolysis, MODS (ingested), MFF (inhaled)	Vineyard sprayer's lung (inhaled); Wilson disease (hepatic and basal ganglia degeneration)	Normal excretion: 25 µg/24 h (urine)	BAL, D-penicillamine, succimer
Iron	Vomiting, GI hemorrhage, cardiac depression, metabolic acidosis	Hepatic cirrhosis	Nontoxic: <300 µg/dL; severe: >500 µg/dL	Deferoxamine
Lead	Nausea, vomiting, encephalopathy (headache, seizures, ataxia, obtundation)	Encephalopathy, anemia, abdominal pain, nephropathy, foot drop/wrist drop	Pediatric: symptoms or [Pb] $\geq$45 µ/dL (blood); Adult: symptoms or [Pb] $\geq$70 µ/dL	BAL, $CaNa_2$ EDTA, succimer
Manganese	MFF (inhaled)	Parkinson-like syndrome, respiratory, neuropsychiatric	No clear reference standard	No accepted chelation regimen; contact a medical toxicologist regarding treatment plan

continued

Table 9.7 Typical Presentation of the Most Commonly Encountered Metals and Their Treatment (Continued)

Metal	Acute Conditions	Chronic Conditions	Toxic Concentration	Treatment
Mercury	Elemental (inhaled): fever, vomiting, diarrhea, ALI; inorganic salts (ingestion): caustic gastroenteritis	Nausea, metallic taste, gingivostomatitis, tremor, neurasthenia, nephrotic syndrome, hypersensitivity (Pink disease)	Background exposure "normal" limits: 10 µg/L (whole blood); 20 µg/L (24-h urine)	BAL, succimer, DMPS (Europe)
Nickel	Dermatitis; nickel carbonyl: myocarditis, ALI, encephalopathy	Occupational (inhaled): pulmonary fibrosis, reduced sperm count, nasopharyngeal tumors	Excessive exposure: ≥8 µg/L (blood); severe poisoning: ≥500 µg/L (8-h urine)	No accepted chelation regimen; contact a medical toxicologist regarding treatment plan
Selenium	Caustic burns, pneumonitis, hypotension	Brittle hair and nails, red skin, paresthesia, hemiplegia	Mild toxicity: [Se] >1 mg/L (serum); serious: >2 mg/L	No accepted chelation regimen; contact a medical toxicologist regarding treatment plan
Silver	Very high doses: hemorrhage, bone marrow suppression, pulmonary edema, hepatorenal necrosis	Argyria: blue-grey discoloration of skin, nails, mucosae	Asymptomatic workers have mean [Ag] of 11 µg/L (serum) and 2.6 µg/L (spot urine)	Selenium, vitamin E (experimental)
Thallium	Early: vomiting, diarrhea, painful neuropathy, coma, autonomic instability, MODS	Late findings: alopecia, Mees lines, residual neurologic symptoms	Toxic: >3 µg/L (blood)	MDAC, Prussian blue
Zinc	MFF (oxide fumes); vomiting, diarrhea, abdominal pain (ingestion)	Copper deficiency: anemia, neurologic degeneration, osteoporosis	Normal range: 0.6–1.1 mg/L (plasma); 10–14 mg/L (red cells)	No accepted chelation regimen; contact a medical toxicologist regarding treatment plan

Source: From http://emedicine.medscape.com/article/814960-overview.
ALI, acute lung injury; ARF, acute renal failure; ATN, acute tubular necrosis; BAL, 2,3-dimercaptopropanol; CaNa$_2$ EDTA, edetate calcium disodium; DMPS, 2,3-dimercapto-1-propane-sulfonic acid; GI, gastrointestinal; LoQTS, long QT syndrome; MDAC, multidose activated charcoal; MFF, metal fume fever; MODS, multiorgan dysfunction syndrome; NAC, N-acetylcysteine.

to tempering. They first undergo oxidation to their respective oxides, which may be followed by decomposition and/or salt formation, depending upon the treatment.

Carbonates—such as calcite ($CaCO_3$), gaspeite [$(Ni,Mg,Fe_2+)CO_3$], magnesite ($MgCO_3$), otavite ($CdCO_3$), rhodochrosite ($MnCO_3$), siderite ($FeCO_3$), smithsonite ($ZnCO_3$), spherocobaltite ($CoCO_3$), cerussite ($PbCO_3$), witherite ($BaCO_3$), and natrite (Na_2CO_3)—decompose to their respective oxides while heating during tempering. The elimination of carbon dioxide gas from the carbonates

Table 9.8 Assessment of the Acute Toxicity of Commonly Used Chinese Mineral Drugs

Drug	Place of Origin	Median Lethal Dose in Mice (g/kg)	Method of Administration	Length of Experiment (days)
Baijiangdan (formula including mercury chloride)	Jiangxi	0.078	i.g.	3
White arsenolite	Jiangxi	0.144	i.g.	3
Red arsenolite	Jiangxi	0.242	i.g.	3
Sublimated sulfur	Hebei	0.266	i.g.	3
Chalcanthite	Handan	0.279	i.g.	3
Calcined alum	Handan	0.060	i.v.	3
Calomel	Xiangtan	2.068	i.g.	3
Alum	Wenzhou	2.153	i.g.	3
Borax	Imported	2.454	i.g.	3
Halite	Qinghai	2.789	i.g.	3
Purple salt rock	Gansu	4.435	i.g.	3
		2.216	i.p.	3
Table salt	Tianjin	4.437	i.g.	3
		2.660	i.v.	3
White sal-ammoniac	Gansu	3.517	i.g.	3
Realgar	Chengde	3.207	i.g.	3
Orpiment	Guizhou	3.83	i.v.	3
Amber	Fushun	0.960	i.v.	3
Pyrite	Wuling	1.920	i.v.	3
Calcined pyrite	Wuling	3.83	i.v.	3
Qiushi (product of urine and sodium chloride)	Anquig	4.437	i.p.	3
Mirabilite	Xian	4.648	i.p.	3
Mirabilite	Chuxiong	6.738	i.p.	3
Borax	Xian	7.395	i.v.	3
Qianfen (product of lead and acetic acid)	Nanjing	13.970	i.v.	3
	Handan	>36.000	i.g.	3
Talc	Hebei	>36.000	i.g.	3
Chlorite-schist	Henan	>36.000	i.g.	3
Litharge (lead monoxide)	Yiyang	6.81	i.v.	3
Pumice	Xiamen	10.00	i.v.	3
Calcined fossil of spirifer	Shanxi	8.25	i.v.	3

continued

Table 9.8 Assessment of the Acute Toxicity of Commonly Used Chinese Mineral Drugs (Continued)

Drug	Place of Origin	Median Lethal Dose in Mice (g/kg)	Method of Administration	Length of Experiment (days)
Calcined pumice	Xiamen	10.00	i.v.	14
Fossil of spirifer	Shanxi	21.50	i.v.	3
Ophicalcite	Hebei	4.22	i.v.	7
Calcined hematite	Hebei	12.10	i.v.	3
Calcined ophicalcite	Hebei	21.50	i.v.	3
Hematite	Hebei	12.90	i.v.	3
Verdigris (basic copper carbonate)	Handan	14.70	i.v.	3
Magnetite	Handan	14.70	i.v.	3
Pumice	Yantai	27.80	i.v.	3
Calcined magnetite	Handan	>36.00	i.g.	3
Stalactite	Guangdong	28.30	i.v.	3
Talc	Yichun	5.62	i.v.	7
Pyrolusite	Chengdu	2.78	i.g.	7
Limonite	Baoding	8.25	i.v.	7
Manufactured HgS	Guangzhou	10.00	i.v.	7
Calcined quartz	Baoding	10.00	i.v.	7
Cinnabar	Sichuan	12.10	i.v.	7
Quartz	Baoding	>36.00	i.g.	7
Kaolin	Luevang	12.90	i.v.	7
Selenite	Xingtai	12.90	i.v.	7
Calamine	Henan	12.90	i.v.	7
Calcined fluorite	Handan	14.70	i.v.	7
Minium	Hunan	16.70	i.v.	7
Calcined stalactite	Handan	16.70	i.v.	7
Calcite	Hubei	16.70	i.v.	7
Vermiculite	Baoding	21.50	i.v.	7
Calcined fossil of *Telohusa* sp.	Handan	21.50	i.v.	7
Dragon's bones	Datong	21.50	i.v.	7
Calcined red ochre	Handan	21.50	i.v.	7
Calcite	Jiangsu	27.80	i.v.	7
Red halloysite	Xinxiang	31.60	i.v.	7
Raw gypsum	Hubei	14.70	i.v.	14
Raw gypsum	Hubei	16.70	i.v.	14
Calcined limonite	Handan	10.00	i.v.	14
Mica	Yunnan	21.50	i.v.	14

Table 9.8 Assessment of the Acute Toxicity of Commonly Used Chinese Mineral Drugs (Continued)

Drug	Place of Origin	Median Lethal Dose in Mice (g/kg)	Method of Administration	Length of Experiment (days)
Dragon's teeth	Shanxi	26.10	i.v.	14
Maifanshi (weathering product of granite)	Neimeng	26.10	i.v.	14

i.g., suspension created and finer particles administered by intragastric gavage; i.p., mineral boiled in water and resulting solution administered intraperitoneally; i.v., mineral boiled in water and resulting solution administered intravenously.

increases the percentage of the metal in the oxides formed. This in turn decreases the quantity of minerals used, increases the efficacy, and also reduces the toxicity.

Hematite, the mineral form of iron(III) oxide (Fe_2O_3); magnetite, one of the two common naturally occurring iron oxides (Fe_3O_4); and limonite [$FeO(OH) \cdot nH_2O$] are treated with vinegar to form ferrous acetate for the treatment of anemia in traditional Chinese medicine (Yu 1995).

9.9 Toxicity

Like any other system of medication, traditional Chinese medicines containing mineral drugs also have potential toxic side effects. Toxicity of the Chinese mineral drugs is mostly due to the presence of heavy metals in the formulations and also due to variations in the composition of various metals in the minerals. The variations in composition are due to geographical availability. As shown in **Table 9.4**, the amounts of calcium and iron in Zhangxian Gansu are 29.22% and 0.99%, respectively, whereas those of Xingtai Hebei are 23.21% and 1.44%. The amounts of potassium and phosphorus are 0.62% and 0.009% in Shexian Hebei, but 0.91% and 0.027% in Xingtai Hebei, respectively (**Table 9.8**).

As shown in **Table 9.5**, the amounts of minor constituents also vary widely in three varieties of Jiangshi. The composition of vanadium varies from 8.05 ppm to 13.17 ppm, chromium varies from 3.42 ppm to 10.26 ppm, molybdenum varies from 0.2 ppm to 0.95 ppm, and boron varies from 1.04 ppm to 8.07 ppm. These wide variations affect dose uniformity. The concentrations of a few elements in the minerals are responsible for the toxicity of the traditional Chinese medicine containing minerals.

The Chinese Standard Association for Industrial Enterprise has suggested five classes of toxicity depending upon the median lethal dose (LD_{50}) of the compound:

- LD_{50} < 10 mg/kg body weight: Hypertoxic
- LD_{50} of 11–100 mg/kg body weight: Highly toxic
- LD_{50} of 101–1,000 mg/kg body weight: Moderately toxic
- LD_{50} of 1,001–10,000 mg/kg body weight: Low toxicity
- LD_{50} > 10,000 mg/kg body weight: Generally recognized as safe

Several compounds, such as arsenolite, cinnabar, chalcanthite, and realgar, contain several elements above moderate toxicity. Cinnabar and calomel contain mercury; minium contains lead oxides; realgar contains white arsenolite; red arsenolite contains arsenic; and chalcanthite contains high levels of copper sulfate. Although these preparations are widely used in traditional Chinese medicine, one has to be very careful when using them, as their toxicity has not yet been fully measured or documented.

References

Ernst, E. 1998. Harmless herbs? A review of the recent literature. *American Journal of Medicine* 104:170–178.

Tan, J.A. 1989. *The Atlas of Endemic Diseases and Their Environments in the People's Republic of China*. Beijing, China: Science Press.

Temple, R. 1986. *The Genius of China: 3,000 Years of Science, Discovery, and Invention*. New York: Simon and Schuster, 133–134.

Yu, W.H., H.D. Foster, and T. Zhang. 1995. Discovering Chinese mineral drugs. *Journal of Orthomolecular Medicine* 10:31.

Yuan, Y.Q. 1956. Treatment of encephalitis B by using traditional Chinese medicine. *China Journal Medicine* 42:110–112.

Zhang, C.S. 1959. Initial result on the effect of borax bacteria. *Journal of Shandong Medical College* 8:42–45.

Metalloenzymes
Relevance in Biological Systems and Potential Applications

Ravi Ramesh Pathak

Contents

10.1 Introduction

Every life form on earth has to undergo a multitude of chemical modifications, which include uptake and alteration of nutrients, synthesis and utilization of energy, cell division and growth, and even death. These chemical modifications are governed by a fundamental principle: One or more chemicals usually react to transform into a new one. To be more specific, a chemical reaction typically involves rearrangement of atoms from the reactant side to the product side: $6CO_2 + 6H_2O$ (+ light energy) $= C_6H_{12}O_6 + 6O_2$ (**Figure 10.1**).

Although most reactions depend on multiple factors, including temperature, pH, and concentrations of reactants, one of the most vital elements for biochemical reactions are enzymes. Chemically, enzymes are mostly globular

CO_2 = carbon dioxide, H_2O = water, light energy is required for the reaction, $C_6H_{12}O_6$ = glucose, O_2 = oxygen

Figure 10.1 *Photosynthesis reaction in which six molecules of water plus six molecules of carbon dioxide produce one molecule of sugar plus six molecules of oxygen.*

proteins that are highly selective catalysts, which significantly accelerate both the rate and specificity of metabolic reactions. German physiologist Wilhelm Kühne (1837–1900) coined the term *enzyme*, which was derived from the Greek word ενζυμον (meaning "in leaven") to describe this process of rapid conversion of substrate to products. Enzymes basically function by lowering the activation energy for a reaction, thereby accelerating the rate of the reaction, which is many orders of magnitudes faster than nonenzymatic reactions.

One of the most important factors that sets enzymes apart from other conventional catalysts is their specificity for a given substrate. An enzyme's activity also depends on temperature, chemical environment (e.g., pH), and the concentration of substrate. With the advent of modern technology, information about the structure and factors that regulate enzymatic function has been studied in great detail. The quest to decipher the true nature of enzymatic regulation stems from the fact that enzymes have been selected under evolutionary pressure to evolve as highly efficient catalysts that have the ability to perform efficiently with a high degree of specificity. Modifying existing enzymes and designing newer enzymes based on existing ones to control desirable reactions that could accentuate economic and medical applications assumes great significance. **Table 10.1** summarizes the modern industrial applications of enzymes (1).

Direct methods such as x-ray crystallography, including the protein conformation, and hydration in solution have contributed immensely to the understanding of enzyme function and structure. However, given the high degree of structural complexity of enzymes, these direct methods fail to truly decipher the "hidden" aspects of reaction mechanisms. These have been resolved to a great extent by using methods such as kinetic studies, substrate analogues, isotope effects, molecular mechanics, and dynamic simulations based on molecular dynamics. It is largely due to the combination of these multiple approaches that certain enzymes were revealed to contain metal ions that are essential to their function. Removal of the ions was found to inactivate the enzymes, whereas addition of these ions to the solution promptly reactivated them. These enzymes were classified as metalloenzymes.

10.2 What are metalloenzymes?

Any enzyme containing tightly bound metal atoms is called a metalloenzyme. Most metalloenzymes have metal ions or metal cofactors embedded within

Table 10.1 Industrial Enzymes and Applications

Industry	Enzyme Class	Application
Detergent (laundry and dish washing)	Protease	Protein stain removal
	Amylase	Starch stain removal
	Lipase	Lipid stain removal
	Cellulase	Cleaning, color clarification, antiredeposition (cotton)
	Mannanase	Manannan stain removal (reappearing stains)
Starch and fuel	Amylase	Starch liquefaction and saccharification
	Amyloglucosidase	Saccharification
	Pullulanase	Saccharification
	Glucose isomerase	Glucose to fructose conversion
	Cyclodextrin glycosyltransferase	Cyclodextrin production
	Xylanase	Viscosity reduction (fuel and starch)
	Protease	Protease (yeast nutrition, fuel)
Food (including dairy)	Protease	Milk clotting, infant formulas (low allergenic), flavor
	Lipase	Cheese flavor
	Lactase	Lactose removal (milk)
	Pectin methyl esterase	Firming fruit-based products
	Pectinase	Fruit-based products
	Transglutaminase	Modify viscoelastic properties
Baking	Amylase	Bread softness and volume, flour adjustment
	Xylanase	Dough conditioning
	Lipase	Dough stability and conditioning (in situ emulsifier)
	Phospholipase	Dough stability and conditioning (in situ emulsifier)
	Glucose oxidase	Dough strengthening
	Lipoxygenase	Dough strengthening, bread whitening
	Protease	Biscuits, cookies
	Transglutaminase	Laminated dough strengths
Animal feed	Phytase	Phytate digestibility, phosphorus release
	Xylanase	Digestibility
	β-Glucanase	Digestibility
Beverage	Pectinase	Depectinization, mashing
	Amylase	Juice treatment, low-calorie beer
	β-Glucanase	Mashing
	Acetolactate decarboxylase	Maturation (beer)
	Laccase	Clarification (juice), flavor (beer), cork stopper treatment
Textile	Cellulase	Denim finishing, cotton softening
	Amylase	Desizing
	Pectate lyase	Scouring

continued

Table 10.1 Industrial Enzymes and Applications (Continued)

Industry	Enzyme Class	Application
	Catalase	Bleach termination
	Laccase	Bleaching
	Peroxidase	Excess dye removal
Pulp and paper	Lipase	Pitch control, contaminant control
	Protease	Biofilm removal
	Amylase	Starch-coating, deinking, drainage improvement
	Xylanase	Bleach boosting
	Cellulase	Deinking, drainage improvement, fiber modification
Fats and oils	Lipase	Transesterification
	Phospholipase	Degumming, lysolecithin production
Organic synthesis	Lipase	Resolution of chiral alcohols and amides
	Acylase	Synthesis of semisynthetic penicillin
	Nitrilase	Synthesis of enantiopure carboxylic acids
Leather	Protease	Unhearing, bating
	Lipase	Depickling
Personal care	Amyloglucosidase	Antimicrobial (combined with glucose oxidase)
	Glucose oxidase	Bleaching, antimicrobial
	Peroxidase	Antimicrobial
Contact lens cleaners	Proteases	Removal of proteins on contact lenses to prevent infections
Rubber industry	Catalase	Generation of oxygen from peroxide to convert latex into foam rubber
Photographic industry	Protease (ficin)	Dissolving gelatin off scrap film, allowing recovery of its silver content

Source: Data from Kirk, O., T.V. Borchert, and C.C. Fuglsang (2002). Industrial enzyme applications. *Curr Opin Biotechnol* 13:345–351.

their structure. These in turn enable protein stabilization and assist the intermolecular and intramolecular bond formation that occurs during enzymatic reactions. Although metalloenzymes have assumed vital significance in modern-day biochemistry, it is interesting to note that the precursors of these metalloenzymes played a critical role in evolution. Wächtershäuser was the first to postulate that surface-bound iron, cobalt, nickel, and other transition metal centers with sulphido, carbonyl, and other ligands were catalysts involved in the synthesis of macromolecular compounds mediated by the process of carbon fixation, which in turn promoted the growth of organic "superstructures" (2). Subsequently, the prebiotic metal centers became parts of peptides and, finally, defined proteins (2, 3). Considering the fact that approximately half of all proteins contain a metal ion and one quarter to one third of all proteins functions are dependent on metals, metalloenzmyes are a diverse and populous group. This diversity makes it very difficult to classify them into different groups, unlike conventional proteins that can be classified based on structural and sequence similarities.

One of the most well accepted grouping methodologies for metalloenzymes relies on bioinorganic motifs, which bases the classification on the type of metal atom involved and the ligands associated with it (**Table 10.2**) (4, 5). Based on the number of bound metal atoms, metalloenzymes can be mononuclear (containing a single metal atom) or polynuclear (containing multiple metal atoms). Additionally, ligands associated with the metal ions are classified as endogenous (polypeptides within the protein) or exogenous (nonprotein component of a conjugated protein).

10.3 Biologically relevant metalloenzymes

One of the unique features of a metalloenzyme is the fact that the metal ion associated with it is located predominantly on or near the active site of the enzyme. This unique positioning allows the metalloenzyme to catalyze biochemical reactions, which are otherwise difficult to regulate by conventional organic chemistry. Based on the type of metal ions associated, metalloenzymes can catalyze a number of biologically critical reactions, as mentioned in **Table 10.2** (6). Some reactions are explained in the following sections to further elucidate their significance in biological systems.

10.3.1 Chlorophyll

Chlorophyll is a vital component of photosynthesis and is composed of magnesium enclosed in a chlorine ring. Discovered in the early 1900s, this was the first time that this element had been detected in living tissue (7). However, the magnesium ion is not vital for photosynthetic function and is replacable by other divalent ions. The structures of different types of chlorophyll are shown in **Figure 10.2**.

10.3.2 Superoxide dismutase

The superoxide dismutase (SOD) enzymes catalyze the dismutation of superoxide into oxygen and hydrogen peroxide (**Figure 10.3**) (8). The superoxide ion O^{2-} is a byproduct of the reduction of molecular oxygen in living organisms. Because it is a free radical, it is a powerful oxidizing agent, which makes it extremely toxic. Human SOD is composed of copper as Cu^{2+} or Cu^+, coordinated tetrahedrally by four histidine residues. It is also comprised of zinc ions that assist in stabilizing the structure, which is activated by a copper chaperone. SOD isozymes may also be composed of iron, manganese, or nickel.

Table 10.2 Function and Structures of Metalloenzymes

Type	Structure	Function	Examples
Mononuclear iron proteins	Amino acid ligands are mostly histidine and/or cysteine (nitrile hydratase) with three cysteine residues and two amide nitrogens.	Electron transfer and oxygen reduction	Aromatic amino acid hydroxylases, aromatic ring cleavage dioxygenases, Fe-superoxide dismutase, lipoxygenases, nitrile hydratase, and Rieske oxygenases
Diiron carboxylate proteins	The enzymes bind a coupled binuclear iron center using four carboxylate and two histidine residues.	Oxygen reduction, storage and transportation, iron storage	Alkene hydroxylase, methane monooxygenase, and phenol hydroxylase; bacterioferritin, ferritin, hemerythrin, myohemerythrin
Hemoproteins	Contain iron porphyrin as prosthetic group; four of the iron ligands are occupied by the porphyrin ring, whereas the other two are used for binding of the protein	Oxygen reduction, electron transfer, iron storage and transport	Catalases, peroxidases, cytochromes, globins
Iron-sulfur proteins	Iron-sulfur clusters containing sulfide-linked di-, tri-, and tetraironcenters are present in variable oxidation states.	Play key role in nearly all respiratory chains (either in bacteria and archaea or in mitochondria of eukaryotic organisms); help in catalysis of stereo-specific isomerization of citrate to isocitrate in the tricarboxylic acid cycle	Aconitase
Type I copper centers	Contain a single copper atom coordinated by two histidine residues and a cysteine residue in a trigonal planar structure and a variable axial amino acid ligand	Oxygen transport or activation processes and electron transport	Ascorbate oxidase, laccase, nitrite reductase, auracyanin, and azurin
Type II copper centers	They are square planar coordinated by N or N/O ligands.	Oxygen transport or activation processes and electron transport	Galactose oxidase
Type III centers	Consist of two copper atoms, each coordinated by three histidine residues	Oxygen transport or activation processes and electron transport	Catechol oxidase, hemocyanins, tyrosinase
Zinc proteins	Zinc ions in catalytic sites are generally coordinated to the side chains of three amino acid residues and a water molecule.	Zinc is either directly involved in forming or breaking chemical bonds, or may act in catalytic reaction with other	Alcohol dehydrogenase (containing both a catalytic and a structural zinc site), metalloproteinases, and carbonic anhydrases with

Table 10.2 Function and Structures of Metalloenzymes (Continued)

Type	Structure	Function	Examples
	In structural zinc sites, the ion is coordinated by four amino acid side chains in a tetrahedral symmetry.	metal ions, or maintains structural integrity of enzymes	catalytic zinc sites; cocatalytic zinc sites are found in superoxide dismutases, phosphatases, amino peptidases, and β-lactamases
Nickel in metalloenzymes	Occurs as Ni-4Fe-5S; coordinated in a tetrapyrrole complex and as dinuclear NiNi or NiFe complexes or a [4Fe-4S]2+ cluster bridged by a cysteine thiolate to a dinuclear NiNi center	Helps activate certain enzymes related to the breakdown or utilization of glucose in the human body	Acetyl-CoA synthase
Manganese in metalloenzymes	Manganese metal centers may be mono-, bi-, tri-, or tetranuclear. Mn-superoxide dismutases exhibit mononuclear centers and the active site manganese is five-coordinate, with the metal ligands arranged in distorted trigonalbipyramidal geometry. Arginase is an enzyme with binuclear centers. The manganese ions in arginase are five- and six-coordinate, with square-pyramidal and (distorted) octahedral coordination symmetries, respectively. The metal centers are interconnected by a solvent-derived hydroxide and carboxylate residues. This metal-bridging hydroxide acts as a nucleophile that attacks the arginine guanidinium carbon.	Acts as a simple Lewis acid catalyst. Mn-superoxide dismutases (SODs) act as antioxidant. SOD decomposes highly reactive superoxide anion (O^{2-}) by a two-step reaction which generates oxygen hydrogen peroxide. Arginase catalyzes the conversion of arginine to urea and ornithine.	Mn-superoxide dismutases, arginase
Molybdenum	Molybdenum is associated with the heterocyclic pterin derivative (molybdopterin) that contains a mononucleate center, coordinated to the thiols of the cofactor except nitrogenase cofactor.	Sulfite oxidase prompts the metabolism of two sulfur-containing amino acids, which are essential to building proteins. Xanthine oxidase prompts nucleotides to break down into uric acid, which supports the antioxidant components of blood plasma.	Sulfite oxidase, xanthine oxidase, and aldehyde oxidase

continued

Table 10.2 Function and Structures of Metalloenzymes (Continued)

Type	Structure	Function	Examples
		Along with aldehyde oxidase, xanthine oxidase also supports the metabolism of drugs and toxins.	
Cobalt	Cobalt is the central metal ion in the tertrapyrrol corrin ring in the coenzyme B12. It is similar to porphyrin rings, with the corrin ring providing four coordination sites.	In the body, cobalt is essential for red blood cell formation. The B12-bearing enzymes catalyze either the transfer of methyl groups between two molecules or isomerization reactions, that is, rearrangements of hydrogen atoms with concomitant exchange of a substituent (e.g., a hydroxyl group) between two adjacent carbon atoms in a molecule. In noncorrin cobalt enzymes, cobalt is attached with amino acids.	Coenzyme B12, noncorrin cobalt enzymes (methionine aminopeptidase 2)
Sodium and potassium	Structures of Na+ and K+ complexes with proteins have a variety of coordination schemes and functions.	Although Na+ and K+ ions are not directly involved in catalytic steps, they activate numerous enzymes.	Na+-activated β-galactosidase or the K+-dependent activity of pyruvate-kinase and glycerol dehydratases
Magnesium	In chlorophyll molecules of photosynthetic reaction centers, magnesium ions are coordinated in a terrapyrrole ring system and to an axial N-histidin, O-aspartate,O-formyl-methionin, O-leucin, or water as ligands.	Mainly involved in production of energy and cardiovascular function; activation and stabilization of enzymes. Enzymes catalyzing phosphorylation of proteins typically use magnesium chelates of adenosine triphosphate (ATP) as co-substrate. The bound Mg^{2+} serves to facilitate nucleophilic attack at the γ-phosphate of the ATP substrate. Phosphate and phosphoryl transfer reactions require Mg^{2+} as an essential cofactor.	Isocitratelyase and glutamine synthetase

Source: Data from Reitner, J., and V. Thiel (2011). *Encyclopedia of Geobiology.* New York, NY: Springer.

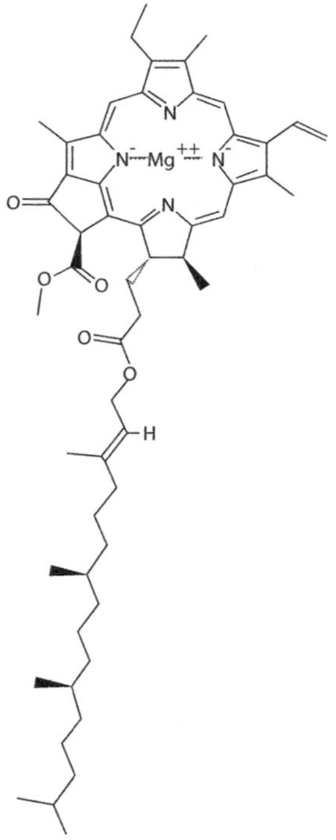

Figure 10.2 *Structure of chlorophyll. (From http://en.wikipedia.org/wiki/Chlorophyll.)*

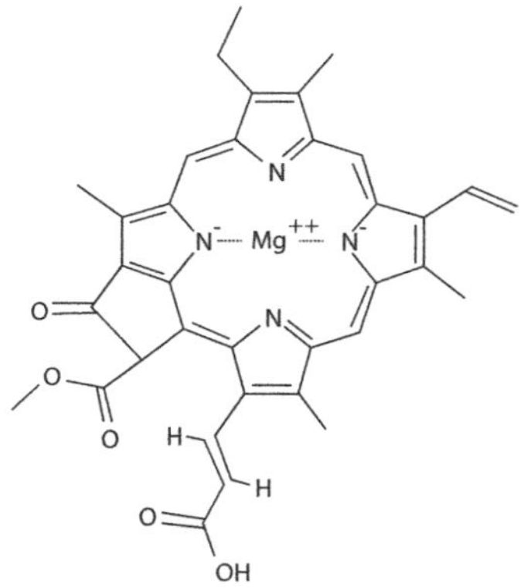

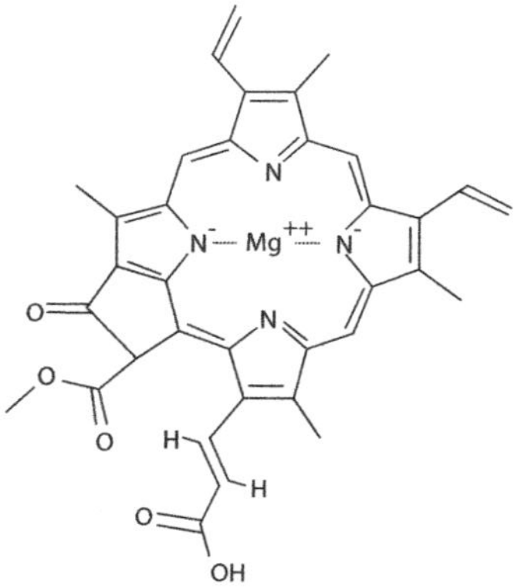

Chlorophyll c1

Chlorophyll c2

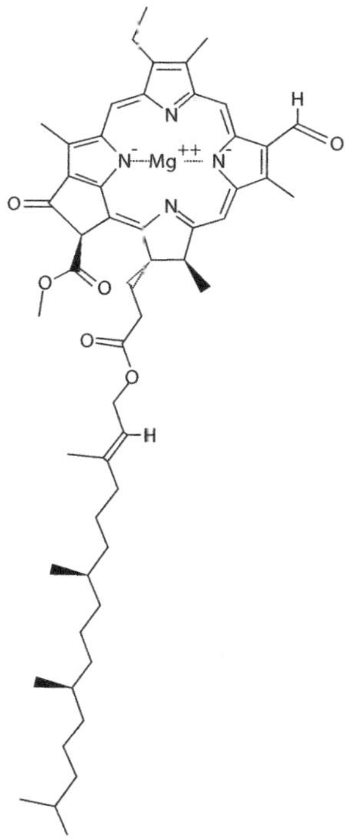

Chlorophyll d

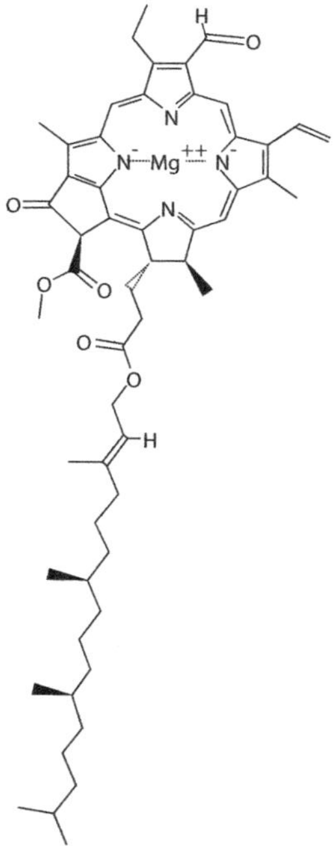

Chlorophyll b

Chlorophyll a

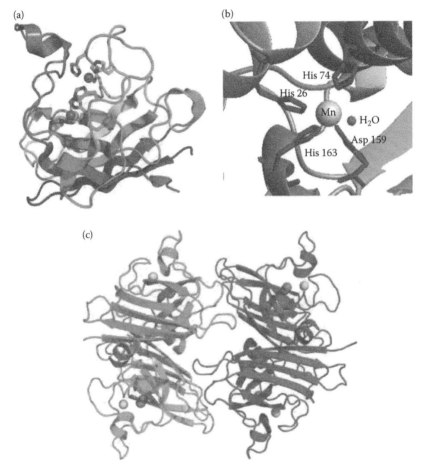

(a)

(b)

His 74

His 26

Mn

H$_2$O

Asp 159

His 163

(c)

Figure 10.3 (See color insert.) *Structure of superoxide dismutases (SODs). (a) SOD1 (soluble). (b) SOD2 (mitochondrial). (c) SOD3 (extracellular).*

10.3.3 Nitrogenase

Nitrogen fixation is one of the most important biological processes in the ecosystem. This energy-intensive reaction is catalyzed by the enzyme nitrogenase, which breaks the highly stable triple bond between the nitrogen atoms. The enzyme is of bacterial origin and is made up of a molybdenum atom at the active site. Iron-sulfur clusters are also present and are involved in transporting the electrons needed to reduce the nitrogen and an abundant energy source. Although the precise structure of the enzyme is yet to be elucidated, it has been postulated to be made up of a MoFe^{7}S^8 cluster, which is able to bind the dinitrogen molecule (**Figure 10.4**) and, presumably, enable the reduction process to begin (9). The electrons are transported by the associated P-cluster, which contains two cubical Fe^{4}S^4 clusters joined by sulfur bridges (10).

10.3.4 Alcohol dehydrogenase

In humans, alcohol dehydrogenase breaks down alcohols that are toxic and participates in the generation of useful aldehyde, ketone, or alcohol groups. Alcohol dehydrogenases are made up of a zinc site that plays a structural role and is crucial for protein stability. The structural zinc site is composed of cysteine ligands (Cys97, Cys100, Cys103, and Cys111 in the amino acid sequence) that are arranged in a quasi-tetrahedron around the Zn ion (**Figure 10.5**). It has been shown that the interaction between zinc and cysteine is governed by primarily an electrostatic contribution, with an additional covalent contribution to the binding (11).

10.3.5 Cytochromes

Redox reactions are vital for biological systems and cytochromes play an important role in these reactions. Cytochromes are made up of iron(II), which is easily oxidized to iron(III) and hence can allow electron transfer. The presence of an iron ion confers an extended versatility to the metal-

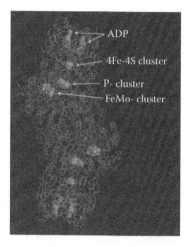

Figure 10.4 Structure of nitrogenase. (From http://www.rcsb.org/pdb/101/ motm.do?momID=26.)

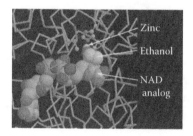

Figure 10.5 Structure of alcohol dehydrogenase. (From http://www.rcsb. org/pdb/101/motm.do?momID=13.)

loenzymes by participating in redox reactions, unlike other enzymes that have a limited set of functional groups found in amino acids (12). The iron atom is mostly contained in the heme group and participates in the mitochondrial electron transport chain (13). **Figure 10.6** shows the structure of the cytochrome.

10.3.6 Transcription factors

Transcription factors are proteins that control which genes are turned on or off in the genome by binding to DNA and other proteins. Naturally, they are critical for gene regulation. Most transcription factors are made up of a structure called the zinc finger, which is a structural module where a region of protein folds around a zinc ion (**Figure 10.7**). The zinc associates with the DNA that the proteins are bound to. However, the cofactor is essential for the stability of the tightly-folded protein chain (14).

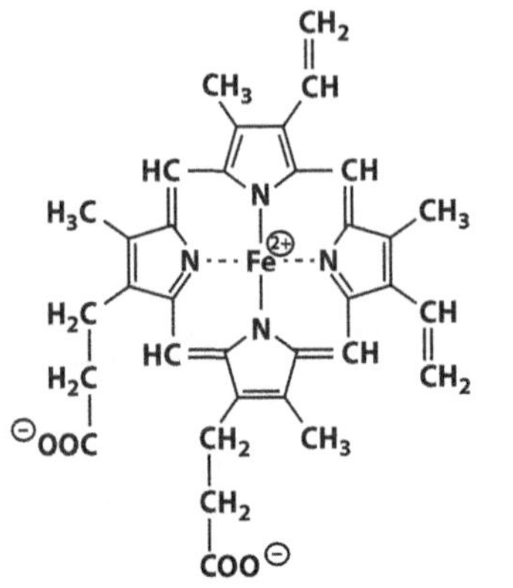

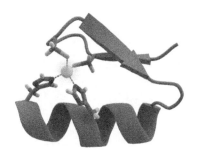

Figure 10.6 *Cytochrome structure.* *(From http://sandwalk.blogspot.com/2007/08/heme-groups.html.)*

Figure 10.7 *Structure of zinc finger.* *(From http://www.scistyle.com.)*

10.4 Metalloenzymes and their association with human diseases

Carbonic anhydrase II was the first zinc metalloenzyme to be discovered in 1940. Since then, upward of 300 enzymes containing zinc spread across six classes of enzymes from multiple species have been discovered (15–17). In recent years, matrix metalloproteinases (MMPs), which are zinc-dependent endopeptidases, have been identified as critical players in a number of human diseases. MMPs belong to a larger family of proteases known as the metzincin superfamily (18), which is distinguished by a highly conserved motif containing three histidines that bind to zinc at the catalytic site and a conserved methionine that sits beneath the active site. The MMP family includes more than 20 related zinc-dependent enzymes, which were initially characterized by their unique ability to degrade extracellular matrix proteins, including collagen, fibronectin, peptidoglycans, and vimentin. The MMPs have a common domain structure composed of the propeptide, the catalytic domain, and the hemopexin-like C-terminal domain,

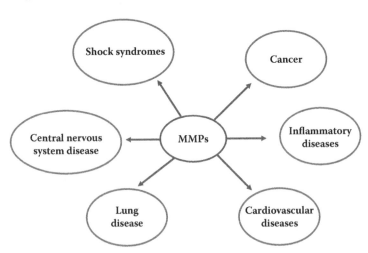

Figure 10.8 *Matrix metalloproteinases and associated diseases.*

which are linked to the catalytic domain by a flexible hinge region. MMPs are classified on the basis of substrate specificity of the MMP and their cellular localization (**Table 10.3**). **Table 10.4** provides a list of human MMPs, their protein lengths, and corresponding Universal Protein Resource identification numbers for reference.

Table 10.3 Classification of Matrix Metalloproteinases (MMPs) Based on Specificity and Localization

Groups	Matrix Metalloproteinases
Collagenases	MMP1
	MMP8
	MMP13
Matrilysin	MMP7
	MMP26
Metalloelastase	MMP12
Gelatinases	MMP2
	MMP9
Stromelysins	MMP3
	MMP10
	MMP11
Enamelysin	MMP20
Others	MMP19
	MMP21
	MMP23A
	MMP23B
	MMP27
	MMP28
Membrane-type	MMP14
	MMP15
	MMP16
	MMP17
	MMP24
	MMP25
Potential inducers of transcription	BCG
	TCF20
	TNF
Inhibitors	TIMP1
	TIMP2
	TIMP3
	TIMP4

BCG, Bacille Calmette-Guérin; TCF, ternary complex factor; TNF, tumor necrosis factor; TIMP, metalloproteinase inhibitor.

Table 10.4 Universal Protein Resource (UniProt) Accession Identifications (IDs) and Protein Lengths for Matrix Metalloproteinases (MMPs)

Gene Names	UniProt ID	Protein Length (Amino Acids)
MMP10, STMY2	P09238	476
MMP11, STMY3	P24347	488
MMP12, HME	P39900	470
MMP13	P45452	471
MMP14	P50281	582
MMP15	P51511	669
MMP16, MMPX2	P51512	607
MMP17, MT4MMP	Q9ULZ9	603
MMP19, MMP18, RASI	Q99542	508
MMP1, CLG	P03956	469
MMP20	O60882	483
MMP21	Q8N119	569
MMP23A, MMP21; MMP23B, MMP21, MMP22	O75900	390
MMP24, MT5MMP	Q9Y5R2	645
MMP25, MMP20, MMPL1, MT6MMP	Q9NPA2	562
MMP26	Q9NRE1	261
MMP27, UNQ2503/PRO5992	Q9H306	513
MMP28, MMP25, UNQ1893/PRO4339	Q9H239	520
MMP2, CLG4A	P08253	660
MMP3, STMY1	P08254	477
MMP7, MPSL1, PUMP1	P09237	267
MMP8, CLG1	P22894	467
MMP9, CLG4B	P14780	707
TCF20, KIAA0292, SPBP	Q9UGU0	1960
TIMP1, CLGI, TIMP	P01033	207
TIMP2	P16035	220
TIMP3	P35625	211
TIMP4	Q99727	224
TNF, TNFA, TNFSF2	P01375	233
MMP28, hCG_1989249	C0H5X0	393
MMP11, hCG_41094	B3KQS8	354
TIMP1, hCG_29141	Q6FGX5	207
TNF, TNFA, hCG_43716	Q5STB3	233
LOC100653515	Q96MC4	621
TNFa	Q9UBM5	15
	Q7KZY0	564
	Q13165	35

Table 10.4 Universal Protein Resource (UniProt) Accession Identifications (IDs) and Protein Lengths for Matrix Metalloproteinases (MMPs) (Continued)

Gene Names	UniProt ID	Protein Length (Amino Acids)
	Q53G96	469
	Q53GF1	267
	Q2PW48	593
	Q53H33	471
MMP13	Q6NWN6	471
	Q7Z5M1	489
	B3KV06	510
	Q6LBE5	121
	Q53G95	338
MMP24	Q86VV6	626
	Q9UHK5	36
	Q53G75	405
	Q99745	240
TIMP1	Q5H9A7	143
MMP1	Q96DZ4	232
	Q5TZP0	469
	Q9Y354	79
	B4DN15	403
TNF	C1K3N5	232
	Q53G97	390
MMP17	Q5U5M0	606
	Q53FK3	469
	Q7Z5M0	383
MMP17	Q8IWC3	606
	Q2EF79	61
	Q71RC1	249
TNFA	Q5RT83	18
	Q6LEP5	11
K222	Q14867	138
	Q8IXG1	20
TIMP1	Q58P21	188
	Q96PI0	13

Because MMPs play a critical role in tissue remodeling, it is hardly surprising that they are associated with various physiological and pathological processes, such as morphogenesis, angiogenesis, tissue repair, and metastasis, to name a few. The following sections discuss the major diseases in which MMPs have been implicated (**Figure 10.8**).

10.4.1 MMPs and cancer

MMPs play a very important role in cancer (19). The migration of tumor cells to blood vessels has been linked to MMP-2 and MMP-9. In addition, MMPs are directly implicated in the stimulation of angiogenesis that promotes tumor progression (20). In addition to this, MMP-9 also performs an important function of releasing fibroblast growth factor and vascular endothelial growth factor, which facilitate the growth of tumors. In humans, elevated MMP levels have been shown to directly contribute to progression of cancer (21), in addition to being identified in dysfunctional apoptosis (22) and altered immune-mediated tumor killing (23).

10.4.2 MMPs in cardiovascular diseases

The normal vascular and heart development depends on continuous alterations in cell–cell adhesion, cell migration, cell proliferation, apoptosis, and remodeling. These are in turn governed by changes in MMP gene expression levels and concomitant activation of pro-MMP factors, which have been rather well documented in the past (24). In addition, aberrant MMP catalytic activity is also associated with atherosclerotic plaque formation and plaque instability (25), vascular smooth muscle cell migration and restenosis (26), development of aortic aneurysm (27), and progressive heart failure in both animal models (28) and humans (29).

10.4.3 MMPs in diseases of the central nervous system

MMPs play a vital role in mediating the disruption of the blood–brain barrier, which is critical for various biological processes. In addition to this, they also regulate extracellular matrix protein destruction and remodeling in response to oxidative stress (30). MMP-8 and MMP-9 are also thought to mediate central nervous tissue injury in bacterial infections, including meningitis (31), multiple sclerosis (32), Alzheimer disease (30), inflammatory myopathies (33), and tumors of the central nervous system, such as gliomas (34).

10.5 Potential applications of metalloenzymes

The advent of biotechnology and advances in genetic engineering have made it possible to design and engineer highly effective enzymes with improved target specificity and efficacy. It is in this regard that artificial metalloenzymes have emerged as promising targets because they offer an attractive possibility to combine properties of homogeneous catalysis and biocatalysis. According to recent estimates, close to 10% of currently approved drugs target one or more metalloenzymes (35), some of which include angiotensin-converting enzyme, carbonic anhydrase, and matrix metalloprotesases (36–39). These drugs are mostly inhibitors that can target the catalytic metal ion group.

Table 10.5 Metalloenzymes Used in Skin Antiaging Formulations

Metalloenzyme	Function	Metal
Tyrosinase	Tyrosine oxidation	Cu, Zn
Superoxide dismutase	Superoxide detoxification	Cu, Zn, Mn
Catalase	Peroxide decomposition	Fe
Matrix metalloproteinases	Protein hydrolysis	Zn
Dopamine hydroxylase	Dopamine conversion	Cu
Amine oxidase	Elastin, collagen synthesis	Cu
Cytochrome oxidase	Oxidation	Cu
Ceruloplasmin	Oxidation	Cu
Glutathione peroxidase	Peroxide detoxification	Se
Glucose tolerance factor	Glucose metabolism	Cr

Source: Data from Gupta, S. (2007). Ayurvedic anti-aging: Ancient medicinal practices have many applications in modern skin care formulations. *Household & Personal Products Industry,* August 1.

MMPs in particular were considered to be an attractive target by the pharmaceutical industry, albeit with a mixed degree of success. Although most drugs have performed satisfactorily in early clinical trials, their applications to humans have been marred with toxicity and off-target effects. Marimastat (BB-2516), a broad-spectrum MMP inhibitor, and cipemastat (Ro 32-3555), an MMP-1 selective inhibitor, are a couple of the drugs that have performed poorly in clinical trials. However, the development of more specific inhibitors and a broader understanding of the basic role of MMPs in pathophysiology will allow development of more effective therapies centered on MMPs.

The importance of trace metals in modern medicine has been recently discovered; however, fields of traditional medicine have long used and applied these principles for the treatment of various diseases. Ayurveda is a 5,000-year-old tradition of medicine that is widely practiced in India and is gaining wider acceptance across the world. For instance, the concept of using zinc in medical treatment has been prevalent in Ayurveda since the eighth century. Formulations that include metalloenzymes have been used for ailments such as diabetes, obesity, and arthritis (40). More recently, the benefits of using formulations with metalloenzymes in skin antiaging formulations have received considerable attention. **Table 10.5** lists the different metalloenzymes that play a crucial role in the human body.

References

1. Kirk, O., T.V. Borchert, and C.C. Fuglsang. 2002. Industrial enzyme applications. *Curr Opin Biotechnol* 13:345–351.
2. Wachtershauser, G. 1990. Evolution of the first metabolic cycles. *Proc Natl Acad Sci U S A* 87:200–204.

3. Russell, M.J., and W. Martin. 2004. The rocky roots of the acetyl-CoA pathway. *Trends Biochem Sci* 29:358–363.
4. Degtyarenko, K. 2000. Bioinorganic motifs: Towards functional classification of metalloproteins. *Bioinformatics* 16:851–864.
5. Degtyarenko, K., and S. Contrino, 2004. COMe: The ontology of bioinorganic proteins. BMC Struct Biol 4:3.
6. Reitner, J., and V. Thiel. 2011. *Encyclopedia of Geobiology.* New York, NY: Springer.
7. Fleming, I. 1967. Absolute configuration and the structure of chlorophyll. *Nature* 216:151–152.
8. Flohe, L., and F. Otting. 1984. Superoxide dismutase assays. *Methods Enzymol* 105:93–104.
9. Kim, J., and D.C. Rees. 1994. Nitrogenase and biological nitrogen fixation. *Biochemistry* 33:389–397.
10. Chan, M.K., J. Kim, and D.C. Rees. 1993. The nitrogenase FeMo-cofactor and P-cluster pair: 2.2 A resolution structures. *Science* 260:792–794.
11. Brandt, E.G., M. Hellgren, T. Brinck, T. Bergman, and O. Edholm. 2009. Molecular dynamics study of zinc binding to cysteines in a peptide mimic of the alcohol dehydrogenase structural zinc site. *Phys Chem Chem Phys* 11:975–983.
12. Messerschmidt, A., R. Huber, T. Poulos, and K. Wieghardt, eds. 2001. *Handbook of Metalloproteins.* Chichester, UK: John Wiley & Sons.
13. Moore, G.R., G.W. Pettigrew, and N.K. Rogers. 1986. Factors influencing redox potentials of electron transfer proteins. *Proc Natl Acad Sci U S A* 83:4998–4999.
14. Berg, J.M. 1990. Zinc finger domains: Hypotheses and current knowledge. *Annu Rev Biophys Biophys Chem* 19:405–421.
15. Christianson, D.W. 1991. Structural biology of zinc. *Adv Protein Chem* 42:281–355.
16. Vallee, B.L., J.E. Coleman, and D.S. Auld. 1991. Zinc fingers, zinc clusters, and zinc twists in DNA-binding protein domains. *Proc Natl Acad Sci U S A* 88:999–1003.
17. Coleman, J.E. 1992. Zinc proteins: Enzymes, storage proteins, transcription factors, and replication proteins. *Annu Rev Biochem* 61:897–946.
18. Brinckerhoff, C.E., and L.M. Matrisian. 2002. Matrix metalloproteinases: A tail of a frog that became a prince. *Nat Rev Mol Cell Biol* 3:207–214.
19. Coussens, L.M., B. Fingleton, and L.M. Matrisian. 2002. Matrix metalloproteinase inhibitors and cancer: trials and tribulations. *Science* 295:2387–2392.
20. Folkman, J. 1996. Endogenous inhibitors of angiogenesis. *Harvey Lect* 92:65–82.
21. Fingleton, B. 2003. Matrix metalloproteinase inhibitors for cancer therapy: The current situation and future prospects. *Expert Opin Ther Targets* 7:385–397.
22. Hanahan, D., and R.A. Weinberg. 2000. The hallmarks of cancer. *Cell* 100:57–70.
23. Sheu, B.C., S.M. Hsu, H.N. Ho, H.C. Lien, S.C. Huang, and R.H. Lin. 2001. A novel role of metalloproteinase in cancer-mediated immunosuppression. *Cancer Res* 61: 237–242.
24. Robbins, J.R., P.G. McGuire, B. Wehrle-Haller, S.L. Rogers. 1999. Diminished matrix metalloproteinase 2 (MMP-2) in ectomesenchyme-derived tissues of the Patch mutant mouse: Regulation of MMP-2 by PDGF and effects on mesenchymal cell migration. *Dev Biol* 212:255–263.
25. Galis, Z.S., G.K. Sukhova, M.W. Lark, and P. Libby. 1994. Increased expression of matrix metalloproteinases and matrix degrading activity in vulnerable regions of human atherosclerotic plaques. *J Clin Invest* 94:2493–2503.
26. Lovdahl, C., J. Thyberg, B. Cercek, et al. 1999. Antisense oligonucleotides to stromelysin mRNA inhibit injury-induced proliferation of arterial smooth muscle cells. *Histol Histopathol* 14:1101–1112.
27. Longo, G.M., W. Xiong, T.C. Greiner, Y. Zhao, N. Fiotti, and B.T. Baxter. 2002. Matrix metalloproteinases 2 and 9 work in concert to produce aortic aneurysms. *J Clin Invest* 110:625–632.
28. Danielsen, C.C., H. Wiggers, and H.R. Andersen. 1998. Increased amounts of collagenase and gelatinase in porcine myocardium following ischemia and reperfusion. *J Mol Cell Cardiol* 30:1431–1442.

29. Yamazaki, T., J.D. Lee, H. Shimizu, H. Uzui, and T. Ueda. 2004. Circulating matrix metalloproteinase-2 is elevated in patients with congestive heart failure. *Eur J Heart Fail* 6:41–45.
30. Yong, V.W., C. Power, P. Forsyth, and D.R. Edwards. 2001. Metalloproteinases in biology and pathology of the nervous system. *Nat Rev Neurosci* 2:502–511.
31. Leppert, D., S.L. Leib, C. Grygar, K.M. Miller, U.B. Schaad, and G.A. Holländer. 2000. Matrix metalloproteinase (MMP)-8 and MMP-9 in cerebrospinal fluid during bacterial meningitis: Association with blood-brain barrier damage and neurological sequelae. *Clin Infect Dis* 31:80–84.
32. Ozenci, V., M. Kouwenhoven, N. Teleshova, M. Pashenkov, S. Fredrikson, and H. Link. 2000. Multiple sclerosis: pro- and anti-inflammatory cytokines and metalloproteinases are affected differentially by treatment with IFN-beta. *J Neuroimmunol* 108:236–243.
33. Yong, V.W., C.A. Krekoski, P.A. Forsyth, R. Bell, and D.R. Edwards. 1998. Matrix metalloproteinases and diseases of the CNS. *Trends Neurosci* 21:75–80.
34. Nakada, M., Y. Okada, and J. Yamashita. 2003. The role of matrix metalloproteinases in glioma invasion. *Front Biosci* 8:e261–e269.
35. Morrison, C. 2009. *The In Vivo Blog. Financings of the Fortnight: Summer Madness.* Available from http://invivoblog.blogspot.com/2009/07/financings-of-fortnight-summer-madness.html.
36. Supuran, C.T. 2010. Carbonic anhydrase inhibitors. *Bioorg Med Chem Lett* 20:3467–3474.
37. Lia, N.-G., Z.-H. Shi, Y.-P. Tang, and J.-A. Duan. 2009. Selective matrix metalloproteinase inhibitors for cancer. *Curr Med Chem* 16:3805–3827.
38. Drag, M., and G.S. Salvesen. 2010. Emerging principles in protease-based drug discovery. *Nat Rev Drug Discov* 9:690–701.
39. Anthony, C.S., G. Masuyer, E.D. Sturrock, and K.R. Acharya. 2012. Structure based drug design of angiotensin-I converting enzyme inhibitors. *Curr Med Chem* 19:845–855.
40. Mishra, L.C. 2004. *Scientific Basis for Ayurvedic Therapies.* Boca Raton, FL: CRC Press.

Application of Nanosilver in Nutraceuticals

Pranab Jyoti Das, Bhaskar Mazumder, and Charles Preuss

Contents

11.1 Introduction

The term *nutraceuticals* is derived from two words: nutrition and pharmaceuticals. It was first coined by Dr. Stephen DeFelice, the founder of the Foundation for Innovation in Medicine (Crawford, New Jersey, USA) in 1989. He described nutraceuticals as "food, or parts of food, that provide medical or health benefits, including the prevention and treatment of disease" (El Sohaimy 2012). Generally, nutraceuticals are foodstuffs that provide health benefits in addition to their basic nutritional value.

The defense mechanism of humans is based on the immune response, which is composed of various cells to form a highly organized, complex, and prompt system that performs molecular disintegration of any invading organism. It needs a constant supply of nutrients or phytochemicals through diet for its survival and functioning. The immune suppression caused by the inflammatory response against foreign attack is an integral act of immune function. Nutraceutical therapy or the supply of active components through diet is essential for recovery from suppression. Today, nutraceutical foods are rich in bioactive and immunomodulating components to govern human health.

In recent decades, the use of different kinds of nutraceuticals has increased worldwide, including dietary supplements and herbal preparations such as vitamins, minerals, ginseng, and ginkgo biloba; functional foods such as oats, bran, prebiotics, omega-3, and milk; and medicinal foods such as transgenic plants and health bars with added medications.

The use of nanometals (e.g., gold, silver, copper, palladium, platinum, molybdenum, titanium, and zinc) as a component of nutraceuticals has been increasing remarkably due to their ability to act as reducing agents, bind proteins and denature enzymes, and act as bactericides in topical formulation (Fung 1996). Silver has been used for thousands of years as a healing and preventive health product. Treating wounds with silver was common from the 1800s to the mid-1900s, when the use of antibiotics took precedence in the medical field. Silver has a long and intriguing history as an antibiotic in human health care. It was developed for use in wound care (Hudspith et al. 2004), bone prostheses (Chu et al. 2002), reconstructive orthopedic surgery (Hardes et al. 2007), cardiac devices (Barde et al. 2006), catheters and surgical appliances (Brosnahan et al. 2004), and antibiotic textiles (Hipler et al. 2006), among others.

The demand for silver nanoparticles is due to its multiple uses in the treatment of traumatic or diabetic wounds, burns, tonsillitis, syphilis (Atiyeh et al. 2007), and chronic infected wounds (Lo et al. 2008), as well as its antiviral (Mehrbod et al. 2009) and antifungal (Kim et al. 2009) properties. Today, silver nanoparticles are used in consumer products and nutraceuticals more than any other metals because of their ability to fight against bacterial growth. Colloidal silver is also consumed as a dietary supplement.

11.2 Formulation of silver nanoparticles

The nanotechnology field is one of the most attractive areas of research in modern science. The advancement of biotechnology has enabled the incorporation of ionizable silver into fabrics for clinical use to reduce the risk of nosocomial infections and for personal hygiene. The antimicrobial action of silver or silver compounds is proportional to the bioactive silver ion (Ag^+) released and its availability to interact with bacterial or fungal cell membranes. Silver metal and inorganic silver compounds ionize in the presence of water, body fluids, or tissue exudates. Silver exhibits low toxicity in humans; when absorbed in the human body, it enters the systemic circulation as a protein complex to be eliminated by the liver and kidneys. Silver metabolism is modulated by induction and binding to metallothioneins. This complex mitigates the cellular toxicity of silver and contributes to tissue repair. Although various approaches are available for the synthesis of silver nanoparticles, there is still need for an economical, commercially viable, and environmentally clean synthesis route to synthesize silver nanoparticles. The various approaches used for the preparation of silver nanoparticles are shown schematically in **Figure 11.1.**

11.2.1 Physical approaches

11.2.1.1 Thermal decomposition method

Crystalline silver nanoparticles can be synthesized by the thermal decomposition of silver oxalate in water and ethylene glycol. Polyvinyl alcohol (PVA) may be employed as a capping agent. Silver oxalate ($Ag_2C_2O_4$) decomposes at approximately 140°C and yields metallic silver and CO_2. Silver oxalate can be prepared by mixing 50 mL of 0.5 M $AgNO_3$ solution with 30 mL of 0.5 M oxalic acid. The white precipitate formed is filtered, washed with distilled water, dried at 60°C, and stored in a dark bottle. To 40 mL of water, the required

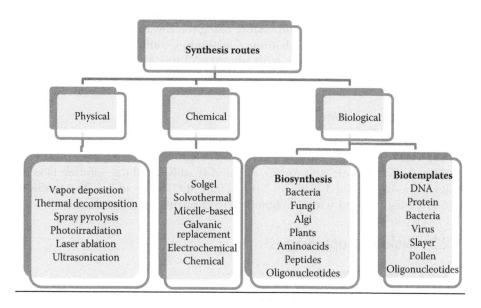

Figure 11.1 (See color insert.) *Schematic diagram of various methods of preparation for silver nanoparticles.*

amount of PVA (molecular weight: 125,000) is added and stirred. After the complete dissolution of PVA, 0.05 g of $Ag_2C_2O_4$ is added, stirred for 10 min, and purged with N_2. This mixture is refluxed in a flow of N_2 gas at 100°C for 3 h in an oil bath. The formation of a yellow colloid will be observed in the reaction mixture. The N_2 gas from the outlet is passed through a 10% solution to confirm the evolution of any CO_2 during the formation of the nanoparticles. Then, it is cooled to room temperature under N_2 atmosphere. The resultant solution is centrifuged for 5 min at 1,000 rpm to separate the yellow silver nanopowder.

11.2.1.2 Ultrasonic spray pyrolysis method

The nanostructured particles can be synthesized by ultrasonic spray pyrolysis using $AgNO_3$ as a precursor. Very fine droplets of the aerosol are obtained in an ultrasonic atomizer Pyrosol. The aerosol is transported by H_2-carrier/ reduction gas via a quartz tube to an electrical heated furnace with a temperature control ±1 °C. Nitrogen with a flow rate of 1 L/min is used for the air removal before the reduction process. Under spray pyrolysis conditions in hydrogen atmosphere and at a flow rate of 1 L/min, the dynamic (continuous) reduction takes place in the quartz tube reactor.

11.2.1.3 Photoirradiation techniques

Different radiations can be used successfully to synthesize silver nanoparticles (AgNPs). AgNPs of well-defined shape and size have been produced by laser irradiation of an aqueous solution of silver salt containing surfactant (Abid et al. 2002). A synthesis procedure using microwave irradiation has also been employed. Microwave radiation of a solution containing carboxymethyl cellulose sodium and silver nitrate produced uniform AgNPs that are stable at room temperature (Chen et al. 2008).

11.2.1.4 Laser ablation method

The AgNPs can also be prepared using gelatin (0.001%) with a mixture of $AgNO_3$ (0.01 M) and 0.01 M cetyltrimethylammonium bromide (CTAB). Initially, 2 mL of gelatin and $AgNO_3$ solution is added and the solution is kept under laser ablation for 1 h. Then, 1 mL of 0.01 M CTAB is added. The solution is continuously stirred with a magnetic stirrer during laser irradiation.

11.2.2 Biological approaches

11.2.2.1 Biosynthetic route

Biological extracts (plant or microorganism) may act as a reducing or capping agent. The bioreduction of metal ions by combinations of biomolecules found in the extracts of certain organisms (e.g., enzymes/proteins, amino acids, polysaccharides, vitamins) is environmentally benign, yet chemically complex. Many studies have reported successful synthesis of silver nanoparticles using organisms (microorganisms and biological systems; Korbekandi et al. 2009; Iravani 2011). Biological methods of nanoparticle synthesis using microorganisms (Nair et al. 2002), enzymes (Willner et al. 2006), or plants or plant extract (Shankar et al. 2004) are possible eco-friendly alternatives to chemical and physical methods. The use of plant extracts for nanoparticle preparation can be advantageous over other biological processes by eliminating the elaborate process of maintaining cell cultures (Shankar et al. 2004). It can also be suitably scaled up for large-scale synthesis of nanoparticles.

It is well known that biological systems can provide a number of metals. Chemical synthesis methods lead to the presence of some toxic chemicals absorbed on the surface, which may have adverse effects in medical applications. Green synthesis provides advancements over chemical and physical methods because it is cost effective, environment friendly, easily scaled up for large-scale synthesis, and does not need high pressure, energy, temperatures, or toxic chemicals. A schematic diagram of the general procedure followed for green synthesis of silver nanoparticles is shown in **Figure 11.2**.

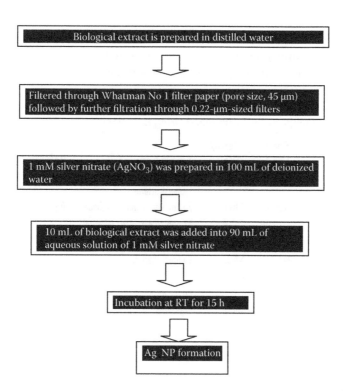

Figure 11.2 *Schematic representation of silver nanoparticles from biological extracts.*

11.2.2.2 Synthesis of silver nanoparticles by bacteria

Stable silver nanoparticles can be synthesized by bioreduction of aqueous silver ions with a culture supernatant of nonpathogenic bacterium (Kalishwaralal et al. 2008). The cell walls of gram-positive bacteria, such as *Bacillus subtilis*, were found to bind with larger quantities of metals than gram-negative bacteria, such as *Escherichia coli* (Beveridge et al. 1985).

11.2.2.3 Synthesis of silver nanoparticles by fungi

Silver nanoparticles are also synthesized extracellularly using *Fusarium oxysporum* with no evidence of flocculation of the particles, even a month after the reaction (Ahmad et al. 2003). The long-term stability of the nanoparticle solution might be due to the stabilization of the silver particles by proteins. The morphology of nanoparticles varies, with generally spherical and occasionally triangular shapes observed in the micrographs.

11.2.2.4 Synthesis of silver nanoparticles by plants

An important branch of the biosynthesis of nanoparticles is the use of plant extracts for biosynthesis reaction. The metals used for the biosynthesis, particle sizes, and shapes are presented in **Table 11.1**. Plant extracts have been used as reducing and stabilizing agents for the biosynthesis of silver nanoparticles in aqueous solutions in ambient conditions (Vilchis-Nestor et al. 2008). Various synthetic and natural polymers such as poly(ethylene glycol), polyvinyl pyrrolidone, starch, heparin, poly-cationic chitosan, glucose, sodium alginate, and

Table 11.1 Bionanoparticles Synthesized Using Botanicals

Plant	Parts Used	Metal/ Alloy	Size/Shape	References
Pelargonium graveolens	Leaves	Silver	16–40 nm/quasi-linear superstructures	Shankar et al. (2003)
Azadirachta indica	Leaves	Silver	20 nm/nearly spherical shape	Tripathy et al. (2010)
Cinnamon zeylanicum	Bark	Silver	31 and 40 nm/quasi-spherical and small, rod-shaped	Sathishkumar et al. (2009)
Aloe vera	Pulp	Silver	25 nm/spherical	Zhang et al. (2010)
Jatropha curcas	Latex	Silver		Bar et al. (2009)
Desmodium triflorum	Leaves	Silver	20 nm/hexagonal and nearly spherical	Philip et al. (2011a, 2011b)
Hibiscus rosa sinensis	Leaves	Silver		Philip (2010)
Medicago sativa	Leaves	Silver	2–20 nm/spherical	Gardea-Torresdey et al. (2003)
Capsicum annuum	Leaves	Silver		Gardea-Torresdey et al. (2005)
Pelargonium graveolens	Leaves	Silver	16–40 nm	Singaravelu et al. (2007)

gum acacia have been reported as reducing and stabilizing agents for biosynthesis of silver nanoparticles.

11.2.3   Chemical approaches

Different reducing agents, such as sodium citrate, ascorbate, sodium borohydride ($NaBH_4$), elemental hydrogen, Tollens reagent, dimethylformamide, and poly(ethylene glycol) block copolymers are used for the reduction of silver ions ($Ag+$) in aqueous or nonaqueous solutions. The aforementioned reducing agents reduce silver ions ($Ag+$) and lead to the formation of metallic silver (Ag^0), which is followed by agglomeration into oligomeric clusters. These clusters eventually lead to formation of metallic colloidal silver particles.

11.2.3.1   Polysaccharide method

In this method, the reduction of silver ions to silver nanoparticles is achieved using polysaccharide as a reducing agent that also acts as a capping agent. For example, a simple method involves starch-tagged silver nanoparticles, where α-D-glucose acts as a reducing agent in a gently heated system. Stable silver nanoparticles can also be synthesized by autoclaving a solution of silver nitrate and starch at 15 psi and 121°C for 5 minutes. AgNPs can also be synthesized by using negatively charged heparin as a reducing or stabilizing agent by heating a solution of silver nitrate or heparin to 70°C for more than 8 hours.

11.2.3.2   Tollens method

The basic Tollens reaction involves the reduction of Tollens reagent by an aldehyde: $Ag(NH_3)_2^+$ (aq.) + RCHO (aq.) = Ag (s) + RCOOH (aq.). In a modified procedure, silver ions are reduced by saccharides in presence of ammonia to form AgNP film and silver hydrosols with nanorange particles. The ammonia concentration and nature of the reductant plays a major role in controlling the size and shape of the AgNPs.

11.2.3.3   Microemulsion technique

Uniform and size-controllable silver nanoparticles can be synthesized using microemulsion techniques. The nanoparticle preparation in two-phase aqueous organic systems is based on the initial spatial separation of reactants (metal precursor and reducing agent) in two immiscible phases. The interface between the two liquids and the intensity of interphase transport between two phases, which is mediated by a quaternary alkyl-ammonium salt, affect the rate of interactions between metal precursors and reducing agents. Metal clusters formed at the interface are stabilized, due to their surface being coated with stabilizer molecules occurring in the non-polar aqueous medium. They are transferred to the organic medium by the interphase transporter (Krutyakov et al. 2008).

11.3   Nutraceutical applications

The Project on Emerging Nanotechnologies (n.d.) lists approximately 240 silver-containing products, including nutraceuticals, clothing, and personal care items, which are in use worldwide. Approximately 35 items from this list could be classified as nutraceuticals that contain nanosilver or colloidal silver. A common use for silver in many of these products is for antimicrobial applications. For example, nanosilver in the form of hydrofiber and nanocrystalline wound dressings is active against gram-positive (e.g., *Enterococcus faecalis*) and gram-negative bacteria (e.g., *Pseudomonas aeroginosa*), including antibiotic-resistant strains in an in vitro model (Percival et al. 2007). Silver nanoparticles are able to increase the antibacterial effects of penicillin G, amoxicillin, erythromycin, clindamycin, and vancomycin against *Staphylococcus aureus* and *E. coli*, as demonstrated in a disk diffusion assay by observing the zone of inhibition (Shahverdi et al. 2007).

Initially, silver nanoparticles are able to penetrate the cell wall of gram-negative bacteria, such as *E. coli*. Furthermore, there are three main mechanisms by which nanosilver exerts its antibacterial activity:

1. Nanosilver in the range of 1–10 nm disrupts permeability and respiration by attaching to the surface of the cell membrane,
2. Nanosilver is able to enter the cell membrane and interact with phosphorus- and sulfur-containing macromolecules, such as DNA and protein, which damages them.
3. Nanosilver can form silver ions, such as Ag^+, which can cause the formation of superoxide radicals that can damage macromolecules in bacteria (Morones et al. 2005; Feng et al. 2000; Hwang et al. 2008).

The various mechanisms of the antibacterial effects of silver nanoparticles include the following:

- Cell death due to uncoupling of oxidative phosphorylation
- Cell death due to induction of free radical formation
- Interference with the respiratory chain at the cytochrome C level
- Interference with components of microbial electron transport system
- Interaction with protein thiol groups and membrane-bound enzymes
- Interaction with phosphorus- and sulfur-containing compounds, such as DNA

A nanocrystalline silver-based dressing demonstrated the fastest and broadest antifungal activity when compared with silver nitrate, silver sulfadiazine, and mafenide acetate (Wright et al. 1999). The nanocrystalline silver-based dressing demonstrated in vitro activity against *Candida albicans, Candida glabrata, Candida tropicalis, Saccharomyces cerevisiae, Aspergillus fumigatus,* and *Mucor.*

Nanosilver (5–20 nm diameter) demonstrated cytoprotective and postin-fected cell antiviral activity in Hut/CCR5 cells (a human T-cell line), which were infected with human immunodeficiency virus (HIV)-1 (Sun et al. 2005). Nanosilver in the range of 1–10 nm prevented the in vitro HIV-1 infection of CD4+ MT-2 cells. The proposed mechanism may be the result of nanosilver binding to the virus gp120, which prevents it from attaching to CD4+ MT-2 cells (Elechiguerra et al. 2005).

When applied topically, nanocrystalline silver cream was effective in treating acute contact dermatitis in a guinea pig model (Bhol et al. 2004). The anti-inflammatory effects of nanocrystalline silver appear to be mediated by the inhibition of matrix metalloproteinases and the induced apoptosis of inflam-matory cells (e.g., neutrophils), as observed in a porcine model of wound healing (Wright et al. 2002). In a thermal injury mouse model, nanosilver-coated dressing treated mice had improved wound healing as compared to animals that received no treatment. The silver nanoparticles caused a reduc-tion in wound inflammation and a modulation of fibrogenic cytokines (Tian et al. 2007).

11.4 Human toxicology

Silver toxicity is a concern in humans. Silver is used in a wide variety of consumer products, such as bactericides, medicines, cosmetics, and drink-ing water, which can lead to an unhealthy exposure of humans to silver (Wijnhoven et al. 2009). In addition, nanosilver can be used in all phases of food production, including production, processing, safety, and packag-ing, which can increase the consumer's nano-silver exposure and the poten-tial for toxicity (Wijnhoven et al. 2009). To better understand the toxicity of nanosilver in humans, the toxicokinetics of nanosilver will be discussed. Toxicokinetics is the study of the time course of absorption, distribution, metabolism, and elimination of a toxicant, which is similar to pharmacoki-netics for drugs (Shen 2010).

The lack of information regarding the toxicity of nanoparticles poses serious problems for human health. Understanding the typical characteristics of fab-ricated nanomaterials and their interactions with biological systems is critical for the safe implementation of these materials in biomedical diagnostics and therapeutics. Therefore, the potential toxicity, biodistribution, and cellular uptake of silver nanoparticles are being studied. Nanoparticles have certain intrinsic properties (e.g., pinocytosis) that make it easy for them to enter into cells. At the cellular level, nanoparticles can be found in various compart-ments and even in the cell nucleus, which contains all the genetic information (Asharani et al. 2009).

Nanoparticles can easily penetrate biological membranes and very small capillaries throughout the body. For example, they can pass through the blood–brain, blood–testes, and blood–placenta barriers). Particle size, surface morphology, and surface area are recognized as important determinants for the toxicity of nanomaterials (Ji et al. 2007). Most of the toxicity studies of silver nanoparticles were undertaken in vitro, with particles ranging from approximately 1 to 100 nm. In vitro and in vivo toxicity studies revealed that silver nanoparticles have the potential ability to cause chromosomal aberrations and DNA damage, enter cells, and cause cellular damage. Silver nanoparticles can bind to different tissues (e.g., proteins and enzymes in mammalian cells) and cause toxic effects, such as adhesive interactions with cellular membranes and production of highly reactive and toxic radicals, including reactive oxygen species, which can cause inflammation and show intensive toxic effects on mitochondrial function. Potential target organs for nanosilver toxicity may involve the liver and immune system. Accumulation and histopathological effects were observed in livers from rats systemically exposed to silver nanoparticles (10–15 nm).

The absorption of nanosilver may occur through oral, dermal, or inhalation routes; it then can travel to the systemic circulation (Wijnhoven et al. 2009). Oral absorption of nanosilver through the gastrointestinal tract would be a common and important route of absorption of nanosilver nutraceuticals. A study in rats found that orally administered nanosilver (approximate average size of 60 nm) for 28 days was absorbed into the systemic circulation (Kim et al. 2008). Argyria (blue-gray hyperpigmentation) in a patient was caused by the oral absorption of colloidal silver, which are liquid-suspended silver particles, mainly in the nanoscale range (Chang et al. 2006).

Inhalation of nanosilver from a spray can be an important route of absorption. The disposition of nanosilver in the respiratory tract depends on factors such as particle size, breathing force, and Brownian diffusion (Wijnhoven et al. 2009). An inhalation study in rats demonstrated that inhalation of nanosilver of approximately 15 nm was absorbed from the lungs into the systemic circulation, suggesting that this could occur in humans as well (Ji et al. 2007). The respiratory system seemed relatively unaffected by exposure to nanosilver in vivo in a 28-day study (Ji et al. 2007). However, the cytotoxic effects of nanosilver particles on alveolar macrophages and alveolar epithelial cells were demonstrated in vitro (Soto et al. 2005, 2007; Park et al. 2010). It has been reported that there is no evidence available to demonstrate that silver is a cause of neurotoxic damage, even though silver deposits have been identified in the region of cutaneous nerves (Lansdown 2007). There is no evidence that nanosilver is allergenic. However, given the many combinations and through possible interactions with other natural and synthetic materials, it is too early to rule out a contribution to the development of allergies (Lansdown 2006). A group of investigators have put forth the hypothesis that nanosilver's toxic effects are proportional to the activity of released ionic silver from silver-containing nanoparticles (Wijnhoven et al. 2009).

11.5   Environmental impacts

Silver has a long history of being associated with food, nutraceuticals, and medicine. Although the probability of environmental toxicity of silver pollution is not so significant, Rodgers et al. (1997) reported that it has a considerable impact on ecological system. Several studies have demonstrated that accumulation of silver in water and soil may cause ecological damage to the environment. Silver is a metal that acts or dissociates as positively charged ions (Ag^+) in aqueous solutions and has a tendency to react with negatively charged ions or ligands that are present in natural water (e.g., fluorides, chlorides, sulfates, hydroxide, carbonate) in order to achieve a stable taste. It also forms a complex with free sulfide dissolved in natural water (Adams and Kramer 1999).

The degree of reaction of nanosilver depends upon the concentration of the negatively charged ion/ligand and also on the strength of silver ion. A small proportion of silver ion always remains in the water as a free ion. Speciation—the distribution of silver between its ionic and its ligand-bound form—is an indicator of how much silver is available to affect the aquatic ecosystem.

Chloride is one of the most abundantly available anions present in natural water. The concentration of chlorides varies widely from freshwater to seawater. In freshwater, chloride concentration is very low; however, due to the high concentration of chloride ion in seawater, silver rapidly reacts with chloride ion, precipitates as silver chloride ($AgCl_2$), and thereby is unavailable for uptake by aquatic organisms (Adams and Kramens 1999). Even when chloride ion concentration reaches approximately 10% of full-strength seawater, one silver ion reacts with multiple numbers of chloride ions to form a complex.

Blaser et al. (2008) reported that the following factors should be considered to get a crude quantification of silver in the environment:

- The nature of the potential sources
- The present number of the sources
- The potential growth of the number of sources
- The concentration of silver in each source
- The potential for dispersal to the environment

However, there is a problem with this estimation: The form of silver that is present—whether it is silver itself or nanosilver (the nanoform is more toxic than the metallic form)—is not clear.

When released to the environment, nanosilver may dissolve, aggregate, remain suspended, or react with other dissolved anionic particles. Once it is exposed to the aquatic environment, it is free to react indefinitely. Few formulations containing nanosilver, on exposure to the environment, dissolve or degrade in acidic pH at nearly normal room temperature. Formulations, especially those containing higher concentrations of silver, have a tendency to agglomerate or agitate to a larger particle in aquatic environments and finally settle down as sediments. This reduces the effectiveness of silver to react.

Factors that should be considered when evaluating environmental risks include the following:

- Sources of the nanosilver
- Concentrations in the environment
- The pathways of nanosilver in the environment
- Bioavailability of nanosilver

Toxicity is determined by the internally accumulated, bioavailable nanosilver in each organism. The effects on ecological structure and function are determined by how many and what kinds of organisms are most affected by nanosilver at the bioavailable concentrations that are present in the environment.

Nanosilver has been found to be toxic at concentrations as low as 0.14 µg/mL to several species of nitrifying bacteria, which play an important role in the environment by converting ammonia in the soil to a form of nitrogen that can be used by plants (Benn and Westerhoff 2008). Nitrifying bacteria are also used by sewage treatment plants (STPs) to convert raw sewage into less harmful products. Nanosilver's toxicity to these organisms has raised concerns that the release of nanosilver to the environment may disrupt the operation of STPs as well as natural processes in the ecosystem that support plant life. However, one recent analysis estimated that concentrations of nanosilver expected to be released from certain textile products currently in use would not likely reach high enough levels to threaten the microbes that are important to STP operation (Mueller et al. 2008).

The formulation and form of a nanoparticle greatly influences the risks that it poses. Silver in different nanoproducts can be in the form of silver ions, silver colloid solutions, or silver nanoparticles. The nanosilver can come in different shapes, have different electrical charges, and be combined with other materials and coated in different ways. Each of these factors, as well as others, affects toxicity and environmental behavior.

11.5.1 Impact on aquatic organisms

Research on nanosilver's toxicity for more complex living organisms is limited. Studies investigated the impacts of exposure to nanosilver on zebrafish embryos (Blaser et al. 2008; Nallathamby et al. 2008). Zebrafish are small minnows that are popular in aquaria and as model organisms in toxicity studies because their embryos are easy to manipulate and study. Two independent studies found that exposure to particles of nanosilver between 5 and 46 nm resulted in increased mortality, heart malformation, and other developmental deformities in zebrafish at concentrations as low as 5 µg/mL.

11.5.2 Impact on fowl

The ability of nanosilver to kill bacteria is being investigated within the agricultural community as a potential replacement for traditional antibiotics. As a result, a few studies of nanosilver's impacts on fowl are being studied (Kulinowski 2008). In an experiment, silver nanoparticles injected into fertilized chicken eggs at a concentration of 10 ppm had no measurable effect on embryo development according to one accepted standard, but they did induce changes in cells important to the immune system. In another study, nanosilver in concentrations up to 25 µg/mL was fed to a species of quail to determine if it could decrease the proportion of harmful bacteria in their intestines, thereby reducing the spread of disease in the animal population.

11.5.3 Impact on humans

Toxicity research of high relevance to human health is very limited. When human skin cells grown in a Petri dish were exposed to nanosilver particles 7–20 nm in size, concentration-dependent changes to cell morphology including abnormal size, shrinkage, and rounded appearance were observed at concentrations above 6.25 µg/mL. Another study describes the result of exposure to nanosilver in a wound dressing used to treat a severe burn victim. After a week of treatment with a wound dressing impregnated with nanosilver, the patient developed reversible signs of liver toxicity and a grayish discoloration of his face similar to that found in patients diagnosed with argyria. The dose received by the patient was not measured but the patient's blood plasma and urine were found to have an elevated concentration of silver (107 and 28 µg/kg, respectively). When the wound dressing was removed, all clinical symptoms returned to normal within 10 months (Kulinowski 2008).

Guidance published by the European Commission stated that a nanoversion of a substance already on the market in non-nano form will require a modification of the registration to include the special properties of the nanoversion, different labeling, and additional risk management measures.

11.6 Conclusion

Although nutraceuticals have significant promise for the promotion of human health and prevention of disease, their use is still not regulated. Quality, safety, long-term adverse effects, and toxicity are primary concerns. The purpose of consuming nutraceuticals is to incorporate them into the body to build overall health and treat disease. However, many important nutraceuticals are not fully absorbed in the human body.

In recent years, more and more nanomaterials—especially nanosilver—found their way into consumer products. Silver has been used for medical applications for hundreds of years. The antibacterial effects have been exploited for the treatment of traumatic or diabetic wounds, burns, tonsillitis, and syphilis.

Ecofriendly bio-organics in plant extracts contain proteins and enzymes that act as both reducing and capping agents, to form stable and shape-controlled silver nanoparticles. The green synthesis of silver nanoparticles involves three main criteria:

1. Selection of solvent medium
2. Selection of an environmentally benign reducing agent
3. Selection of nontoxic substances for the silver nanoparticle's stability

Silver nanoparticles are not water soluble; therefore, silver colloids will not release silver ions into the environment. Silver nanoparticles do not last as nanoparticles in nature for very long. Instead, they grow into the harmless clumps of silver metal that have existed in nature since the beginning of our planet.

Silver nanoparticle-based diagnostics and therapeutics hold great promise because multiple functions can be built onto the particles. The potential applicability of these silver nanoparticles is simple, sensitive, and selective for the versatile applications related to diagnostics and therapeutics. The use of silver nanoparticles is safe for consumer health and environment.

References

Abid, J.P., A.W. Wark, P.F. Brevet, and H.H. Girault. 2002. Preparation of silver nanoparticles in solution from a silver salt by laser irradiation. *Chem Commun* 7:792–793.

Adams, N.W.H., and J.R. Kramer. (1999). Silver speciation in wastewater effluent, surface waters and pore waters. *Environ Toxicol Chem* 18:2667–2673.

Ahmad, A., P. Mukherjee, S. Senapati, et al. 2003. Extracellular biosynthesis of silver nanoparticles using the fungus *Fusarium oxysporum*. *Colloids Surf B* 28:313–318.

Asharani, P.V., G. Low Kah Mun, M.P. Hande, and S. Valiyaveettil S. 2009. Cytotoxicity and genotoxicity of silver nanoparticles in human cells. *ACS Nano* 3:279e90.

Atiyeh, B.S., M. Costagliola, S.N. Hayek, and S.A. Dibo. 2007. Effect of silver on burn wound infection control and healing: Review of the literature. *Burns* 33:139–148.

Bar, H., D.K. Bhui, G.P. Sahoo, P. Sarkar, S.P. De, and A. Misra. 2009. Green synthesis of silver nanoparticles using latex of Jatrophacurcas. *Coll Surf A Physicochem Eng Asp* 339:134–139.

Barde, P.B., G.D. Jindal, R. Singh, and K.K. Deepak. 2006. New method of electrode placement for determination of cardiac output using impedence cardiography. *Indian J Physiol Pharmacol* 50:234–240.

Benn, T.M., and P. Westerhoff, P. 2008. Nanoparticle silver released into water from commercially available sock fabrics. *Environ Sci Technol* 42:4133–4139.

Beveridge, T.J., and W.S. Fyfe. 1985. Metal fixation by bacterial cell walls. *Can J Earth Sci* 22:1893–1898.

Bhol, K., J. Alroy, and P. Schechter. 2004. Anti-inflammatory effect of topical nanocrystalline silver cream on allergic contact dermatitis in a guinea pig model. *Clin Exp Dermatol* 29:282–287.

Blaser, S.A., M. Scheringer, M. Macleod, and K. Hungerbuhler. 2008. Estimation of cumulative aquatic exposure and risk due to silver: Contribution of nanofunctionalized plastics and textiles. *Sci Total Environ* 390:396–409.

Chang, A., V. Khosravi, and B. Egbert. 2006. A case of argyria after colloidal silver ingestion. *J Cutan Pathol* 33:809–811.

Chen, J., K. Wang, J. Xin, and Y. Jin. 2008. Microwave-assisted green synthesis of silver nanoparticles by carboxymethyl cellulose sodium and silver nitrate. *Mater Chem Phys* 108:421–424.

Chu, C.H., E.C. Lo, and H.C. Lin. 2002. Effectiveness of silver diamine fluoride varnish in arresting dentin caries in Chinese preschool children. *J Dent Res* 81:767–770.

Elechiguerra, J., J. Burt, J. Morones, et al. 2005. Interaction of silver nanoparticles with HIV-1. *Journal of Nanobiotechnology* 3:6.

Feng, Q.L., J. Wu, G.Q. Chen, et al. 2000. A mechanistic study of the antibacterial effect of silver ions on *Escherichia coli* and *Staphylococcus aureus*. *J Biomed Mater Res* 52:662–680.

Fung, M.C., and D.L. Bowen. 1996. Silver products for medical indications: Risk-benefit assessment. *J Toxicol Clin Toxicol* 34:119–126.

Gardea-Torresdey, J.L., E. Gomez, J.R. Peralta-Videa, J.G. Parsons, H. Troiani, and M. Jose-Yacaman. 2003. Alfalfa sprouts: A natural source for the synthesis of silver nanoparticles. *Langmuir* 19:1357–1361.

Hardes, J., H. Ahrens, C. Gebert, et al. 2007. Lack of toxicological side-effects in silver-coated megaprostheses in humans. *Biomaterials* 28:2869–2875.

Hipler, U.C., P. Elsner, and J.W. Fluhr. 2006. A new silver loaded cellulosic fibre with antifungal and antibacterial properties. *Curr Probl Dermatol* 33:165–178.

Hudspith, J., and S. Rayatt. 2004. First aid and treatment of minor burns. *Brit Med J* 328:1487.

Hwang, E.T., J.H. Lee, Y.J. Chae, et al. 2008. Analysis of the toxic mode of action of silver nanoparticles using stress-specific bioluminescent bacteria. *Small* 4:746–750.

Iravani, S. 2011. Green synthesis of metal nanoparticles using plants. *Green Chem* 13:2638–2650.

Ji, J.H., J.H. Jung, S.S. Kim, et al. 2007. Twenty-eight-day inhalation toxicity study of silver nanoparticles in Sprague-Dawley rats. *Inhalation Toxicol* 19:857–871.

Kalishwaralal, K., V. Deepak, S. Ramkumarpandian, et al. 2008. Extracellular biosynthesis of silver nanoparticles by the culture supernatant of *Bacillus licheniformis*. *Mater Lett* 62:4411–4413.

Kim, K.-J. et al. 2009. Antifungal activity and mode of action of silver nano-particles on *Candida albicans*. *Biometals* 22:235–242.

Kim, Y.S., J.S. Kim, H.S. Cho, et al. 2008. Twenty-eight-day oral toxicity, genotoxicity, and gender-related tissue distribution of silver nanoparticles in Sprague-Dawley rats. *Inhalation Toxicol* 20:575–583.

Korbekandi, H., S. Iravani, and S. Abbasi. 2009. Production of nanoparticles using organisms. *Crit Rev Biotechnol* 29:279–306.

Krutyakov, Y.A., A.A. Kudrinskiy, A.Y. Olenin, and G.V. Lisichkin. 2008. Synthesis and properties of silver nanoparticles: Advances and prospects. *Russ Chem Rev* 77:233–257.

Kulinowski, K.M. 2008. Environmental impacts of nanosilver: An ICON backgrounder. Available from http://cohesion.rice.edu/centersandinst/icon/resources.cfm?doc_id=12722.

Lansdown, A. 2006. Silver in health care: Antimicrobial effects and safety in use. *Curr Probl Dermatol* 33:17–34.

Lansdown, A.B. 2007. Critical observations on the neurotoxicity of silver. *Crit Rev Toxicol*, 37:237–250.

Lo, S.-F., Hayter, M., and C.J. Chang. 2008. A systematic review of silver-releasing dressings in the management of infected chronic wounds. *J Clin Nurs* 17:1973–1985.

Mehrbod, P., N. Motamed, and M. Tabatabaian. 2009. In vitro antiviral effect of "nanosilver" on influenza virus. *DARU J Pharm Sci* 17:88–93.

Morones, J.R., J.L. Elechiguerra, A. Camacho, et al. 2005. The bactericidal effect of silver nanoparticles. *Nanotechnology* 16:2346–2353.

Mueller, N.C., and B. Nowack. 2008. Exposure modeling of engineered nanoparticles in the environment. *Environ Sci Technol* 42:4447–4453.

Nair, B., and T. Pradeep. 2002. Coalescence of nanoclusters and formulation of submicron crystallites assisted by *Lactobacillus* strains. *Cryst Growth Des* 2:293–298.

Nallathamby, P.D., K.J. Lee, and X.H.-L. Xu. 2008. Design of single and stable nanoparticle photonics for in vivo dynamics imaging of nanoenvironments of zebra fish embryonic fluids. *ACS Nano* 2:1371–1380.

Park, E.-J., E. Bae., and J. Yi. 2010. Repeated-dose toxicity and inflammatory responses in mice by oral administration of silver nanoparticles. *Environ Toxicol Pharmacol* 30:162–168.

Percival, S.L., P.G. Bowler, and J. Dolman. 2007. Antimicrobial activity of silver-containing dressings on wound microorganisms using an *in vitro* biofilm model. *Int Wound J* 4:186–191.

Philip, D. 2010. Green synthesis of gold and silver nanoparticles using Hibiscus rosa sinensis. *Physica* E 42:1417–1424.

Philip, D. 2011a. Mangifera Indica leaf-assisted biosynthesis of well-dispersed silver nananoparticles. *Spectrochimica Acta Part A* 78:327–331.

Philip, D., S. Unnib, A. Aswathy, and V.K. Vidhua. 2011b. Murraya Koenigii leaf-assisted rapid green synthesis of silver and gold nanoparticles. *Spectrochimica Acta Part A* 78:899–904.

Project on Emerging Nanotechnologies. n.d. Available at http://www.nanotechproject.org/process/assets/files/7039/silver_database_fauss_sept2_final.pdf.

Sathishkumar, M., K. Sneha, W.S. Won, C.-W. Cho, S. Kim, and Y.-S. Yun. 2009. Cynamon zeylanicum bark extract and powder-mediated green synthesis of nanocrystalline silver particles and its bactericidal activity. *Coll Surf B Biointer* 73:332–338.

Shahverdi, A.R., A. Fakhimi, H.R. Shahverdi, and S. Minaian. 2007. Synthesis and effect of silver nanoparticles on the antibacterial activity of different antibiotics against *Staphylococcus aureus* and *Escherichia coli*. *Nanomedicine* 3:168–171.

Shankar, S., A. Ahmad, and M. Sastry. 2003. Geranium leaf assisted biosynthesis of silver nanoparticles. *Biotechno Prog* 19:1627–1631.

Shankar, S.S., A. Rai, A. Ahmad, and M. Sastry. 2004. Rapid synthesis of Au, Ag, and bimetallic Au core–Ag shell nanoparticles using Neem (*Azadirachta indica*) leaf broth. *J Colloid Interf Sci* 275:496–502.

Shen, D. 2010. Toxicokinetics. In *Casarett and Doull's Essentials of Toxicology*, 2nd ed., ed. C. Klassen and J. Watkins III. McGraw-Hill Medical/Jaypee Brothers Medical Publishers: New Delhi, India.

Sohaimy, S.A.E. 2012. Functional foods and nutraceuticals—modern approach to food science. *World Appl Sci J* 20:691–708.

Soto, K.F., A. Carrasco, T.G. Powell, et al. 2005. Comparative in vitro cytotoxicity assessment of some manufactured nanoparticulate materials characterized by transmission electron microscopy. *J Nanopart Res* 7:145–169.

Soto, K., K.M. Garza, and L.E. Murr. 2007. Cytotoxic effects of aggregated nanomaterials. *Acta Biomater* 3:351–358.

Sun, R., R. Chen, N. Chung, et al. 2005. Silver nanoparticles fabricated in Hepes buffer exhibit cytoprotective activities toward HIV-1 infected cells. *Chem Commun* 28:5059–5061.

Tian, J., K. Wong, C.M. Ho, et al. 2007. Topical delivery of silver nanoparticles promotes wound healing. *ChemMedChem* 2:129–136.

Tripathy, A., M.A. Raichur, N. Chandrasekaran, T.C. Prathna, and A. Mukherjee. 2010. Process variables in biomimetic synthesis of silver nanoparticles by aqueous extract of *Azadirachta indica* (Neem) leaves. *J Nano Res* 12:237–246.

Vilchis-Nestor, A.R., V. Sánchez-Mendieta, M.A. Camacho-López, et al. 2008. Solventless synthesis and optical properties of Au and Ag nanoparticles using *Camellia sinensis* extract. *Material Lett* 62:3103–3105.

Wijnhoven, S.W.P., W.J.G.M. Peijnenburg, C.A. Herberts, et al. 2009. Nano-silver: A review of available data and knowledge gaps in human and environmental risk assessment. *Nanotoxicology* 3:109–138.

Willner, I., R. Baron, and B. Willner. 2006. Growing metal nanoparticles by enzymes. *Adv Mater* 18:1109–1120.

Wright, J.B., L. Kan, A.G. Buret, et al. 2002. Early healing events in a porcine model of contaminated wounds: Effects of nanocrystalline silver on matrix metalloproteinases, cell apoptosis, and healing. *Wound Rep Reg* 10:141–151.

Wright, J.B., K. Lam, D. Hansen, and R.E. Burrell. 1999. Efficacy of topical silver against fungal burn wound pathogens. *Am J Infect Control* 27:344–350.

Zhang, Y., D. Yang, Y. Kong, X. Wang, O. Pandoli, and G. Gao. 2010. Synergetic antibacterial effects of silver nanoparticles: Aloe vera prepared via a green method. *Nano Biomed Eng* 2:252–257.

Regulatory Pathways and Intellectual Property Rights for Metallonutraceuticals

Vishal Kataria, Jayant Lokhande, Vrinda Vedi, and Yashwant Pathak

Contents

12.1   Introduction

The term *nutraceuticals* is a linguistic combination of *nutrient* and *pharmaceutical.* A nutraceutical is defined as a substance that may be considered a food or part of a food and provides medical or health benefits, including the prevention and treatment of disease. Nutraceuticals belong to a diverse product category with various synonyms and are used internationally. Metallonutraceuticals may create an environment for new products that promise novel solutions to health-related issues. Metallonutraceuticals will play an important role in future therapeutic developments. The greatest challenges concern public policy and regulation. The research and development of products providing health benefits should be encouraged, while truthful, nonmisleading communications about these products protect public health and maintain public confidence (1).

Metallonutraceuticals are clearly not drugs, which are potential pharmacologically active substances that potentiate, antagonize, or otherwise modify any physiological or metabolic function. On the other hand, a nutraceutical is a nutrient that not only maintains, supports, and normalizes any physiologic or metabolic function; it also potentiates, antagonizes, or otherwise modifies physiologic or metabolic functions (2). The approach to regulating and marketing nutraceuticals is notably heterogeneous on the global level. This is largely due to the challenges in classifying these products, absence of a suitable regulatory category for these hybrid products, and varying views on what is considered sufficient scientific substantiation to determine functionality (3).

The rational use of metallonutraceuticals is based on objective evaluation of the clinical evidence as well as subjective evaluation of the risks, benefits, economic costs, and potential drug interactions. The pharmaceutical industry is known for its high costs of research and development associated with drug development, as well as the use of patents to protect the discoveries from this research; therefore, this industry is associated with high product margins. The food industry is noted for its low margins and the commoditization of its inputs and, in some cases, its products. Metallonutraceuticals will fall somewhere in between the two. Nutraceuticals are higher priced with greater margins than conventional foods, thus generating a great incentive for companies to enter this market (4). Metallonutraceutical products represent an excellent growth opportunity, but companies will have to take appropriate actions to develop, preserve, and protect their intellectual property rights in order to stay competitive.

Ayurveda ras shastra (Indian iatrochemistry) remains the origin of metallonutraceuticals. It is nothing but reinventing the wheel. In recent years, another innovative concept has given birth to nutrition genomics (NUGO), which combines state-of-the-art technologies such as biotechnology and genomics. It is not incorrect to say that NUGO practices reverse engineering of traditional knowledge in an innovative.

12.2 Intellectual property rights

In general, introducing a new concept or idea in the form of a product or process (or both) for industrial use is an innovation/invention. Novelty in the form of a product and/or process can be claimed in the form of a legal document, in which the owner of that novel product and/or process can obtain legal rights, called *intellectual property rights* (IPRs). An IPR is an act of intellect that has different dimensions. IPRs are recognized worldwide, although they are territorial in nature. The IPR is a symbol of prosperity, growth, and economy. Industrially, it is a technological development for newer products, processes, devices, designs, literacy, and artistic works.

12.3 Forms of intellectual property

Intellectual property (IP) is an asset of an owner. The rights obtained by the owner for said IP asset are IPRs. IPRs can be sought by a variety of forms, including the following:

- A *patent* provides exclusive rights for a limited period to a product and/or process that qualifies for novelty or inventiveness in an industrial application.
- A *trademark* is a visual symbol or sign in the form of a word or a label used to identify and distinguish from other similar goods or services from others.
- A *copyright* provides exclusive rights in literacy, art, and music.
- *Industrial design* provides protection to designs.
- *Geographical indication* is used for registering and seeking protection for goods pertaining to geographical location/region.
- *Layout designs (topography) of integrated circuits* provides protection to semiconductor integrated circuits and matters connected therewith.
- *Undisclosed information/trade secrets* provide protection for confidential information that has commercial value, which remains with the owner and is not disclosed to anyone, including the government (composition, formulae, business model, process details, drawings, and information on inventions on which patent applications are not filed or that do not fall under patent protection category).
- *Plant varieties and farmers' rights* recognizes and protects the rights of the farmers for their contribution made at any time in conserving, improving, and making available plant genetic resources for development of new plant varieties.

Internationally recognized conventions, including the General Agreement on Trade and Tariff, World Trade Organization, and Trade Related Aspects of Intellectual Property Rights (TRIPS) agreements, define the scope and magnitude of IPR in various fields. Most countries have an IPR policy based on the TRIPS agreement. The Convention of Biodiversity, in conjunction with the Biological Diversity Act and National Biodiversity Act, address access to traditional knowledge, informed consent, and benefit sharing.

12.4 Provisions and scope of patentability for metallonutraceuticals

Metallonutraceuticals are neither a normal food nor a pharmaceutical. Rather, they link food and pharmaceuticals. Like pharmaceuticals, metallonutraceuticals play an important role in the health care of humans or animals, being nutritive and/or therapeutic. Metallonutraceuticals have been categorized under a specific product category. When an innovation is claimed, it raises the issue of claims and rights to own the innovation. It is necessary to have certain rules and regulations for seeking protection. Transparency and stringency matter where life is concerned.

Every country has its own IP laws to protect IP forms. A patent law adds value and brings in technological advances. Patents provide territorial monopoly rights for an invention for a defined period, as well as provide incentives and generate money for the owner. It is an offense to use, sell, or manufacture a patented product or process without the permission of the patent owner. If any party or individual is interested in using, selling, or manufacturing the patented invention or taking over the invention (patent) or technology, then said party or interested person can apply for a procuring license or opt for a technology transfer between the parties by defining the terms and conditions.

A patent granted to the owner is territorial in nature; hence, each country has its own patent system and statutes defining rules and regulations. For any invention, it is essential to determine that the invention does not fall in the domain of one or more sections and clauses therein of the respective patent laws. *Invention* means a new product or process involving an inventive step that is capable of industrial applications. Novel metallonutraceutical compositions, formulations, or manufacturing processes should be (but are not always) patentable. Invention differs from innovation though derived from an idea that is new/novel. Consider the following scenarios:

1. Idea into invention + exploitation into new useful form = patentable *and* patentable invention = rights to claim + own rights for certain time period = patent
2. Idea into innovation + exploitation of new ideas in useful form = not patentable *and* innovation = no rights to claim = no patent

A metallonutraceutical product or process invention that does not fall into the nonpatentable domain and is indeed a patentable invention meets the following conditions:

- As per the clause of primary or intended use or commercial exploitation of which would be contrary to public order or morality or which causes serious prejudice to human, animal, or plant life or health or to the environment are not patentable.
- Mere discovery of a scientific principle or the formulation or discovery of any living thing or nonliving substance occurring in nature is not patentable.

These clauses eliminate the chances of patentability for inventions whose primary, intended use or commercial exploitation are contrary to public order, morality, or cause serious prejudice to humans, animals, plants, health and/or the environment or discover the scientific principle, formulation, or discovery of any living thing (of biological origin) or nonliving substance (of chemical origin) that is occurring in nature.

Further metallonutraceutical inventions are not patentable wherein the following conditions are true:

- Mere discovery of a new form of a known substance, which does not result in the enhancement of the known efficacy of that substance or the mere discovery of any new property or new use for a known substance or of the mere use of a known process, machine, or apparatus, unless such known process results in a new product or employs at least one new reactant.
- Any substance obtained by a mere admixture, resulting only in aggregation of the properties of the components thereof or a process producing such substance.
- Any process, for the medicinal, surgical, curative, prophylactic, diagnostic, therapeutic, or other treatment of human beings or any process for a similar treatment of animals to render them free of disease or to increase their economic value or that of their products.

Metallonutraceutical inventions are to be read in context with the chemical nature, form, composition, formulation, and preparation process. Interpretation with respect to product/process invention (or both) is therefore crucial.

Countries such as India have vast knowledge and practices (e.g., Ayurveda, Siddha, and Unani) using herbal or herbomineral compositions, formulas, and preparations that have commercial value and potential. Many individuals (e.g., Vaidya, Hakims) in India practice an ancient system that is herbal or herbomineral based and has nutritive and/or therapeutic value. Their knowledge remains as confidential undisclosed information/trade secrets. For such nonpatentable information, other forms of legal protection (trade secret) can be sought to avoid illegal access.

12.5 Data protection: Is there a need?

The composition and formulation of a metallonutraceutical include one or more natural/biological resources and may be based on traditional or indigenous knowledge claiming health benefits. This knowledge is either publicly known or practiced. Because India is the source of this knowledge, such data protection can open avenues at national and international levels.

Metallonutraceuticals can be natural, semisynthetic, or synthetic in nature. They are herbal or herbomineral based, for which traditional knowledge (Ayurveda and similar) may be involved for health benefits. Some metallonutraceuticals may be novel and some may not. Also, being prophylactic in nature, they may claim to alleviate certain diseases or disorders. Therefore, metallonutraceuticals do fall under nonpatentable clauses related to not new, natural products (minerals, plants, or animals) and/or traditional knowledge and drugs. Hence, clauses related to pharmaceuticals, natural resources, or herbal treatments (biological and mineral) are equally important to consider. Furthermore, these clauses are encased in moral, ethical, and safety issues.

12.6 Protocol compliance

Pharmaceutical companies are focusing on strategies to develop and discover natural drugs (5). Various herbal preparations are currently being marketed. The methods adopted to manufacture them do not satisfy safety norms and proof of scientific data. Furthermore, incorrect procedures can result in toxic preparations with metal contamination. Ayurveda *ras shastra* preparations follow purification methods and processes in a defined way to convert a toxic form to a nontoxic consumable.

Table 12.1 provides an overview of the regulatory acts from different countries. Governmental agencies must insist upon transparency and stringency in regulatory frameworks and policies, especially with products and processes that claim health benefits (nutritive and/or therapeutic). To obtain clearance in global markets, governments should require manufacturers and dealers to comply with good manufacturing practices (GMPs), good clinical practices, good laboratory practices, and good harvesting practices using state-of-the-art technologies.

Other important concerns include the labeling of product information (including a list of ingredients), resources, safety norms, dosages, expiration dates, and authentic claims (therapeutic or otherwise) to avoid any undue delay in permitting certification of these products in national and international markets. By addressing these issues, protection can be sought for age-old practiced compositions, formulas, and processes that are now value-added products and processes.

12.7 Nutraceuticals: International scenario

There is a need for clearly delineated classifications of metallonutraceuticals to formulate regulatory frameworks more distinctly, as these products fall under different categories than drugs and normal food. Although nutraceuticals have been defined to some extent, metallonutraceuticals have not yet been clearly defined. The international guidelines from the World Health Organization for the manufacture, quality control, and evaluation of botanicals should be considered when drafting these classifications.

The World Health Organization has defined recognizable regulatory guidelines for the following purposes:

1. To evaluate the quality, safety, and efficacy of botanical medicines
2. To assist national regulatory authorities, scientific organizations, and manufacturers in undertaking an assessment of the documentation/ submissions/dossiers for botanical medicines
3. To assess the quality of botanical medicines
4. To assess the quality of botanical materials to ensure the quality of medicinal plant products by using modern techniques and applying suitable standards

Table 12.1 Nutraceutical Regulation and Legislation by Country

Country	Legislation	Regulatory Issue Covered
Australia	*Therapeutic Goods Act, 1990:* Traditional claims for herbal remedies are allowed, providing that general advertising requirements are complied with and such claims are justified by literature references.	Develops food standards to cover the food industry in Australia and New Zealand. Modifications made in the act are available in Parliamentary Counsel. Ensures food for sale is safe and suitable for human consumption. Regulates food safety for food businesses.
European Union	*Evaluation of Medicinal Products:* General guidelines to determine a uniform set of specifications for botanical preparations manufactured and sold in Europe.	Establish a science-based approach for concepts in functional food science. Applies to good manufacturing practices (GMPs), foods, and food ingredients. Authorizes probiotics used as additives. Medicinal claims are made based on traditional use of herbs. Establishes rules in labeling, presentation, and the advertising of foods. Establishes implementation of rules for health claims in Regulation (EC) No 1924/2006. Authorizes food that reduces disease risk and improves children's health.
India	*The Drugs and Cosmetics Act (1940):* Regulates drugs; amended in 1964 to include Ayurveda and Unani under its purview and cosmetics. *Prevention of Food Adulteration Act (1954):* Regulates packaged foods. *Promulgated GMP regulations for traditional systems of medicines:* From part of the Drugs and Cosmetics Act (1940), this came into force in 2000 to improve the quality and standard of Ayurvedic, Siddha, and Unani drugs in pharmacies. *Food Safety and Standard Act (FSSA; 2006):* To integrate and streamline the many regulations covering nutraceuticals, foods, and dietary supplements; old PFA was replaced; likely to be implemented by the end of the year 2010.	Regulates the manufacture, sale, or import of novel foods, GMF, irradiated food, organic food, food for special dietary uses, functional food, nutraceuticals, and health supplements. Serves as a single reference point for all matters relating to food safety and standards. Places more emphasis on science-based and participatory decisions. Drafted rules and regulations under the FSSA.
Japan	*Foods for Specified Health Uses (FOSHU; 1993):* Permitted health claims legally for selected functional food. *New regulatory system (2001):* Regulated foods with health claims (FHC) using a system of foods with nutrient function claims. *Introduction of the new subsystems of FOSHU (2005):* Standardized FOSHU; qualified FOSHU and disease risk reduction claims for FOSHU (Ohama, H., H. Ikeda, and H. Moriyama. 2006. Health foods and foods with health claims in Japan. *Toxicology* 221:95–111).	Focuses on health claims for specific products. Category of products was expanded to include capsules and tablets. Restricted to the specified nutrients having nutritional function claims in FHC.

continued

Country	Legislation	Regulatory Issue Covered
Canada	Canadian Food and Drugs Act and Regulation, 1953 Food Directorate of the Health Protection Branch of Health Canada, 1996 Canadian Food and Drugs Act, 2001 Natural Health Product Directorate, 2003	Presented the definition of food. Nutraceutical generally sold in medicinal forms are not usually associated with food. Described foods with health benefits beyond basic nutrition. Defined nutraceuticals.
United States	*Food and Drugs Act (1906)*: Regulates dietary supplements under a different set of regulations; responsible for ensuring the safety and accurate labeling of nearly all food products in the United States. *Federal Food, Drug, and Cosmetic Act (1938)*: Required the manufacturer to prove the safety of a drug before it could be marketed; requires that manufacturers and distributors who wish to market dietary supplements that contain new dietary ingredients notify the Food and Drug Administration about these ingredients. *Nutrition Labeling and Education Act (1990)*: Required consistent, scientifically based labeling for almost all processed foods. *Dietary Supplement Health and Education Act (1994)*: Watershed legislation to regulate the manufacture and marketing of nutraceuticals. *Food and Drug Administration Modernization Act (1997)*: Made available additional options to the manufacturers of nutraceuticals.	Nutritional labeling of most foods is regulated by the Food and Drug Administration. Described the role of a nutrient or a dietary ingredient in the normal structure or function of the human body. The Federal Food, Drug, and Cosmetic Act relates to the regulation of food, drugs, devices, and biological products. Ensures safe U.S. food supply by preventing contamination.

Sources: Data from Singh, J., and S. Sinha. 2012. Classification, regulatory acts and applications of nutraceuticals for health. *J Pharm Bio Sci* 2:177–187 and http://www.pharmainfo.net/justvishal/publications/regulatory-issues-herbal-products-review.

5. To perform pharmacological evaluation, including certain norms, such as bitterness value and hemolytic activity; safety of the botanicals in terms of pesticidal residue, arsenic, and heavy metal content; microbial load; and radioactive contaminants for safety of the botanical materials

6. To standardize methodology on the research and evaluation of traditional medicine

7. To define good agricultural and collection practices for medicinal plants

12.8   Conclusion

The following important concerns about metallonutraceutical regulations illustrate the need for international harmonization:

- No appropriate regulatory framework exists for nutraceuticals, especially metallonutraceuticals.
- There is no effective regulatory system or marketing regulation for metallonutraceuticals yet.
- No regulatory system for these products exists because they do not fall under any specific law of the land.
- Nutraceuticals and food supplements are classified as neither food nor drugs for licensing purposes.
- A lack of clarity in regulation has resulted in a growing number of products in this category with false and unsubstantiated medical claims.

References

1. Palthur, M.P., S.S.S. Palthur, and S.K. Chitta. 2010. Nutraceuticals: Concept and regulatory scenario. *International Journal of Pharmacy and Pharmaceutical Sciences* 2:14.
2. Hardy, G. 2000. Nutraceuticals and functional foods: Introduction and meaning. Nutrition 16:688–689.
3. Kotilainen, L., R. Rajalahti, C. Ragasa, and E. Pehu. 2006. *Health-Enhancing Foods: Opportunities for Strengthening the Sector in Developing Countries.* Washington, DC: World Bank.
4. Scott Wolfe Management. 2002. *Potential Benefits of Functional Foods and Nutraceuticals to the Agri-Food Industry in Canada.* Ottawa, Ontario, Canada: Agriculture and Agri-Food Canada.
5. Seidl, P.R. 2002. Pharmaceuticals from natural products: Current trends. *Anais da Academia Brasileira de Ciências* 74:145–150.

Gold Nanoparticles

A Promising Nanometallic Drug Delivery System with Many Therapeutic Applications

Anastasia Groshev, Danielle Dantuma, Vaibhav Alandikar, Samuel Rapaka, Yashwant Pathak, and Vijaykumar Sutariya

Contents

13.1 What are gold nanoparticles?

13.1.1 Historical significance

Gold is a noble, highly unreactive element. Gold has played an important role in the historical arena, with gold artifacts preserving the course of history without deterioration and tarnishing. The use of bulk gold in jewelry, monetary coins, electronics, and other bulk forms is widespread. In the molecular form, gold has been used as a catalyst (Chen, 2004; Valden, 1998; Green, 2011).

Recently, however, more attention has been paid to the noble metal nanoparticles (NPs) and colloids due to their unique properties—even though gold colloids have been used for much longer. Artists have been working with colloidal suspensions to create colors for stained glass since the seventeenth century (Daniel, 2004). Then, in 1857, Faraday reported the successful formation of gold colloidal solution in one of his early published works on gold nanoparticles, describing its "beautiful ruby" color and the optical properties of the dried colloidal solutions on the film (Faraday, 1857). Other uses and synthesis methods may predate the era of peer-reviewed literature.

The beautiful range of colors that noble metal nanoparticles exhibit are due to their surface properties (**Figure 13.1**). In his 1908 work based on the laws of theoretical thermodynamics, Mie showed that the color properties of gold colloidal solutions are a result of the absorption and scattering of light by nanoparticles (Maxwell, 1865; Mie, 1908). It is well known that materials on the nanoscale demonstrate properties somewhere between those of bulk material and atoms, as initially predicted by physicists (Chen, 1999; Dreaden, 2011; El-Sayed, 2001). In the case of gold, its metallic properties allow for formation of a highly dynamic electron cloud around the particle. These electrons (in the 6s orbital of the elemental gold) collectively oscillate in response to the incoming photons of light, resulting in a surface plasmon band (SPB) of broad absorption around 520 nm; they are responsible for the ruby appearance of gold colloidal solution (Mulvaney, 1996; Venkatesh, 1983). Specifically due to this SPB, gold nanoparticles (AuNPs) are significant in the fields of spectroscopy and photography (Kam, 1983).

The SPB of an AuNP depends on many factors, including the nanoparticle's makeup, synthesis method, and interactions with the environment. Adjustment of these components allows for shifts in absorbance and, therefore, results in applications in the whole spectrum of areas where absorbance properties are important, such as imaging, fluorescence tagging, and diagnostic contrast. An adjustment in a nanoparticle's size effects the absorbance band, resulting in a blue shift with a decrease in size (**Figure 13.1**) (Logunov, 1997). A nanoparticle's shape and dimensions also considerably influence the SPB. A change in the shell's thickness in a spherical nanoparticle or a change in the amount of gold present in the nanocage will result in different absorption ranges (**Figure 13.2**) (West, 2003; Skrabalak, 2008). Furthermore, temperature has also been shown to contribute to SPB shifts. The solvent used in an AuNP's suspension results in further adjustments to the SPB. For instance, solvents with high dielectric constants result in a red shift of the SPB (Malikova, 2002).

These colorimetric and fluorescence properties of gold nanoparticles have been widely used (Hwang, 2002; Hu, 2001; Wang, 2002; Thomas, 2000; Xu, Yanagi, 1999; Makarova, 1999; Dulkeith, 2002; Gu, 2003; Dubertret, 2001). Their interactions with light energy warrant wide use of this advanced photonic technology in the field of biotechnology, including nanobioimaging (Skrabalak, 2008; Tsung, 2013), drug delivery (Geldenhuys, 2011), fluorescence

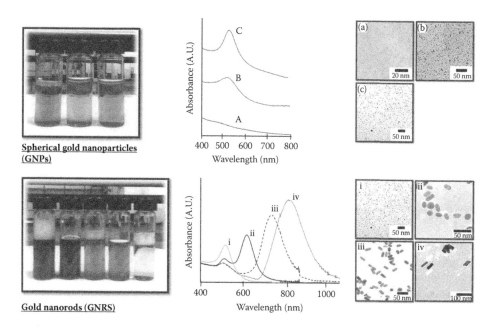

Figure 13.1 *Effect of the size and shape of gold nanoparticles on the surface plasmon band (SPB) property. Spherical nanoparticles of less than 2 nm do not support SPBs due to their size. However, gold nanorods exhibit two peaks of absorbance due to their shape. (From Alkilany, A.M., S.E. Lohse, and C.J. Murphy. 2013. The gold standard: Gold nanoparticle libraries to understand the nano-bio interface.* Accounts of Chemical Research *46:650–661. With permission.)*

tagging, radiation and hyperthermia therapy, imaging (Sokolov, 2004; Yelin, 2003; Yelin, 2002), diagnostic contrast (Min, 2013; Ventura, 2012), targeting to subcellar components (Mandal, 2009; Ma, 2013), and enhancement of the pharmacodynamic properties of active pharmacological ingredients (APIs) (Song, 2013; Du, 2013; Liu, Wang, 2012). Furthermore, the sensitivity of the SPB to environmental conditions, such as temperature and solvent, has made gold NPs very valuable in the field of sensors and biology. For instance, the fluorescence capacities of gold NPs have been extensively studied (Hwang, 2002; Hu, 2001; Wang, 2002; Thomas, 2000; Xu, 1999; Makarova, 1999; Dulkeith, 2002; Gu, 2003; Dubertret, 2001).

In medicine, the use of gold colloids dates back to the Middle Ages, when they were used for the diagnosis of syphilis and treatment of various diseases using soluble gold (Antonii, 1618; Knuckels, 1676). These practices (although they did not provide a cure) were continued well into Renaissance. The antimicrobial properties of gold nanoparticles have been demonstrated (Ahmad, 2013; Pender, 2013). Since it was discovered in the 1950s that the particles can be bound with peptides without loss of activity, the applications of AuNPs in the areas of bioimaging, immunodiagnostics, and histopathology flourished,

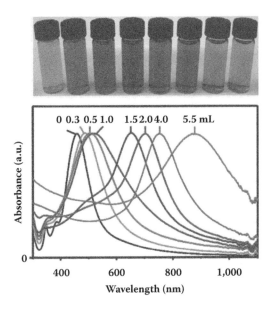

Figure 13.2 *Gold colloids or nanosupensions of various colors. Suspensions containing nanocages prepared by reacting various amounts of HAlCl4 solutions with nanocubes are shown. The difference of the surface properties of the nanocages due to the various amounts of gold present at the surface of the particle results in absorbance at different wavelengths, which is responsible for the various colorful presentations. (From Skrabalak, S.E, J. Chen, Y. Sun, et al. 2008. Gold nanocages: Synthesis, properties, and applications.* Accounts of Chemical Research *41:1587–1595. With permission.)*

with successful studies demonstrating use of gold nanoparticles in DNA diagnostics (Mirkin, 1996).

In 1964, the potential of AuNPs in cancer treatment was demonstrated by Rubin and Levitt. More recently, since the start of the twenty-first century, research publications on synthesis and applications have experienced exponential growth (**Figure 13.3**), demonstrating exciting potential for various applications in diagnostics and drug delivery. However, most of the developed technologies based on AuNPs are still in clinical trials, such as the use of nanoparticles in an analysis of a patient's breath for early detection of head and neck cancers and as immunosensors for the detection of adrenal cortical hormone (Hakim, 2011; Tang, 2005). A particularly exciting phase I clinical trial study was completed by Kharlamov (2012), in which targeting gold nanoparticles used stem cells to treat atherosclerosis plaques. The mechanism of this application was based on the plasmonic resonance of the AuNPs and their capacity to break up plaques after laser irradiation. The study found a significant decrease in atheroma volume in AuNPs-treated patients with an increase of the lumen diameter, as well as significantly lower risks of cardiovascular incidents and death.

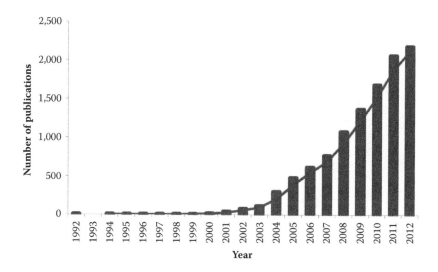

Figure 13.3 *Growth in the number of publications in the area of gold nanoparticles synthesis and applications. The graph represents the number of publications pertaining to gold nanoparticles in the PubMed database per indicated year. The number of publications grew exponentially following the National Nanotechnology Initiative by President Clinton in 2000.*

Although gold nanoparticles have been developed for a plethora of medical applications, most of the medical applications fall into two categories—namely, diagnostic imaging and targeted drug delivery—with many successful designs, such as contrast agents for early cancer detection (Sokolov, 2004).

13.1.2 Synthesis methods

Although the first synthesis of colloidal gold solutions occurred before the age of peer-reviewed literature, the first published work on gold nanoparticle synthesis by Faraday in 1857 paved the way. In this work, Faraday reduced the aqueous solution of chloraurate $AuCl_4^-$ using phosphorus in CS_2. Even today, many of the current methods follow essentially the same strategy, in which solvated gold is reduced in the presence of surface-capping ligands to prevent aggregation of the particles.

Before the advent of the electron microscope in 1932, the structure of AuNPs was unclear (Knoll, 1932). However, new breakthroughs soon followed, with Turkevich publishing the first work on structure and nanoparticle images in 1951 (**Figure 13.4**) (Turkevich, Stevenson, 1951). The Turkevich method for synthesis of AuNP colloids used hydroxylamine hydrochloride and sodium citrate as two nucleating agents in a gold sol. Once manufactured, the nanoparticles were studied using a nephelometer or an electron microscope. The overall reaction of gold chloride and hydroxylamine was first order, with

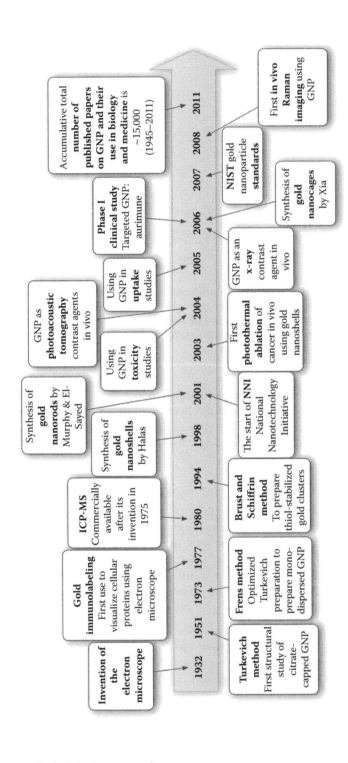

Figure 13.4 *The timeline of development and major breakthroughs in the history of gold nanoparticles. (From Alkilany, A.M., S.E. Lobse, and C.J. Murphy. 2013. The gold standard: Gold nanoparticle libraries to understand the nano-bio interface. Accounts of Chemical Research 46:650–661. With permission.)*

zero order between gold chloride and sodium citrate: $HAuCl_4 + NH_2OH \rightarrow$ Au + 4HCl + NO (Turkevich, Hubbell, 1951).

It is now known that the mechanism of the method proposed by Turkevich occurs via nucleation reaction, where dicarboxyacetone (DCA) intermediate formed by citrate and chloroauric acid (Dreaden, 2012). Further complexation of Au with DCA results in reduction of Au with degration of the complex to form AuNPs. The polymer then degraded to form AuNPs. After the growth of the polymer (organic–inorganic interface) was large enough, an autoassembly or autoreorganization occurs to allow for the formation of AuNPs.

In 1973, the Turkevich methods for citrate-mediated reduction synthesis were extensively studied by Frens, who successfully produced monodisperse spherical gold nanoparticles with diameters ranging from 16 to 150 nm (Frens, 1973). The Frens method perfected the earlier Turkevich method by preparing gold sols using the citrate method, but also introducing sodium citrate to boiling silver nitrate, centrifugation, and redispersion of silver sol in 1% sodium citrate (Frens, 1972). This method allowed for the coagulation of the AuNPs, which were then easily extracted. Frens also proposed that smaller particles lead to more stable colloids (Hostetler, 1996; Templeton, 1998; Hostetler, 1999).

The citrate reduction method has been further improved by others, who regulated the parameters by varying the citrate salt and using an amphiphilic surfactant as a stabilizer. In 1981, Schmid et al. reported a method for phosphine-stabilized AuNPs with a more-controlled molecular weight, smaller size (1.4 ± 0.4 nm), and very narrow polydistribution range. Later on in 2005, Hussain et al. improved the method by reducing the sodium borohydride solution in the presence of alkyl thioethers in a substitution reaction. This method yielded AuNPs that were less than 5 nm.

In 1994, Brust et al. took a different approach to nanoparticle synthesis, experimenting with thiol-stabilized gold clusters in a two-phase system of toluene-dissolved gold chloride, using p-mercaptophenol as a phase transfer reagent along with hydrogen tetrachloroaurate trihydrate (Brust, 1995; Brust, 1994). The above components were mixed in methanol, and acetic acid and sodium borohydrate were introduced to prevent deprotonation. The solution was then stirred and kept under much milder, isothermal conditions ($50°C$). After washing with and evaporating the diethyl ether to remove excess reagents, the solutions were kept under low pressure to obtain AuNPs. The infrared spectrum revealed that the AuNPs incorporated the p-mercaptophenol as part of their structure. This method, which produced AuNPs of approximately 2 nm, later allowed for further assembly of larger AuNPs. The stability of the NPs was also significantly improved, making them more yielding to manipulation without precipitation out of the product.

Using this so-called Brust-Schiffrin method as the basis for further development of adjustable parameters, it has been repeatedly demonstrated that the size of AuNPs can be conveniently regulated by adjusting thol/gold mole ratios, cooling the solution, using bulky ligands, or quenching the reaction (Chen, 1999; Hostetler, 1996; Templeton, 2000). After this single-solvent approach was developed, the method was widely used for the synthesis of AuNPs (also called monolayer-protected clusters) to contain functionalized thiols (Brust, 1994; Templeton, 1998; Hostetler, 1999; Templeton, 2000; Hostetler, 1998). The versatility of the AuNPs that can by synthesized allow for many applications in various fields.

In 1997, Demaille et al. demonstrated that AuNPs could be assembled into ultramicroelectrodes, with the parameters perfectly suitable for use as a probe in scanning electron microscopy (Demaille, 1997). In 1998, Chen et al. used 10-nm and 30-nm particles for improvement of glucose oxidase activity (Chen, 1998). The attachment of glucose oxidase to 10-nm AuNPs allowed for the best improvement in enzyme activity due to the quantum size effect. While researching the optical properties of the AuNPs dependent on the physical parameters such as shape of the AuNPs, Zhou et. al. introduced a new synthesis method by seeding of the nanoparticles (1994). Nanoparticles with a thin gold shell were produced by reacting synthesized Au_2S with NaS; when the Au–S bond is reduced, the particle gradually becomes AuNP. This method allowed for preparation of 4- or 5-nm-diameter quantum dots and nanometer-sized gold nanoparticles in all dimensions. Using this method, the size of the quantum dots can be manipulated by variations of seeds in the metal salt content.

Perrault and Chan (2009) later used a similar method when seeking to increase the particle size beyond 50 nm, which was a limitation for previous methods. In this method, the particle seeds of gold chloride were reduced by hydroxyquinone (HQ), with a correlation between seed size and final AuNP particle size. HQ, as a reducing agent, improved monodispersity and lowered new nucleation, resulting in improved-quality particles of 200 nm in diameter.

With the start of new millennia, new improvements in synthesis and applications of AuNPs soon followed. In 2000, Weare et al. improved upon the Brust-Schiffrin method, showing that a range of sizes (1.4–10 nm) of phosphine-stabilized nanoparticles can be synthesized by ligand exchange under less rigorous reaction conditions. Furthermore, the digestive ripening method was developed by Prasad et al. in 2002, in which the heating of a colloidal suspension with alkanethiols near the boiling point allowed for synthesis of AuNPs with much reduced average size (4.5 nm) and narrow polydispersity (Prasad, 2003). Once cooled to room temperature, these AuNPs would organize in two-dimensional (2D) and three-dimensional (3D) superlattices with hexagonal packing, which then further led to their study. Subsequently, a novel decomposition method of $AuCl(PPh_3)$ was developed by Khomutov (2002), where decomposition of an insoluble metal-organic precursor with amphiphile monolayer resulted in 2D growth of nanoparticles at the gas–liquid interphase upon reduction in the monolayer (Khomutov, 2011).

The form of any particle is dictated by its lowest energy state. Most methods, until more recently, were only successful in manufacturing spherical AuNPs, which were easy to detect using transmission electron microscopy but served a limited purpose. With the advent of different technologies, vibrational spectroscopy, and the newer methods of thermogravimetric analysis and polymer-based synthesis, different shapes of AuNPs are being produced regularly (Dreaden, 2011; Alexandridis, 2011; Langille, 2012; Alkilany, 2012). These new shapes include, but are not limited to, nanorods (Wu, 2013), nanocubes (Guan, 2012; Sau, 2004), nanocages (Mackey, 2013), nanodendrites (Wang, 2012), star-shaped AuNPs (Melnikau, 2013), and multicomponent nanoassemblies (Khaletskaya, 2013), nanodumbells (Huang, 2006), nanoshells, nanoprisms, and other high-index faceted shapes (Kuo, 2004). Most of the methods attain these structures via control of reaction kinetics in the process of synthesis, which are accomplished primarily by varying ratios of the reagents, introducing silver ions (which cause an increase in the number of facets), or adding higher halides. For details concerning the methods of the synthesis of various shapes and high-index faceted AuNPs, see the extensive review by Langille et al. (2012).

The optical properties of AuNPs are a fascinating area of study that may drive the visually cued scientist to search for the treasures that lie beneath their watery solutions. When exposed to radiation, the plasmon effects on different AuNP shapes, sizes, and resonant frequencies allow for different wavelengths of light to be refracted and emitted, which are observed as different colors. Due to the high sensitivity of AuNPs to dielectric constants and their high surface-area-to-volume ratios, perturbation of colloids or bulk particles yields different colors. These properties, along with Rayleigh scattering, are responsible for the relationship between particle size, shape, and visible color. Nanorods change color with increasing aspect ratios, nanospheres turn more violet with increasing size, and nanocages turn more violet with decreasing gold percentages (**Figure 13.5**). The spheres are affected by a single plasmon band; thus, changes in particle size result in less dramatic changes in color. However, nanorods, which are affected by two plasmon bands, show greater contrasts in color with increasing particle size (Alkilany, 2013; Fischer et al., 2013).

Other methods of synthesis of AuNPs using a source of energy to induce growth of the nanoparticles have been demonstrated, including laser photolysis and radiolysis, which allow tight control over the parameters of the nanoparticles (Bronstein, 1999; Henglein, 1998; Dawson, 2000; Gachard, 1998). Sakamoto et al. (2006) demonstrated successful laser-induced formation of AuNPs, in which the laser flash results in the production of radicals, inducing growth of the nanoparticles in polyvinyl alcohol solution. A similar mechanism was used by Meyre et al. (2008), where γ-irradiation was used in the synthesis of AuNPs aligned in lamellar sheets.

Research on the structural organization of AuNPs using polymers arose from the need to stabilize colloidal gold solutions in order to make them suitable for uses in semiconductor, biological, pharmaceutical, and medical fields. Earlier

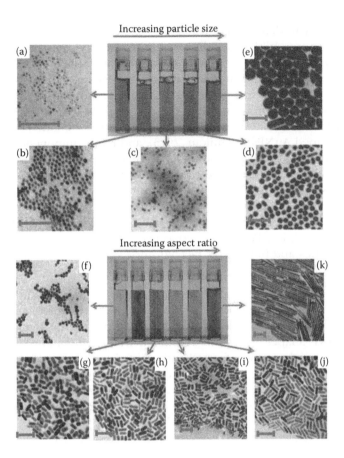

Figure 13.5 *Optical properties of gold nanoparticles with variations of physical parameters. An increase in nanosphere size results in a more violet color (a–d) and an increase of the aspect ratio results in a color change (f–j). (Figure 13.5a is reprinted from Murphy C.J., A.M. Gole, J.W. Stone, et al. 2008. Gold nanoparticles in biology: Beyond toxicity to cellular imaging.* Accounts of Chemical Research *41:1721–1730. With permission.)*

studies by Gittins demonstrated layer-by-layer polymer deposition using the covalent attachment of anionic thiols for polyelectrolyte modification, thus allowing for the construction of very stable multilayers (Gittins, 2001). The benefits of using polymers in AuNP assembly include inhibition of agglomeration of the nanoparticles, in situ synthesis, and biocompatibility. In addition, the polymerization matrix around the nanoparticles has a great capacity for functionalization.

Another synthesis approach based on the use of templates, such as silica, was demonstrated by Pol et al. (2003), who successfully assembled small 5-nm AuNPs using silica submicrospheres. Deposition of gold on the surface of a silica bead was facilitated by ultrasound, using a relatively simple process that allows scale-up of the procedure and reproducibility. Other reports of successful assembly of AuNPs within the pores of silica have been published (Pol, 2003; Chen, 2001; Chen, Cai, Liang, and Zhang, 2001).

Although organic synthesis presents many exciting opportunities for applications in various areas ranging from computers to pharmacotherapeutics, concerns about the health safety of AuNPs remain. Chauhan et al. (2011) proposed a method of synthesis of nanoparticles using a biologically inspired model in a *Candida albicans*-mediated synthesis of gold nanoparticles. In this study, a cytosolic extract of *C. albicans* was added to aqueous $HAuCl_4$. After an incubation period of 24 hours, 60-nm and 20-nm particles were produced, depending on the amount of the cytosolic extract added. Furthermore, the group demonstrated the capacity of these nanoparticles to be used for the detection of liver cancer. Other fungus-mediated methods were also reported; however, these methods were rather time consuming (Sharma, 2012; Veeraapandian, 2013; Deplanche, 2013). In an attempt to expedite biologically mediated synthesis, Thakker et al. (2013) used *Fusarium oxysporum* f. sp. *cubense* JT1 for successful accomplishment of 22-nm AuNPs with antimicrobial activity from the organism capping agents. Synthesis of the nanoparticles was completed in 60 minutes via oxidation of $HAuCl_4$ by an organism-produced radical species.

In summary, in order to attain specific parameters of nanoparticles, various methods are available. The versatility of the methods, however, demands careful design to meet the specific demands of the applications in mind. Specifically, in the area of medical research, concerns about health safety of the developed therapeutic agents are a primary issue. Because the AuNPs exhibit high variability in properties based on factors that pertain to both the design of the NP and its environment, gold nanoparticles present exciting ranges of possibilities for therapeutic application of targeted action with low off-target and potentially toxic activity. Necessary tight regulation of the nanoparticle parameters can potentially be accomplished by various coatings and functionalization.

13.1.3   Coatings and functionalization of AuNPs

Gold nanoparticles have vast possibilities for applications in the medical field. However, surface functionalization is required for the particles to be targeted to a specific area of the body for interaction with a particular type of cell or molecule, as well as for delivering stable particles.

13.1.3.1   Stability enhancement

Usually, electrostatic repulsion between charged ligands on the surface of hydrophilic nanoparticles contributes to their stability (R.A. Sperling, 2010). However, interactions between nanoparticles can change dramatically in different biological environments, causing precipitation of the nanoparticles. For example, the high osmolarity in blood serum diminishes the electric field between nanoparticles, which leads to aggregation via attractive forces

(Timo Laakonen, 2006). Furthermore, the charge of the nanoparticles may change or be eliminated, depending on the pH and isoelectric point of the solution. At physiological pH, AuNPs can cause aggregation of proteins to the surface of the AuNPs and to each other, which leads to nanoparticle-induced protein misfolding that can potentially induce toxicity to the cells (Dongmaio, 2009; Jenifer, 2012).

To address the stability issues of nanosuspensions, NPs can be supplemented with a complementary ligand coating, such as sulfur-containing compounds (Mathias, 1994) or amino acids (PR. Selvakannan, 2003; Shaheena, 2013), to prevent complications in various environments. The solvent, particle size, and material of the core must be considered when determining which ligand molecules are sufficient to stabilize a particular nanoparticle (R.A. Sperling, 2010). When dealing with aqueous solutions, ligands with a strong charge have a prolonged stabilizing effect and continue to stabilize the particle in high salt concentrations. However, ligands that deliver steric stabilization have a better stabilizing effect in high salt concentrations than those with a strong charge (Takeshi, 2005). In some cases, a combination of electrostatic and steric stabilization may be used (Victoria, 2001; Gerhard, 2002).

AuNPs that were stabilized electrosterically show an initial decrease in stability due to the loss of electrostatic stabilization; however, at higher salt concentrations, the stability of the nanoparticles dramatically increased (Maryann, 1992). This becomes important in medical applications of nanoparticles when delivery to the bloodstream is expected; in the biological system, blood serum exhibits high osmolality due to normal buffer systems and dissolved proteins in the blood.

13.1.3.2   Sulfur-containing ligands

The abundance of sulfur in cells allows for AuNPs to have broad bioapplications, such as efficient drug-delivery platforms (Marie-Christine, 2011; Zhifei, 2013), antibiotic detection (Jamee Bresee, 2010), and microRNA detection (Shao-Yi, 2012). Also, citrate-stabilized AuNPs functionalized with thiolated polyethylene glycol (PEG) can be used as contrast agents for in vivo x-ray computed tomography imaging (Dongkyu, 2007). Furthermore, sulfur-containing ligands have a great capacity to stabilize AuNPs due to the strong bond (47 kcal/mole) between gold and sulfur. This was first demonstrated in 1993, when Giersig and Mulvaney used alkanethiol ligands for the stabilization of gold nanoparticles. Later published, the Brust–Schiffrin method of AuNPs synthesis provides temperature- and air-stable AuNPs with thiol ligands, controlled size, and decreased dispersity in size, which can also withstand repeated isolations and can be redissolved without decomposition or permanent aggregation (Brust et al. 1994).

Many other sulfur-containing ligands are successful at stabilizing AuNPs. For example, dithiols (Christof Baur, 1999), trithiols (Niti, 2010), xanthates

(O. Tzhayik, 2002), disulfides (Lon, 1998; Tetsu Yonezawa, 2000), thio-acetates (Zhang, 2009), dithiocarbamates (Vickers, 2006), trithiolates (Wojczykowski, 2006), thioctic acid (Volkert, 2011), and resorcinarene tetrathiols (Balasubramanian, 2002) have been used to stabilize AuNPs; however, ligands are application specific. For example, thiols used for catalysis (Tomoya, 2012) are more successful at stabilizing AuNPs than disulfides (Lon, 1998). Thioethers do not bind AuNPs strongly; however, it was found that polythioethers do bind strongly, which is important in preventing disassembly (Xue, 2001). Reversible AuNP assemblies can be produced by using tetradentate thiethers (Daniel, 2004). Also, resorcinarene tetrathiols are effective at stabilizing mid-nanometer-sized (16–87 nm) AuNPs in organic solvents by improving dispersion and stability (Balasubramanian, 2002).

13.1.3.3 Lysine stabilization

The production of water-dispersible nanoparticles has broad implications, especially in the areas of biolabeling (Gittins, 2002) and biosensing (Cumberland, 2001). Although thiolate ligands are great at stabilizing AuNPs, arylthiolate ligands containing alcohol, carboxylic acid, or amine terminal groups do not allow the AuNPs to be soluble in aqueous solutions (Selvakannan, 2003). AuNPs capped with the amino acid lysine are an alternative to thiol ligands; lysine electrostatically stabilizes the particles in aqueous solutions, thus preventing aggregation in biological environments. Also, lysine renders the AuNPs stable in air; therefore, these nanoparticles can be produced as a dry powder by evaporating the aqueous solvent. This powder is very stable in air and is easily dispersed in water. Therefore, AuNPs capped with lysine could potentially be applied to intravenous and inhalation therapies. Future studies are required for these possible applications.

Although the specific interaction between lysine and the AuNPs is not completely understood, the lysine attached to the AuNPs was found to be more stable than lysine alone. However, these AuNPs aggregate into large superstructures at pH 10, which can be dispersed again if the pH drops.

13.1.3.4 Alkalothermophilic actinomycetes

Biologically mediated synthesis of nanomaterials has been of increased interest in order to avoid the use of toxic chemicals that are often generated during chemical synthesis. AuNPs have been synthesized by bacterial cells by incubating the cells with Au^{3+} ions (Southam, 1996; Mewada, 2012) or by alfalfa plants if solid media containing Au ions is supplied (Gardea-Tarresdey, 2002).

A study by Ahmad et al. identified the alkalothermophilic actinomycete *Thermomonospora* sp. as a candidate for the production of AuNPs via a chemical reaction of biomass with aqueous chloroaurate ions (2003). This method efficiently produced AuNPs with monodisperisty and an average size of 8 nm. However, the reduction of $AuCl_4^-$ is very time consuming—the reaction is complete after approximately 120 hours. The AuNPs produced by *Thermomonospora* sp. remained stable for more than 6 months and showed little aggregation because a proteinaceous capping agent, likely secreted by the biomass, was present in the solution. The structure of this capping agent could be established in future studies to use for synthesis and the stabilization of AuNPs. Furthermore, living organisms can be genetically modified to produce stable AuNPs via recombinant DNA technology and protein engineering.

13.1.3.5 Polymers, microemulsions, reversed micelles, surfactants, membranes, and polyelectrolytes: Stabilization and functionalization properties

The use of polymers to stabilize AuNPs has many benefits, such as longer stability, the ability to adjust solubility, increased amphiphilicity, high surface density, and processibility (Juan Shan, 2007). AuNPs stabilized with polymers are currently being used for catalysis (Masatake Haruta, 2001; Avelino Corma, 2006), optics (Bardhan, 2011; Schatz, 2007), and biology (Sperling, 2008). Polymers are an attractive option for coating because they are easily incorporated in the synthesis of metal nanoparticles, such as AuNPs (Huhn, 2013).

In addition, reversed micelle solutions can be produced by using surfactants. These solutions are thermodynamically stable, transparent, and isotropic. Surfactant molecules stabilize the reverse micelles at the interface of water and oil; the polar hydrophilic core provides a template for AuNP synthesis and it also prevents aggregation. The size and shape of reverse micelles produced by copolymer surfactants can be easily modified by changing the composition or structure of the copolymer and the length of the fundamental blocks (Penxiang Zhao, 2013). The stability of the AuNPs can be modified by adjusting the length of the coronal blocks (longer blocks increase stability) and amount of NaBH4 used for reduction (Filali, 2005).

Functionalization of the nanoparticles allows for a variety of applications. AuNPs have been used for specific detection of polynucleotides (Elghanian, 1997), as contrast agents for fluorescent and x-ray dual-modality imaging (Zhang, 2012), as scaffolds for 2D and 3D building (Shenhar, 2003; Luo, 2011), as bioprobes for real-time imaging (Leduc, 2013) and bimetallic catalysts (Villa, 2010), and many other applications.

13.2 Therapeutic applications of gold nanoparticles

13.2.1 Diagnostic applications

Gold nanoparticles have many diagnostic applications, which can be inexpensive and noninvasive. Colloidal gold is easily synthesized in large quantities and high quality by various methods that can yield a variety of gold nanoparticle structures, including nanorods, hollow nanoparticles, and silica/gold nanoshells (Huang, 2010). Gold nanoparticles become extremely useful in molecular imaging, especially in cancer, because of their increased radioactive properties of absorption and scattering. These properties allow for enhanced targeted molecular imaging and improved contrast of diagnostic techniques.

Gold nanoparticles have also been of use in improving important diagnostic techniques, such as computed tomography (CT) (Reuveni, 2011). Current CT techniques are limited in the areas of targeted molecular imagery and imaging time. Unlike magnetic resonance imaging, CT is not considered for its molecular targeting modality because of a lack of specific contrast agents. The contrast agents currently being used are iodine-containing agents that absorb the x-rays well but do not specifically bind to any one biological component; they also are readily expelled by the kidneys, which is the cause of short imaging times.

Gold nanoparticles can focus on targeted tumor antigens and thus provide higher contrast in CT imaging (Geso, 2007; Hainfeld, 2006). Attenuation coefficients differ based on atomic number; gold has a much higher atomic number than iodine and thus provides a sharper contrast. Gold provides about 2.7 times greater contrast per unit weight and is absorbed by soft tissue instead of bone, which lowers the dose of radiation exposure. The large molecular weight also makes the nanoparticles less readily expelled; thus, gold has a longer blood retention time, making it possible to use intravenous injection instead of the more invasive catheterization.

Currently, gold agents are not being used in clinical settings, but they have been patented, including an innovative stabilized gold nanoparticle contrast agent (WO 3009005752 A1). The future of nanoparticle technology aims for earlier detection and staging and microtumor identification in cancer diagnoses. Improvements in targeted molecular imaging have already been demonstrated in squamous cell carcinoma (SCC), which is characterized by the overexpression of the A9 (alpha6-beta4 integrin) antigen, whose expression level is also correlated to the maturity of the tumor (i.e., ability to metastasize) (Reuveni, 2011). In this study by Rueveni et al., AuNPs were manufactured and conjugated with UM-A9 antibodies specific to SCC tumor cells (2011). CT showed a significant contrast in comparison to normal surrounding cells due to the high density of gold nanorods attached to the tumor cells. The ability to conjugate gold nanoparticles with particular antibodies, as well as for ligands to bind to overexpressed antigens and receptors, improves the ability of CT as a targeted molecular imaging modality.

In addition to molecular imaging, gold nanoparticles have been used in an array of sensors that can distinguish the breath of healthy patients from patients with lung cancer based on the presence of volatile organic compounds (VOCs), which serve as cancer biomolecules (Peng, 2009). This is based on the concept that tumorigenic cells are complemented by gene and protein defects in the tightly packed mosaic cell membrane, triggering oxidative stress and peroxidation of the proteins and phospholipids, thus leading to the emission of VOCs (Barash, 2012). Gas chromatography studies have shown several of these compounds are in concentrations of 1–20 ppb in healthy patients, whereas they are present in extreme levels of 10–200 ppb in patients with lung cancer (Peng, 2009).

Currently, lung cancer diagnosis involves an invasive procedure to retrieve a tissue specimen, and it is highly limited due to small nodule size and challenging accessibility. Breath analysis is not a new procedure, but the functionalized gold nanoparticles allow molecules from volatile organic compounds to be detected without pretreating the exhaled breath so that the relative concentration of the cancer biomarkers increases to a detectable level. The sensor can be composed of chemiresistive layers of gold nanoparticles in chloroform solution on top of a semicircular microelectronic transducer (Barash, 2012). The gold nanoparticles are coated with a variety of organic ligands, which react to the different VOCs that are present in breath. Different optical separation of gold nanoparticles was seen in samples from healthy patients as well as between the different subsets of lung cancer (small cell and non-small cell), which was indicated by the change in resistance of the chemiresistor. In addition to distinguishing the breath of patients with lung cancer from healthy patients, these sensors were also able to distinguish between the histologically different types of lung cancers based on subtle differences in the metabolites present in the headspace of the cancer cells.

Another use of nanoparticles involves their use in aggregation-based immunoassays along with micrometer-sized latex particles to cultivate a home pregnancy test (Sadauskas, 2009; Thanh, 2002). In this particular assay, both the nanoparticles and the latex particles are derivatized with beta-human chorionic gonadotropin (HCG), which is normally released by pregnant women shortly after conception. When this assay is mixed with a urine sample, the particles agglomerate to form very notable pink aggregates. The indicator thus turns pink when HCG levels are high, indicating pregnancy. This application of gold nanoparticles is widely used on the market in popular home pregnancy tests.

Gold nanoparticles are at the forefront of cancer research due to their tunable optical properties and compatibility with a wide variety of ligands and antibodies whose biological interactions can provide targeted exposure (Huan, 2010). These properties make gold nanoparticles highly valuable in many diagnostic applications. Their high attenuation coefficient and properties of absorption and scattering make them an efficient contrast dye as well as a module for targeted molecular imaging. Their use in biosensors can provide an inexpensive noninvasive tool that can provide in-depth lung cancer screening. Gold nanoparticles are already used in many home pregnancy tests, which are inexpensive and easy to use.

13.2.2 Therapeutic applications

Nanoparticles can be a thousand times smaller than a human cell, offering unprecedented interactions with extracellular and intracellular molecules, which could be the future of cancer treatment. The multifunctionality of gold nanoparticles makes them highly applicable. The nanoparticles' ability to be conjugated with organic ligands and antibodies makes them a valuable vehicle for targeted cancer therapy. These conjugated nanoparticles can interact with tumor cells that overexpress certain receptors and antigens to deliver drugs effectively. In addition to organic ligands and antibodies, these gold nanoparticles can be coated with a variety of natural and biological compounds to overcome biological barriers (Kudgus, 2011).

Photodynamic therapy (PDT) is a technique that begins with the intravenous injection of photosensitizers—hydrophobic agents usually located in the plasma membrane of cells (Cheng, 2008). These photosensitizers can then be excited by light of a specific wavelength, which will generate highly reactive oxygen species within the cells, thus inducing apoptosis and cell death. One of the challenges of PDT is low solubility of the APIs, resulting in low bioavailability and tissue saturation; however, a certain degree of hydrophobicity is required for penetration of the cell membranes. PEGylated gold nanoparticles can be used as a drug delivery vehicle to help PDT APIs to solubilize in the blood and rapidly reach the organelles of the cells. Studies have shown that when PEGylated gold nanoparticles are conjugated with the PDT drug, the administration of therapy to cells is nearly twice as fast as when using the PDT drug alone.

Although the concentrations of systemically administered gold nanoparticles are highest in the liver and spleen after intravenous administration, a considerable amount is detected in the brain, indicating that 15-nm gold nanoparticles are able to penetrate the blood–brain barrier (Schroeder, 1998). This has many applications for drug delivery to the brain because drugs are often excluded from the brain based on their chemical properties, such as molecular weight and hydrophobicity.

The area of ocular drug delivery has been a challenge due to the natural barriers of the eye, such as the blood–aqueous and blood–retinal barriers, for any one route of administration. In a study by Guadana et al., intravenously administered gold nanoparticles passed through the blood–retinal barrier and were evenly distributed in the ocular layers in C57BL/6 mice (2010). These nanoparticles exhibited no cytotoxicity; cells containing the nanoparticles did not demonstrate structural abnormalities or cell death, indicating that small (20-nm) gold nanoparticles may be an acceptable form of drug delivery in vivo.

Gold nanoparticles can be used in a variety of therapeutic applications providing targeted drug delivery due to their ability to overcome biological barriers. The targeting of these gold nanoparticles takes advantage of cancer cells' tendencies to overexpress certain ligands. Gold nanoparticles can thus be engineered to contain a ligand with strong binding affinity for these overexpressed receptors to provide targeted drug delivery. One such use of targeted therapy is the use of tea-extracted compounds for drug delivery to the prostate in prostate cancer. The compound in tea is a part of polyphenolic flavonoids, called epigallocatechin gallate (EGCG), and is also the subject of research dealing with the anticarcinogenic effects of drinking tea. Current radiation prostate cancer treatments include brachytherapy, intensity-modulated radiation therapy, and proton therapy, among others. Most of these treatments are unsuccessful for aggressive forms of prostate cancer because of their inability to penetrate tumor cells, low radiation retention around tumor, and other factors causing ineffective dose delivery.

The phytochemical EGCG in tea is an effective reducing agent. It also reacts with gold salt to form nanoparticles with a rough coating of the prostate tumor-specific phytochemical on the surface (Shukla, 2012). EGCG selectively binds to Laminin67R receptors, which are overexpressed in prostate tumor cells, thus allowing for more effective drug delivery due to a smaller delivery medium and increased localization provided by the tea-extracted compound. The feasible synthesis of these nanoparticles is a completely green process involving the combination of an aqueous solution of sodium tetrachloroaurate ($NaAuCl_4$) and an aqueous solution of Darjeeling tea leaves, incubated at 25°C for 30 minutes. The tea-integrated nanoparticles have been shown to provide effective delivery in both prostate (PC3) and breast (MCF5) cancer cells due to the similarity of overexpressed receptors.

Based on the surface plasmon resonance property of gold nanoparticles, they can convert light energy into heat to induce hypothermia, in which the cell temperature increases to more than 40°C (Qu, 2012). This photothermal therapy uses heat to cause cell death and apoptosis of tumor cells, providing a safe therapeutic technique for cancer treatment. Efficient photothermal therapy relies on specific targeting of tumor cells, so that only minimal heat and damage is applied to surrounding normal cells. Antibody-conjugated gold nanoparticles differ in cellular uptake among different cell types, depending on what overexpressed receptors are present on the surface of the tumor. In Hodgkin cells (L-428), the CD30 receptors are overexpressed. Gold nanoparticles were conjugated with BerH2 antibodies, which specifically bind to the CD30 receptors that are overexpressed in Hodgkin cells—effectively killing them with the application of laser irradiation.

For applications in targeted drug delivery, it is important to understand how the charged nanoparticles interact with double-stranded DNA. Even weakly charged cationic nanoparticles induce DNA bending and strand separation (Ralisback, 2012). The agglomeration of hydrophobic gold nanoparticles leads to the intercalation of the two strands, which disrupts the base pairing. Furthermore, the charged groups on the nanoparticles may hold onto the negatively charged strands, bending them or unzipping them. On the other hand, AuNPs can be engineered by tuning the charge and hydrophobicity to induce a structural change in DNA, which may improve the success of transfections (Xu, 2013). A study by Tandon et al. (2013) demonstrated the application of this interaction of the AuNP with DNA in inhibiting fibrosis by introduction of the *BMP7* gene into rabbit keratocytes in vivo.

As demonstrated in this chapter by a few examples of AuNP applications, AuNPs have promising potential for the early diagnosis and treatment of cancer, diagnostic imaging, and targeted delivery. This nanotechnology allows physiological barriers to be overcome, thus offering a host of new opportunities and targeting of the treatment in order to obtain therapeutic effects in the most challenging disorders, such as cancer, Alzheimer disease, retinal diseases, and others.

13.3   Challenges and future perspectives

AuNPs present exceptional opportunities for the next-generation of pharmaceuticals. However, scaling up the synthesis may present a challenge in the future due to the need for tight control of the method's parameters, which may be costly. Nevertheless, an especially challenging area is the determination of the potential toxicity implications and the health safety of pharmacological systems containing AuNPs. Extensive in vitro and in vivo studies have been performed to understand the pharmacokinetics and toxicological implications (Chuang, 2013; Di Guglielmo, 2012; Alkilany, 2010; Lewinski, 2008, Dreaden, 2012). The studies tend to indicate low toxicity; however, smaller-sized particles of 2 nm or less may have increased toxicity because of their higher reactive surfaces (Alkilany, 2010; Turner, 2008). To overcome this challenge, various coatings, such as hyaluronic acid or PEG (Di Guglielmo, 2012; Zhang, 2011), may be implemented, although some of the correlation between toxicity and size may still exist. This may be due to difference of biointeractions of smaller nanoparticles, as biodistribution tends to vary with size.

Zhou et al. demonstrated in their studies that small AuNPs (2 nm) form larger aggregates in the blood and may not be cleared by the kidneys as easily (2011). Furthermore, the size of the nanoparticles also has an effect on their metabolism and distribution in the body. The elimination of nanoparticles of less than 5.5 nm in diameter may also be filtrated and excreted by the kidneys (Jain, 2006). On other hand, most of the larger nanoparticles (5–30 nm) accumulate in the liver and spleen due to the reticuloendothelial system responsible for capturing and eliminating foreign antigens (Khlebtsov, 2011; Zhang, 2011; Longmire, 2008). These particles may be slowly eliminated by liver and excreted (Cho, 2010); however, more studies are required to determine this possibility. However, the metabolism of the nanoparticles and its potential toxicity may change based on the route of exposure. For example, in a study by Schulz et al. (2012), AuNPs of 2, 20, and 200 nm did not lead to significant toxicity, as assessed by DNA damage.

Specific AuNP properties, such as charge and surface functionalization ligands, may further complicate the estimation of toxicity. For instance, positively charged AuNPs are believed to penetrate membranes more effectively, with potentially increased toxicity (Ghosh, 2008; Kim, 2010). This effect may be due to the interactions of the charged nanoparticles with the biological environment with formation of protein corona. Specifically, Torrano et al. demonstrated that the capping agents are more important in determining the interaction of the AuNPs with the membrane. However, toxicity of the AuNPs may arise from any component of the system, such as the capping agents, degradation products, organic solvents or chemicals used in the synthesis process, or any other impurities. To understand the toxicity of these components, numerous studies have been conducted, demonstrating that toxicity is specific to factors other than the core of the nanoparticles themselves (Hauck, 2008; Alkilany, 2009).

In sum, the future of AuNPs in medicine and pharmacy depends on a successful design of nanoparticles, with consideration of the specific parameters of the core and surface, to achieve biocompatibility with therapeutic effect. Further research on synthesis and purification methods is required to ascertain health safety. Given the high versatility of the synthesis methods and excellent capacity for functionalization, we expect that more pharmaceutical systems based on AuNPs will be developed to address the current needs of medical field by offering innovative treatments for the most challenging medical issues.

13.4 Conclusion

As a precious metal, gold has been historically highly esteemed. With the rise of the nanotechnology field, the role of gold has been reinvented in a whole new way. There is immense and exciting potential for numerous applications in the pharmacological and medical fields, allowing for ways to overcome many challenges, such as the penetration of biological barriers, targeted delivery, and off-site toxicity.

The versatility of the methods of synthesis and functionalization of nanoparticles may present more potential applications than are currently documented in the literature. However, it is immensely important to ascertain the safety of these therapeutic applications. Some of the safety concerns arise in the areas of immunological toxicity due to normal responses of the body, as well as cytotoxicity due to modification of genes and proteins as a result of the high reactivity of the gold nanoparticles. Functionalization and careful design of the coatings of gold nanoparticles may alleviate some of those risks. Nevertheless, the need for long-term toxicity studies is essential in order to ascertain safe and therapeutic treatments. To address these safety concerns, the National Cancer Institute has established the Nanotechnology

Characterization Laboratory to standardize toxicity characterization methods (McNeil, 2011). As the field develops further, we expect the laboratory will address the safety concerns for clinical trials by conducting both in vitro and in vivo studies.

Along the lines of toxicity, biocompatibility and safe elimination are among the prime issues in the field of nanotechnology. Specifically, in order to use gold nanoparticles for therapeutic applications, their safe processing and elimination are essential. Careful design of nanoparticles and their surfaces is required for appropriate metabolism from the time of exposure to elimination. Therefore, knowledge about physiological processes and interactions on the nanoscale is essential.

Furthermore, because many synthesis methods require tight control of reaction conditions, the synthesis of the AuNPs is difficult and expensive to scale up. Some of the recent methods usng biological organisms present a promising path towards more economical and efficient production of AuNPs, which perhaps may take advantage of the current methods of industrial manufacturing of bacterial products; however, further investigation in the field is required.

In conclusion, materialization of these exciting opportunities presented by nanotechnology would require the joint effort of researchers in chemistry, physics, biology, engineering, and perhaps other disciplines in designing and manufacturing nanostructures, understanding their activity, and ensuring safety for therapeutic applications in a clinical setting.

References

Ahmad, A., S. Senapati, M.I. Khan, R. Kumar, and M. Sastry. 2003. Extracellular biosynthesis of monodisperse gold nanoparticles by a novel extremophilic actinomycete, *Thermomonospora* Sp. *Langmuir* 19:3550–3553.

Ahmad, T., I.A. Wani, N. Manzoor, J. Ahmed, and A.M. Asiri. 2013. Biosynthesis, structural characterization and antimicrobial activity of gold and silver nanoparticles. *Colloids and Surfaces B: Biointerfaces* 107C:227–234.

Alexandridis, P. 2011. Gold nanoparticle synthesis, morphology control, and stabilization facilitated by functional polymers. *Chemical Engineering & Technology* 34:15–28.

Alkilany, A.M., S.E. Lohse, and C.J. Murphy. 2012. The gold standard: Gold nanoparticle libraries to understand the nano-bio interface. *Accounts of Chemical Research* 46:650–661.

Alkilany, A.M., and C.J. Murphy. 2010. Toxicity and cellular uptake of gold nanoparticles: What we have learned so far? *Journal of Nanoparticle Research* 12:2313–2333.

Alkilany, A.M., P.K. Nagaria, C.R. Hexel, T.J. Shaw, C.J. Murphy, and M.D. Wyatt. 2009. Cellular uptake and cytotoxicity of gold nanorods: Molecular origin of cytotoxicity and surface effects. *Small* 5:701–708.

Antonii, F. Panacea aurea-auro potabile. Hamburg: Ex Bibliopolio Frobeniano 1618:250.

Ashmi, M. G. Oza, S. Pandey, and M Sharon. 2012. Extracellular synthesis of gold nanoparticles using *Pseudomonas denitrificans* and comprehending its stability. *Journal of Microbiology and Biotechnology Research* 2:493–499.

Avelino, C., and P. Serna. 2006. Chemoselective hydrogenation of nitro compounds with supported gold catalysts. *Science* 313:332–334.

Balasubramanian, R., B. Kim, S.L. Tripp, X. Wang, M. Lieberman, and A. Wei. 2002. Dispersion and stability studies of resorcinarene-encapsulated gold nanoparticles. *Langmuir* 18:3676–3681.

Barash, O., N. Peled, U. Tisch, P.A. Bunn, Jr., F.R. Hirsch, and H. Haick. 2012. Classification of lung cancer histology by gold nanoparticle sensors. *Nanomedicine* 8:580–589.

Bardhan, R., S. Lal, A. Joshi, and N. J. Halas. 2011. Theranostic nanoshells: From probe design to imaging and treatment of cancer. *Accounts Chemical Research* 44:936–946.

Baur, C., R. Bugacov, B.E. Koel, et al. 1999. Linking and manipulation of gold multi-nanoparticle structures using dithiols and scanning force microscopy. *The Journal of Physical Chemistry B* 103:3647–3650.

Bresee, J., K.E. Maier, C. Melander, and D.L. Feldheim. 2010. Identification of antibiotics using small molecule variable ligand display on gold nanoparticles. *Chemical Communications* 46:7516–7518.

Bronstein, L., D. Chernyshov, P. Valetsky, N. Tkachenko, H. Lemmetyinen, J. Hartmann, and S. Forster. 1999. Laser photolysis formation of gold colloids in block copolymer micelles. *Langmuir* 15:83–91.

Brown, L.O., and J.E. Hutchison. 2001. Formation and electron diffraction studies of ordered 2-D and 3-D superlattices of amine-stabilized gold nanocrystals. *Journal of Physical Chemistry B* 105:8911–8916.

Brust, M., J. Fink, D. Bethell, D.J. Schiffrin, and C. Kiely. 1995. Synthesis and reactions of functionalized gold nanoparticles. *Journal of the Chemical Society–Chemical Communications* 16:1655–1656.

Brust, M., M. Walker, D. Bethell, D.J. Schiffrin, and R. Whyman. 1994. Synthesis of thiol-derivatized gold nanoparticles in a 2-phase liquid-liquid system. *Journal of the Chemical Society–Chemical Communications* 7:801–802.

Chauhan, A., S. Zubair, S. Tufail, et al. 2011. Fungus-mediated biological synthesis of gold nanoparticles: Potential in detection of liver cancer. *Int J Nanomedicine* 6:2305–2319.

Chen, M.S., and D.W. Goodman. 2004. The structure of catalytically active gold on titania. *Science* 306:252–255.

Chen, S.W. 1999. 4-Hydroxythiophenol-protected gold nanoclusters in aqueous media. *Langmuir* 15:7551–7557.

Chen, T.I., P.H. Chang, Y.C. Chang, and T.F. Guo. 2013. Lighting up ultraviolet fluorescence from chicken albumen through plasmon resonance energy transfer of gold nanoparticles. *Scientific Reports* 3:1505.

Chen, W., W.P. Cai, C.H. Liang, and L.D. Zhang. 2001. Synthesis of gold nanoparticles dispersed within pores of mesoporous silica induced by ultrasonic irradiation and its characterization. *Materials Research Bulletin* 36:335–342.

Chen, W., W.P. Cai, L. Zhang, G.Z. Wang, and L.D. Zhang. 2001. Sonochemical processes and formation of gold nanoparticles within pores of mesoporous silica. *Journal of Colloid and Interface Science* 238:291–295.

Chen, X.Y., J.R. Li, X.C. Li, and L. Jiang. 1998. A new step to the mechanism of the enhancement effect of gold nanoparticles on glucose oxidase. *Biochemical and Biophysical Research Communications* 245:352–355.

Cheng, Y., A.C. Samia, J.D. Meyers, I. Panagopoulos, B.W. Fei, and C. Burda. 2008. Highly efficient drug delivery with gold nanoparticle vectors for in vivo photodynamic therapy of cancer. *Journal of the American Chemical Society* 130:10643–10647.

Cho, W.S., M. Cho, J. Jeong, et al. 2010. Size-dependent tissue kinetics of peg-coated gold nanoparticles. *Toxicology and Applied Pharmacology* 245:116–123.

Chuang, S.M., Y.H. Lee, R.Y. Liang, et al. 2013. Extensive evaluations of the cytotoxic effects of gold nanoparticles. *Biochim Biophys Acta* 1830:4960–4973.

Cumberland, S.L., and G.F. Strouse. 2002. Analysis of the nature of oxyanion adsorption on gold nanomaterial surfaces. *Langmuir* 18:269–276.

Daniel, M.C., and D. Astruc. 2004. Gold nanoparticles: Assembly, supramolecular chemistry, quantum-size-related properties, and applications toward biology, catalysis, and nanotechnology. *Chemical Reviews* 104:293–346.

Daniel, M.C., M.E. Grow, H. Pan, et al. 2011. Gold nanoparticle-cored poly(propyleneimine) dendrimers as a new platform for multifunctional drug delivery systems. *New Journal of Chemistry* 35:2366–2374.

Dawson, A., and P.V. Kamat. 2000. Complexation of gold nanoparticles with radiolytically generated thiocyanate radicals ((Scn)$_2$·⁻). *Journal of Physical Chemistry B* 104:11842–11846.

Demaille, C., M. Brust, M. Tsionsky, and A.J. Bard. 1997. Fabrication and characterization of self-assembled spherical gold ultramicroelectrodes. *Anal Chem* 69:2323–2328.

Deplanche, K., M.L. Merroun, M. Casadesus, et al. 2012. Microbial synthesis of core/shell gold/palladium nanoparticles for applications in green chemistry. *Journal of the Royal Society Interface* 9:1705–1712.

Di Felice, R., and A. Selloni. 2004. Adsorption modes of cysteine on Au(111): Thiolate, amino-thiolate, disulfide. *The Journal of Chemical Physics* 120:4906–4914.

Di Guglielmo, C., J. De Lapuente, C. Porredon, D. Ramos-Lopez, J. Sendra, and M. Borras. 2012. In vitro safety toxicology data for evaluation of gold nanoparticles—chronic cytotoxicity, genotoxicity and uptake. *Journal of Nanoscience and Nanotechnology* 12:6185–6191.

Dieluweit, S., D. Pum, and U.B. Sleytr. 1998. Formation of a gold superlattice on an s-layer with square lattice symmetry. *Supramolecular Science* 5:15–19.

Dreaden, E.C., A.M. Alkilany, X.H. Huang, C.J. Murphy, and M.A. El-Sayed. 2012. The golden age: Gold nanoparticles for biomedicine. *Chemical Society Reviews* 41:2740–2779.

Dreaden, E.C., M.A. Mackey, X.H. Huang, B. Kang, and M.A. El-Sayed. 2011. Beating cancer in multiple ways using nanogold. *Chemical Society Reviews* 40:3391–3404.

Du, L., S. Suo, G. Wang, et al. 2013. Mechanism and cellular kinetic studies of the enhancement of antioxidant activity by using surface-functionalized gold nanoparticles. *Chemistry* 19:1281–1287.

Dubertret, B., M. Calame, and A.J. Libchaber. 2001. Single-mismatch detection using gold-quenched fluorescent oligonucleotides. *Nature Biotechnology* 19:365–370.

Dulkeith, E., A.C. Morteani, T. Niedereichholz, et al. 2002. Fluorescence quenching of dye molecules near gold nanoparticles: Radiative and nonradiative effects. *Physical Review Letters* 89:1–4.

Einarson, M.B., and J.C. Berg. 1992. Effect of salt on polymer solvency: Implications for dispersion stability. *Langmuir* 8:2611–2615.

Elghanian, R., J.J. Storhoff, R.C. Mucic, R.L. Letsinger, and C.A. Mirkin. 1997. Selective colorimetric detection of polynucleotides based on the distance-dependent optical properties of gold nanoparticles. *Science* 1997:277:1078–1081.

El-Sayed, M.A. 2001. Some interesting properties of metals confined in time and nanometer space of different shapes. *Accounts of Chemical Research* 34:257–264.

Faraday, M. 1857. The Bakerian Lecture: Experimental relations of gold (and other metals) to light. *Philosophical Transactions of the Royal Society of London* 147:145–181.

Filali, M., M.A.R. Meier, U.S. Schubert, and J.-F. Gohy. 2005. Star-block copolymers as templates for the preparation of stable gold nanoparticles. *Langmuir* 21:7995–8000.

Fischer, S., F. Hallermann, T. Eichelkraut, et al. 2013. Plasmon enhanced upconversion luminescence near gold nanoparticles—Simulation and analysis of the interactions: Errata. *Optics Express* 21:10606–10611.

Frens, G. 1972. Particle-size and sol stability in metal colloids. *Kolloid-Zeitschrift and Zeitschrift Fur Polymere* 250:736.

Frens, G. 1973. Controlled nucleation for regulation of particle-size in monodisperse gold suspensions. *Nature Physical Science* 241:20–22.

Fritz, G., V. Schadler, N. Willenbacher, N.J. Wagner. 2002. Electrosteric stabilization of colloidal dispersions. *Langmuir* 18:6381–6390.

Gachard, E., H. Remita, J. Khatouri, B. Keita, L. Nadjo, and J. Belloni. 1998. Radiation-induced and chemical formation of gold clusters. *New Journal of Chemistry* 22:1257–1265.

Gagner, J.E. X. Qian, M.M. Lopez, J.S. Dordick, and R.W. Siegel. 2012. Effect of gold nanoparticle structure on the conformation and function of adsorbed proteins. *Biomaterials* 33:8503–8516.

Gardea-Torresdey, J.L., J.G. Parsons, E. Gomez, et al. 2002. Formation and growth of Au nanoparticles inside live alfalfa plants. *Nano Letters* 2:397–401.

Garg, N., A. Mohanty, N. Lazarus, et al. 2010. Robust gold nanoparticles stabilized by trithiol for application in chemiresistive sensors. *Nanotechnology* 21:40.

Gaudana, R., H.K. Ananthula, A. Parenky, and A.K. Mitra. 2010. Ocular drug delivery. *AAPS Journal* 12:348–360.

Geldenhuys, W., T. Mbimba, T. Bui, K. Harrison, and V. Sutariya. 2011. Brain-targeted delivery of paclitaxel using glutathione-coated nanoparticles for brain cancers. *Journal of Drug Targeting* 19:837–845.

Geso, M. 2007. Gold nanoparticles: A new x-ray contrast agent. *British Journal of Radiology* 80:64–65.

Ghosh, P.S., C.K. Kim, G. Han, N.S. Forbes, and V.M. Rotello. 2008. Efficient gene delivery vectors by tuning the surface charge density of amino acid-functionalized gold nanoparticles. *ACS Nano* 2:2213–2218.

Giersig, M., and P. Mulvaney. 1993. Preparation of ordered colloid monolayers by electrophoretic deposition. *Langmuir* 9:3408–3413.

Gittins, D.I., and F. Caruso. 2001. Tailoring the polyelectrolyte coating of metal nanoparticles. *Journal of Physical Chemistry B* 105:6846–6852.

Gittins, D.I., and F. Caruso. 2002. Biological and physical applications of water-based metal nanoparticles synthesised in organic solution. *ChemPhysChem* 3:110–113.

Green, I.X., W.J. Tang, M. Neurock, and J.T. Yates. 2011. Spectroscopic observation of dual catalytic sites during oxidation of Co on a Au/TiO$_2$ catalyst. *Science* 333:736–739.

Gu, T., J.K. Whitesell, and M.A. Fox. 2003. Energy transfer from a surface-bound arene to the gold core in omega-fluorenyl-alkane-1-thiolate monolayer-protected gold clusters. *Chemistry of Materials* 15:1358–1366.

Gaudana, R., H.K. Ananthula, A. Parenky, et al. 2010. Ocular drug delivery. *American Association of Pharmaceutical Scientists Journal* 12:348–360.

Guan, Z., S. Li, P.B. Cheng, N. Zhou, N. Gao, and Q.H. Xu. 2012. Band-selective coupling-induced enhancement of two-photon photoluminescence in gold nanocubes and its application as turn-on fluorescent probes for cysteine and glutathione. *ACS Appl Mater Interfaces* 4:5711–5716.

Hainfeld, J.F., D.N. Slatkin, T.M. Focella, and H.M. Smilowitz. 2006. Gold nanoparticles: A new x-ray contrast agent. *British Journal of Radiology* 79:248–253.

Hakim, M., S. Billan, U. Tisch, G. Peng, I. Dvrokind, O. Marom, et al. Diagnosis of head-and-neck cancer from exhaled breath. *British Journal of Cancer* 104:1649–1655.

Haruta, M., and M. Daté. 2001. Advances in the catalysis of Au nanoparticles. *Applied Catalysis A* 222:427–437.

Hauck, T.S., A.A. Ghazani, and W.C.W. Chan. 2008. Assessing the effect of surface chemistry on gold nanorod uptake, toxicity, and gene expression in mammalian cells. *Small* 4:153–159.

Henglein, A., and D. Meisel. 1998. Radiolytic control of the size of colloidal gold nanoparticles. *Langmuir* 14:7392–7396.

Hostetler, M.J., S.J. Green, J.J. Stokes, and R.W. Murray. 1996. Monolayers in three dimensions: Synthesis and electrochemistry of omega-functionalized alkanethiolate-stabilized gold cluster compounds. *Journal of the American Chemical Society* 118:4212–4213.

Hostetler, M.J., A.C. Templeton, and R.W. Murray. 1999. Dynamics of place-exchange reactions on monolayer-protected gold cluster molecules. *Langmuir* 15:3782–3789.

Hostetler, M.J., J.E. Wingate, C.J. Zhong, et al. 1998. Alkanethiolate gold cluster molecules with core diameters from 1.5 to 5.2 nm: Core and monolayer properties as a function of core size. *Langmuir* 14:17–30.

Hou, S.-Y., Y.-L. Hsiao, M.-S. Lin, C.-C. Yen, and C.-S. Chang. 2012. MicroRNA detection using lateral flow nucleic acid strips with gold nanoparticles. *Talanta* 99:375–379.

Hu, J., J. Zhang, F. Liu, K. Kittredge, J.K. Whitesell, and M.A. Fox. 2001. Competitive photochemical reactivity in a self-assembled monolayer on a colloidal gold cluster. *Journal of the American Chemical Society* 123:1464–1470.

Huang, C.J., P.H. Chiu, Y.H. Wang, and C.F. Yang. 2006. Synthesis of the gold nanodumbbells by electrochemical method. *Journal of Colloid and Interface Science* 303:430–436.

Huang, X., and M.A. El-Sayed. 2010. Gold nanoparticles: Optical properties and implementations in cancer diagnosis and photothermal therapy. *Journal of Advanced Research* 1:13–28.

Hühn, D., K. Kantner, C. Geidel, et al. 2013. Polymer-coated nanoparticles interacting with proteins and cells: Focusing on the sign of the net charge. *ACS Nano* 7:3253–3263.

Hussain, I., S. Graham, Z. Wang, et al. 2005. Size-controlled synthesis of near-monodisperse gold nanoparticles in the 1–4 nm range using polymeric stabilizers. *Journal of the American Chemical Society* 127:16398–16399.

Hwang, Y.N., D.H. Jeong, H.J. Shin, et al. 2002. Femtosecond emission studies on gold nanoparticles. *Journal of Physical Chemistry B* 106:7581–7584.

Jain, P.K, K.S. Lee, I.H. El-Sayed, and M.A. El-Sayed. 2006. Calculated absorption and scattering properties of gold nanoparticles of different size, shape, and composition: Applications in biological imaging and biomedicine. *The Journal of Physical Chemistry B* 110:7238–7248.

Jana, N.R., L. Gearheart, and C.J. Murphy. 2001. Evidence for seed-mediated nucleation in the chemical reduction of gold salts to gold nanoparticles. *Chemistry of Materials* 13:2313–2322.

Kam, Z. 1983. Absorption and scattering of light by small particles. *Nature* 306:625–625.

Kanehara, M., Y. Oumi, T. Sano, and T. Teranishi. 2003. Formation of low-symmetric 2d superlattices of gold nanoparticles through surface modification by acid-base interaction. *Journal of the American Chemical Society* 125:8708–8709.

Khaletskaya, K., J. Reboul, M. Meilikhov, et al. 2013. Integration of porous coordination polymers and gold nanorods into core-shell mesoscopic composites toward light-induced molecular release. *Journal of the American Chemical Society* 135:10998–11005.

Kharlamov, A. 2012. *Plasmonic Nanophotothermic Therapy of Atherosclerosis*. Bethesda, MD: National Institutes of Health.

Khlebtsov, N., and L. Dykman. 2011. Biodistribution and toxicity of engineered gold nanoparticles: A review of in vitro and in vivo studies. *Chemical Society Reviews* 40:1647–1671.

Khomutov, G.B. 2002. Two-dimensional synthesis of anisotropic nanoparticles. *Colloids and Surfaces A* 202:243–267.

Kim, B., G. Han, B.J. Toley, C.K. Kim, V.M. Rotello, and N.S. Forbes. 2010. Tuning payload delivery in tumour cylindroids using gold nanoparticles. *Nature Nanotechnology* 5:465–472.

Kim, D., S. Park, J.H. Lee, Y.Y. Jeong, and S. Jon. 2007. Antibiofouling polymer-coated gold nanoparticles as a contrast agent for in vivo x-ray computed tomography imaging. *Journal of the American Chemical Society* 129:7661–7665.

Knoll, M., and E. Ruska. 1932. The electron microscope [in German]. *Zeitschrift Fur Physik* 78:318–339.

Kudgus, R.A., R. Bhattacharya, and P. Mukherjee. 2011. Cancer nanotechnology: Emerging role of gold nanoconjugates. *Anticancer Agents Med Chem* 11:965–973.

Kuo, C.H., T.F. Chiang, L.J. Chen, and M.H. Huang. 2004. Synthesis of highly faceted pentagonal- and hexagonal-shaped gold nanoparticles with controlled sizes by sodium dodecyl sulfate. *Langmuir* 20:7820–7824.

Laaksonen, T., P. Ahonen, C. Johans, and K. Kontturi. 2006. Stability and electrostatics of mercaptoundecanoic acid-capped gold nanoparticles with varying counterion size. *ChemPhysChem* 7:2143–2149.

Langille, M.R., M.L. Personick, J. Zhang, and C.A. Mirkin. 2012. Defining rules for the shape evolution of gold nanoparticles. *Journal of the American Chemical Society* 134:14542–14554.

Leduc, C. S. Si, J. Gautier, et al. 2013. Highly specific gold nanoprobe for live-cell single-molecule imaging. *Nano Letters* 13:1489–1494.

Lewinski, N., V. Colvin, and R. Drezek. 2008. Cytotoxicity of nanoparticles. *Small* 4:26–49.

Li, W., L.H. Huo, D.M. Wang, et al. 2000. Self-assembled multilayers of alternating gold nanoparticles and dithiols: Approaching to superlattice. *Colloids and Surfaces A* 175:217–223.

Li, X.-M., M.R. de Jong, K. Inoue, S. Shinkai, J. Huskens, D.N. Reinhoudt. 2001. Formation of gold colloids using thioether derivatives as stabilizing ligands. *Journal of Materials Chemistry* 11:1919–1923.

Lin, X.M., C.M. Sorensen, and K.J. Klabunde. 1999. Ligand-induced gold nanocrystal super-lattice formation in colloidal solution. *Chemistry of Materials* 11:198–202.

Liu, Z., W. Li, F. Wang, et al. enhancement of lipopolysaccharide-induced nitric oxide and interleukin-6 production by pegylated gold nanoparticles in raw cells. *Nanoscale* 4:7135–7142.

Logunov, S.L., T.S. Ahmadi, M.A. El-Sayed, J.T. Khoury, and R.L. Whetten. 1997. Electron dynamics of passivated gold nanocrystals probed by subpicosecond transient absorption spectroscopy. *Journal of Physical Chemistry B* 101:3713–3719.

Longmire, M., P.L. Choyke, and H. Kobayashi. 2008. Clearance properties of nano-sized particles and molecules as imaging agents: Considerations and caveats. *Nanomedicine* 3:703–717.

Luo, J., M. Dong, F. Lin, et al. 2011. Three-dimensional network polyamidoamine dendrimer-Au nanocomposite for the construction of a mediator-free horseradish peroxidase biosensor. *Analyst* 136:4500–4506.

Ma, X., X. Wang, M. Zhou, and H. Fei. 2013. A mitochondria-targeting gold-peptide nano-assembly for enhanced cancer-cell killing. *Advanced Healthcare Materials* 12:1638–1643.

Mackey, M.A., F. Saira, M.A. Mahmoud, and M.A. El-Sayed. 2013. Inducing cancer cell death by targeting its nucleus: Solid gold nanospheres versus hollow gold nanocages. *Bioconjug Chem* 24:897–906.

Makarova, O.V., A.E. Ostafin, H. Miyoshi, J.R. Norris, and D. Meisel. 1999. Adsorption and encapsulation of fluorescent probes in nanoparticles. *Journal of Physical Chemistry B* 103:9080–9084.

Malikova, N., I. Pastoriza-Santos, M. Schierhorn, N.A. Kotov, and L.M. Liz-Marzan. 2002. Layer-by-layer assembled mixed spherical and planar gold nanoparticles: Control of interparticle interactions. *Langmuir* 18:3694–3697.

Mallick, K., Z.L. Wang, and T. Pal. 2001. Seed-mediated successive growth of gold particles accomplished by UV irradiation: A photochemical approach for size-controlled synthesis. *Journal of Photochemistry and Photobiology A* 140:75–80.

Mandal, D., A. Maran, M.J. Yaszemski, M.E. Bolander, and G. Sarkar. 2009. Cellular uptake of gold nanoparticles directly cross-linked with carrier peptides by osteosarcoma cells. *Journal of Materials Science–Materials in Medicine* 20:347–350.

Maxwell, J.C. 1865. A dynamical theory of the electromagnetic field. *Philosophical Transactions of the Royal Society of London* 155:459–512.

McNeil, S.E. 2011. *Characterization of Nanoparticles Intended for Drug Delivery.* New York, NY: Humana Press.

Melnikau, D., D. Savateeva, A. Susha, A.L. Rogach, and Y.P. Rakovich. 2013. Strong plasmon-exciton coupling in a hybrid system of gold nanostars and J-aggregates. *Nanoscale Res Lett* 8:134.

Meltzer, S., R. Resch, B.E. Koel, et al. 2001. Fabrication of nanostructures by hydroxylamine seeding of gold nanoparticle templates. *Langmuir* 17:1713–1718.

Meyre, M.E., M. Treguer-Delapierre, and C. Faure. 2008. Radiation-induced synthesis of gold nanoparticles within lamellar phases. Formation of aligned colloidal gold by radiolysis. *Langmuir* 24:4421–4425.

Mie, G. 1908. Articles on the optical characteristics of turbid tubes, especially colloidal metal solutions [in German]. *Annalen Der Physik* 25:377–445.

Min, J., H. Jung, H.H. Shin, G. Cho, H. Cho, and S. Kang. 2013. Implementation of P22 viral capsids as intravascular magnetic resonance T contrast conjugates via site-selective attachment of Gd(Iii)-chelating agents. *Biomacromolecules* 14:2332–2339.

Mirkin, C.A., R.L. Letsinger, R.C. Mucic, and J.J. Storhoff. 1996. A DNA-based method for rationally assembling nanoparticles into macroscopic materials. *Nature* 382:607–609.

Mulvaney, P. 1996. Surface plasmon spectroscopy of nanosized metal particles. *Langmuir* 12:788–800.

Parween, S., A. Ali, and V.S. Chauhan. 2013. Non-natural amino acids containing peptide capped gold nanoparticles for drug delivery application. *ACS Applied Materials and Interfaces* 5:6484–6493.

Pender, D.S., L.M. Vangala, V.D. Badwaik, et al. 2013. Bactericidal activity of starch-encapsulated gold nanoparticles. *Frontiers in Bioscience* 18:993–1002.

Peng, G., U. Tisch, O. Adams, et al. 2009. Diagnosing lung cancer in exhaled breath using gold nanoparticles. *Nature Nanotechnology* 4:669–673.

Perrault, S.D., and W.C. Chan. Synthesis and surface modification of highly monodispersed, spherical gold nanoparticles of 50-200 nm. [In Eng]. *Journal of the American Chemical Society* 131, no. 47 (Dec 2 2009): 17042-3.

Pol, V.G., A. Gedanken, and J. Calderon-Moreno. 2003. Deposition of gold nanoparticles on silica spheres: A sonochemical approach. *Chemistry of Materials* 15:1111–1118.

Porter, L.A., D. Ji, S.L. Westcott, et al. 1998. Gold and silver nanoparticles functionalized by the adsorption of dialkyl disulfides. *Langmuir* 14:7378–7386.

Prasad, B.L.V., S.I. Stoeva, C.M. Sorensen, and K.J. Klabunde. 2003. Digestive-ripening agents for gold nanoparticles: alternatives to thiols. *Chemistry of Materials* 15:935–942.

Prasad, B.L.V., S.I. Stoeva, C.M. Sorensen, and K.J. Klabunde. 2002. Digestive ripening of thiolated gold nanoparticles: the effect of alkyl chain length. *Langmuir* 18:7515–7520.

Qu, X.C., C.P. Yao, J. Wang, Z. Li, and Z.X. Zhang. 2012. Anti-Cd30-targeted gold nanoparticles for photothermal therapy of L-428 Hodgkin's cell. *International Journal of Nanomedicine* 7:6095–6103.

Railsback, J.G., A. Singh, R.C. Pearce, et al. 2012. Weakly charged cationic nanoparticles induce DNA bending and strand separation. *Advanced Materials* 24:4261.

Reuveni, T., M. Motiei, Z. Romman, A. Popovtzer, and R. Popovtzer. 2011. Targeted gold nanoparticles enable molecular CT imaging of cancer: An in vivo study. *International Journal of Nanomedicine* 6:2859–2864.

Rubin, P., and S.H. Levitt. 1964. The response of disseminated reticulum cell sarcoma to the intravenous injection of colloidal radioactive gold. *Journal of Nuclear Medicine* 5:581–594.

Sadauskas, E., N.R. Jacobsen, G. Danscher, et al. 2009. Biodistribution of gold nanoparticles in mouse lung following intratracheal instillation. *Chemistry Central Journal* 3:16.

Sakamoto, M., T. Tachikawa, M. Fujitsuka, and T. Majima. 2006. Acceleration of laser-induced formation of gold nanoparticles in a poly(vinyl alcohol) film. *Langmuir* 22:6361–6366.

Sakura, T., T. Takahashi, K. Kataoka, and Y. Nagasaki. 2005. One-pot preparation of mono-dispersed and physiologically stabilized gold colloid. *Colloid and Polymer Science* 284:97–101.

Sarathy, K.V., G.U. Kulkarni, and C.N.R. Rao. 1997. A novel method of preparing thiol-derivatised nanoparticles of gold, platinum and silver forming superstructures. *Chemical Communications* 6:537–538.

Sau, T.K., and C.J. Murphy. 2004. Room temperature, high-yield synthesis of multiple shapes of gold nanoparticles in aqueous solution. *Journal of the American Chemical Society* 126:8648–8649.

Sau, T.K., A. Pal, N.R. Jana, Z.L. Wang, and T. Pal. 2001. Size controlled synthesis of gold nanoparticles using photochemically prepared seed particles. *Journal of Nanoparticle Research* 3:257–261.

Schatz, G.C. 2007. Using theory and computation to model nanoscale properties. *Proceedings of the National Academy of Science U S A* 104:6885–6892.

Schmid, G., R. Pfeil, R. Boese, et al. 1981. $Au_{55}[P(C_6H_5)_3]_{12}Cl_6$—A gold cluster of unusual size. *Chemische Berichte* 114:3634–3642.

Schroeder, U., P. Sommerfeld, S. Ulrich, and B.A. Sabel. 1998. Nanoparticle technology for delivery of drugs across the blood-brain barrier. *Journal of Pharmaceutical Sciences* 87:1305–1307.

Schulz, M., L. Ma-Hock, S. Brill, et al. 2012. Investigation on the genotoxicity of different sizes of gold nanoparticles administered to the lungs of rats. *Mutation Research* 745:51–57.

Selvakannan, P.R., S. Mandal, S. Phadtare, R. Pasricha, and M. Sastry. 2003. Capping of gold nanoparticles by the amino acid lysine renders them water-dispersible. *Langmuir* 19:3545–3549.

Shan, J., and H. Tenhu. 2007. Recent advances in polymer protected gold nanoparticles: Synthesis, properties and applications. *Chemical Communications* 44:4580–4598.

Sharma, N., A.K. Pinnaka, M. Raje, A. Fnu, M.S. Bhattacharyya, and A.R. Choudhury. 2012. Exploitation of marine bacteria for production of gold nanoparticles. *Microbiol Cell Factories* 11:86.

Shenhar, R., and V.M. Rotello. 2003. Nanoparticles: Scaffolds and building blocks. *Accounts of Chemical Research* 36:549–561.

Shimizu, T., T. Teranishi, S. Hasegawa, and M. Miyake. 2003. Size evolution of alkanethiol-protected gold nanoparticles by heat treatment in the solid state. *Journal of Physical Chemistry B* 107:2719–2724.

Shukla, R., N. Chanda, A. Zambre, et al. 2012. Laminin receptor specific therapeutic gold nanoparticles ((AuNP)-Au-198-EGCG) show efficacy in treating prostate cancer. *Proc Natl Acad Sci U S A* 109:12426–12431.

Skrabalak, S.E, J. Chen, Y. Sun, et al. 2008. Gold nanocages: Synthesis, properties, and applications. *Accounts of Chemical Research* 41:1587–1595.

Sokolov, K., M. Follen, J. Aaron, et al. 2003. Real-time vital optical imaging of precancer using anti-epidermal growth factor receptor antibodies conjugated to gold nanoparticles. *Cancer Research* 63:1999–2004.

Song, J., Z. Fang, C. Wang, et al. 2013. Photolabile plasmonic vesicles assembled from amphiphilic gold nanoparticles for remote-controlled traceable drug delivery. *Nanoscale* 5:5816–5824.

Southam, G., and T.J. Beveridge. 1996. The occurrence of sulfur and phosphorus within bacterially derived crystalline and pseudocrystalline octahedral gold formed in vitro. *Geochimica et Cosmochimica Acta* 60:4369–4376.

Sperling, R.A., P.R. Gil, F. Zhang, M. Zanella, and W.J. Parak. 2008. Biological applications of gold nanoparticles. *Chemical Society Reviews* 37:1896–1908.

Sperling, R.A., and W.J. Parak. 2010. Surface modification, functionalization and bioconjugation of colloidal inorganic nanoparticles. *Philosophical Transactions of the Royal Society A* 368:1333–1383.

Stenkamp, V.S., P. McGuiggan, and J.C. Berg. 2001. Restabilization of electrosterically stabilized colloids in high salt media. *Langmuir* 17:637–651.

Tandon, A., A. Sharma, J.T. Rodier, A.M. Klibanov, F.G. Rieger, and R.R. Mohan. 2013. Bmp7 gene transfer via gold nanoparticles into stroma inhibits corneal fibrosis in vivo. *PLoS One* 8:e66434.

Tang, D.Q., D.Y. Tang, and D.P. Tang. 2005. Construction of a novel immunoassay for the relationship between anxiety and the development of a primary immune response to adrenal cortical hormone. *Bioprocess Biosyst Eng* 27:135–141.

Templeton, A.C., M.J. Hostetler, C.T. Kraft, and R.W. Murray. 1998. Reactivity of monolayer-protected gold cluster molecules: Steric effects. *Journal of the American Chemical Society* 120:1906–1911.

Templeton, A.C., M.P. Wuelfing, and R.W. Murray. 2000. Monolayer protected cluster molecules. *Accounts of Chemical Research* 33:27–36.

Thakker, J.N., P. Dalwadi, and P.C. Dhandhukia. 2013. Biosynthesis of gold nanoparticles using *Fusarium oxysporum* F. sp. cubense Jt1, a plant pathogenic fungus. *ISRN Biotechnology* 2013:5.

Thanh, N.T., and Z. Rosenzweig. 2002. Development of an aggregation-based immunoassay for anti-protein a using gold nanoparticles. *Analytical Chemistry* 74:1624–1628.

Thomas, K.G., and P.V. Kamat. 2000. Making gold nanoparticles glow: enhanced emission from a surface-bound fluoroprobe. *Journal of the American Chemical Society* 122:2655–2656.

Taguchi, T., K. Isozaki, and K. Miki. 2012. Enhanced catalytic activity of self-assembled-monolayer-capped gold nanoparticles. *Advanced Materials* 24:6462–6467.

Torrano, A.A., Â.S. Pereira, O.N. Oliveira, Jr, and A. Barros-Timmons. 2013. Probing the interaction of oppositely charged gold nanoparticles with DPPG and DPPC Langmuir monolayers as cell membrane models. *Colloids Surf B Biointerfaces* 108:120–126.

Turkevich, J., and H.H. Hubbell. 1951. Low angle x-ray diffraction of colloidal gold and carbon black. *Journal of the American Chemical Society* 73:1–7.

Turkevich, J., P.C. Stevenson, and J. Hillier. 1951. A study of the nucleation and growth processes in the synthesis of colloidal gold. *Discussions of the Faraday Society* 11:55.

Turner, M., V.B. Golovko, O.P.H. Vaughan, et al. 2008. Selective oxidation with dioxygen by gold nanoparticle catalysts derived from 55-atom clusters. *Nature* 454:981–983.

Tzhayik, O., P. Sawant, S. Efrima, E. Kovalev, and J.T. Klug. 2002. Xanthate capping of silver, copper, and gold colloids. *Langmuir* 18:3364–3369.

Valden, M., X. Lai, and D.W. Goodman. 1998. Onset of catalytic activity of gold clusters on titania with the appearance of nonmetallic properties. *Science* 281:1647–1650.

Veeraapandian, S., S.N. Sawant, and M. Doble. 2012. Antibacterial and antioxidant activity of protein capped silver and gold nanoparticles synthesized with *Escherichia coli*. *Journal of Biomedical Nanotechnology* 8:140–148.

Venkatesh, P., and C.R. Prasad. 1983. The scattering of light and other electromagnetic-radiation. *Applied Optics* 22:645–645.

Ventura, M., Y. Sun, V. Rusu, et al. 2013. Dual contrast agent for computed tomography and magnetic resonance hard tissue imaging. *Tissue Engineering Part C* 19:405–416.

Vickers, M.S., J. Cookson, P.D. Beer, P.T. Bishop, and B. Thiebaut. 2006. Dithiocarbamate ligand stabilised gold nanoparticles. *Journal of Materials Chemistry* 16:209–215.

Villa, A., D. Wang, D. Su, G.M. Veith, and L. Prati. 2010. Using supported Au nanoparticles as starting material for preparing uniform Au/Pd bimetallic catalysts. *Physical Chemistry Chemical Physics* 12:2183–2189.

Volkert, A.A., V. Subramaniam, M.R. Ivanov, A.M. Goodman, and A.J. Haes. 2011. Salt-mediated self-assembly of thioctic acid on gold nanoparticles. *ACS Nano* 5:4570–4580.

Wang, H., Z. Sun, Y. Yang, and D. Su. 2013. The growth and enhanced catalytic performance of AU–PD core-shell nanodendrites. *Nanoscale* 5:139–142.

Wang, T.X., D.Q. Zhang, W. Xu, J.L. Yang, R. Han, and D.B. Zhu. 2002. Preparation, characterization, and photophysical properties of alkanethiols with pyrene units-capped gold nanoparticles: Unusual fluorescence enhancement for the aged solutions of these gold nanoparticles. *Langmuir* 18:1840–1848.

Wang, Z., L. Jia, and M.-H. Li. 2013. Gold nanoparticles decorated by amphiphilic block copolymer as efficient system for drug delivery. *Journal of Biomedical Nanotechnology* 9:61–68.

Weare, W.W., S.M. Reed, M.G. Warner, and J.E. Hutchison. 2000. Improved synthesis of small (D(core) approximate to 1.5 nm) phosphine-stabilized gold nanoparticles. *Journal of the American Chemical Society* 122:12890–12891.

West, J.L., and N.J. Halas. 2003. Engineered nanomaterials for biophotonics applications: Improving sensing, imaging, and therapeutics. *Annual Review of Biomedical Engineering* 5:285–292.

Wojczykowski, K., D. Meißner, P. Jutzi, et al. 2006. Reliable stabilization and functionalization of nanoparticles through tridentate thiolate ligands. *Chemical Communications* 35:3693–3695.

Wu, X., J.-Y. Chen, A. Brech, et al. 2013. The use of femto-second lasers to trigger powerful explosions of gold nanorods to destroy cancer cells. *Biomaterials* 34:6157–6162.

Xu, C., D. Yang, L. Mei, et al. 2013. Encapsulating gold nanoparticles or nanorods in graphene oxide shells as a novel gene vector. *ACS Applied Materials and Interfaces* 5:2715–2724.

Xu, P., and H. Yanagi. 1999. Fluorescence patterning in dye-doped sol-gel films by generation of gold nanoparticles. *Chemistry of Materials* 11:2626–2628.

Yelin, D., D. Oron, E. Korkotian, M. Segal, and Y. Silberberg. 2002. Third-harmonic microscopy with a titanium-sapphire laser. *Applied Physics B* 74:S97–S101.

Yelin, D., D. Oron, S. Thiberge, E. Moses, and Y. Silberberg. 2003. Multiphoton plasmon-resonance microscopy. *Optics Express* 11:1385–1391.

Yonezawa, T., K. Yasui, and N. Kimizuka. 2000. Controlled formation of smaller gold nanoparticles by the use of four-chained disulfide stabilizer. *Langmuir* 17:271–273.

Zhang, A., Y. Tu, S. Qin, et al. 2012. Gold nanoclusters as contrast agents for fluorescent and x-ray dual-modality imaging. *Journal of Colloid and Interface Science* 372:239–244.

Zhang, D., O. Neumann, H. Wang, et al. 2009. Gold nanoparticles can induce the formation of protein-based aggregates at physiological pH. *Nano Letters* 9:666–671.

Zhang, S., G. Leem, and T.R. Lee. 2009. Monolayer-protected gold nanoparticles prepared using long-chain alkanethioacetates. *Langmuir* 25:13855–13860.

Zhang, X.D., D. Wu, X. Shen, et al. 2011. Size-dependent in vivo toxicity of PEG-coated gold nanoparticles. *International Journal of Nanomedicine* 6:2071–2081.

Zhao, P., N. Li, and D. Astruc. 2013. State of the art in gold nanoparticle synthesis. *Coordination Chemistry Reviews* 257:638–665.

Zhou, C., M. Long, Y. Qin, X. Sun, and J. Zheng. 2011. Luminescent gold nanoparticles with efficient renal clearance. *Angewandte Chemie* 123:3226–3230.

Zhou, H.S., I.I. Honma, H. Komiyama, and J.W. Haus. 1994. Controlled synthesis and quantum-size effect in gold-coated nanoparticles. *Physical Review B* 50:12052–12056.

Index

Thermionic emission, 110
Thioflavin-T (ThT), 169, 174
Thiolate ligands, 303
Three-dimensional nanostructures, 81
ThT. *See* Thioflavin-T
Tiksna loha. See Iron
Tin
    metallurgical properties of, 57–58
    nutraceutical and therapeutic properties
        of, 73–74
Titania, 190
Titanium dioxide nanoparticles, 171
    in dentistry, 152
    dermal protectant applications, 151
TMOs. *See* Transition metal oxides
Tollens method, 269
Toxicity
    calcium, 19–20
    of Chinese traditional medicine,
        239–242
    chromium, 25
    copper, 25–26
    environmental, 162
    iron, 21
    issues of NPs, 175
    magnesium, 21–22
    manganese, 26
    phosphorus, 22
    potassium, 23
    and safety of nanomaterials, 210–211
    selenium, 27
    silver, 271–272, 274
    sodium, 24
    zinc, 29
Toxic minerals, effect of, 230
Trace elements, 10, 16
Trace metals
    in biological systems, 138–140
    in modern medicine, 259
Trace minerals, health benefits of, 230,
    231
Tracer method, 129
Trade Related Aspects of Intellectual
    Property Rights (TRIPS)
    agreements, 283
Traditional Chinese medicine (TCM),
    225–226, 234
Transcription factors, 253
Transcytosis, 166
Transition metal oxides (TMOs), 191
Transmission electron microscopy (TEM),
    112, 198, 199
Transporter proteins, 213
Turkevich method
    for synthesis of AuNP colloids, 295

for citrate-mediated reduction synthesis,
    297
2D gel protein profiling system, 117
Two-dimensional nanostructures, 80

U

Ultrafine powder, molten metal convert
    to, 40
Ultrasonic spray pyrolysis method, 266
Ultraviolet (UV) rays, 151
Ultraviolet photoelectron spectroscopy
    (UPS), 109
Ultraviolet-visible (UV-vis) spectroscopy,
    104, 108–109
Universal Protein Resource (UniProt)
    Accession Identifications (IDs),
    256–257
UPS. *See* Ultraviolet photoelectron
    spectroscopy
Urdhvapatana, 38
Urinary excretion, 129
U.S. Food and Drug Administration (FDA),
    124

V

Valuka yantra, 50
Vanga. See Tin
Vapor-phase synthesis, 136–137
Varitara (water buoyancy), 43
Varta loha
    metallurgical properties of, 61
    nutraceutical and therapeutic properties
        of, 76–77
Vasoactive intestinal neuroprotective
    peptide, 169
Verapamil, therapeutic effects of, 132
Vipaka, 65
Virya, 65
Vishesha shodhana (purification), 39, 58,
    60
Volatile organic compounds (VOCs), 305,
    306
von Laue, Max, 103
Vrddha jarana, 46

W

Water-dispersible nanoparticles, 303
Water-soluble C60 carboxylic acid, 172
Weiss's hypothesis, 204
Wet-chemistry synthesis, 194–198
Wheat germ agglutinin (WGA), 169
Wilson's disease, 25–26, 168

Woolen filters, 63–64
Work function electron tunneling,
    109–110
World Trade Organization agreements, 283

X

XPS. *See* X-ray photoelectron spectroscopy
X-ray crystallography, 244
X-ray generation, 111
X-ray photoelectron spectroscopy (XPS),
    108–109
X-ray powder diffraction (XRD), 103,
    200–201

Y

Yasada. See Zinc

yin-yang theory, 226–229
Yttrium NPs, 171

Z

Zero-dimensional nanostructures, 80
Zinc
    bioavailability, 130
    in biological systems, 139
    drug interaction with, 132
    metallurgical properties of, 58–59
    nanoparticles, 142
        dermal protectant applications, 151
        nanometal antibacterial applications,
            149
    nutraceutical and therapeutic properties
        of, 74–75
    in nutritional physiology, 28–29